The Analysis of Chemically Reacting Systems: A Stochastic Approach

Topics in Chemical Engineering

A series edited by R. Hughes, University of Salford, UK

Volume 1 HEAT AND MASS TRANSFER IN PACKED BEDS
by N. Wakao and S. Kaguei

Volume 2 THREE-PHASE CATALYTIC REACTORS
by P. A. Ramachandran and R. V. Chaudhari

Volume 3 DRYING: PRINCIPLES, APPLICATIONS AND DESIGN
by C. Strumillo and T. Kudra

Volume 4 THE ANALYSIS OF CHEMICALLY REACTING SYSTEMS: A STOCHASTIC APPROACH
by L. K. Doraiswamy and B. D. Kulkarni

Additional volumes in preparation

ISSN: 0277-5883

This book is part of a series. The publisher will accept continuation orders which may be cancelled at any time and which provide for automatic billing and shipping of each title in the series upon publication. Please write for details.

THE ANALYSIS OF CHEMICALLY REACTING SYSTEMS:

A Stochastic Approach

By L. K. Doraiswamy and B. D. Kulkarni
National Chemical Laboratory, Pune, India

GORDON AND BREACH SCIENCE PUBLISHERS
New York London Paris Montreux Tokyo

Gordon and Breach Science Publishers

Post Office Box 786
Cooper Station
New York, New York 10276
United States of America

Post Office Box 197
London WC2E 9PX
England

58, rue Lhomond
75005 Paris
France

14-9, Okubo 3-chome
Shinjuku-ku, Tokyo 160
Japan

Library of Congress Cataloging-in-Publication Data

Doraiswamy, L. J. (Laxmangudi Krishnamurthy)
The analysis of chemically reacting systems.

(Topics in chemical engineering; v. 4)
Bibliography: p.
Includes index.
1. Chemical reactions -- Mathematical models.
2. Chemical reactions -- Statistical methods.
I. Kulkarni, B. D. II. Title. III. Series.
QD501.D586 1987 541.3′94 87-269
ISBN 0-677-21670-X
ISSN 0277-5883

Introduction to the Series

The subject matter of chemical engineering covers a very wide spectrum of learning, and the number of subject areas encompassed in both undergraduate and graduate courses is inevitably increasing each year. This wide variety of subjects makes it difficult to cover the whole subject matter of chemical engineering in a single book. The present series is therefore planned as a number of books covering areas of chemical engineering that, although important, are not treated at any length in graduate and postgraduate standard texts. Additionally, the series will incorporate recent research material that has reached the stage where an overall survey is appropriate, and where sufficient information is available to merit publication in book form for the benefit of the profession as a whole.

Inevitably, with a series such as this, constant revision is necessary if the value of the texts for both teaching and research purposes is to be maintained. I would be grateful to individuals for criticisms and for suggestions for future editions.

R. HUGHES

Dedicated to

Rajalakshmi Doraiswamy

"Experience has shown, and a true philosophy will always show, that a vast, perhaps the larger, portion of the truth arises from the seemingly irrelevant."

Edgar Allan Poe

Contents

Preface

Over the years, chemical engineering science has grown through infusions of new physico-chemical thoughts and mathematical methods in the formulation, analysis and solution of problems. One tacit assumption is that all systems can be treated at the macroscopic level. This often leads to time integration of a set of (usually) coupled differential equations. Time evolution occurs as a continuous process that is wholly predictable. The chemical reaction engineering literature to date is based almost entirely on the macroscopic analysis of problems. The books of Levenspiel, Carberry, Froment and Bischoff, Hill, Hlaváček, and Doraiswamy and Sharma, for instance, fall in this category. Even more specialised books, such as the treatment of reaction–diffusion in pellets by Aris and the book on chemical reactor theory edited by Lapidus and Armundson, have based their treatments on macroscopic (deterministic) formulations.

While conventional macroscopic analysis has yielded very useful information regarding the behaviour of systems, its routine application to all types of situations is suspect. For instance, the phenomenon of chemical instability cannot be based on a simple macroscopic level of description, and requires a more refined microscopic formulation. The microscopic approach, which in its rigorous form is very difficult to use, can be considerably simplified by the usual method of averaging (over all possible variables of the system) to eliminate the irrelevant fast variables. It should be noted, however, that these eliminated variables contain some information about the system, and their neglect may mean some loss in the exact description. An intermediate description, such as the one attempted in this book, aims at incorporating this information in suitable ways – the simplest of which is to give the eliminated variables a random character. This necessitates the use of probabilistic considerations, and the literature in the physical and chemical sciences (particularly chemical physics) and in electrical engineering has progressed considerably in using this approach.

In the field of chemical physics, with which we are more immediately concerned, books such as *Handbook of Stochastic Methods* by Gardiner, *Stochastic Processes in Physics and Chemistry* by van Kampen, *Synergetics* edited by Haken, *The Fokker–Planck Equation* by Risken and the excellent reviews published in *Advances in Chemical Physics* (edited by Prigogine and others) are noteworthy. There is, however, a clear lack of effort in adapting these various physico-chemical concepts to chemical engineering, although a few papers have appeared in this area in the past few years, such as those of L. T. Fan. Seinfeld and Lapidus' *Mathematical Methods in Chemical Engineering* gives a lucid exposition of the mathematical methods involved in stochastic analysis. The present book represents perhaps the first attempt to put together all the available information in a manner that can provide a rational framework for further research by chemical engineers in this challenging area.

Yet the implications of the book are much wider than the title would indicate. Traditionally, chemical engineers have been concerned with systems of large dimensions, such as chemical plants and process machinery. Although over the years chemical engineering has tended to be more and more science oriented, bringing within its orbit analysis of happenings at the micro level, the very concept of "micro" has undergone a remarkable change in the past few years. We look at chemical reactions today in terms of their particulate behaviour, and at certain transport processes in terms of sub-micron particle systems. It is becoming increasingly clear that dramatic enhancements in mass transfer can be obtained in the presence of sub-micron size particles. The application of stochastic methods for such systems is inevitable, but unfortunately conventional chemical engineering curricula do not cover these adequately.

We realised at the outset the need to outline the basic principle of probability theory before undertaking any rigorous study of its application to chemically reacting systems. Thus Part I of this book is devoted to a presentation of the principles and methods of statistics insofar as they relate to the developments that follow. The rest of the book is divided into three parts, one pertaining to the role of internal fluctuations, the second to the role of external fluctuations and the last to the combined influence of the two. It is suggested that this book be read in conjunction with books on similar subjects using the more common deterministic methods.

In preparing this monograph we have been greatly helped by our students S. S. Tambe, N. S. Dabke and Y. S. Srinivas, and we are heavily in debt to them. We also thank Dr Ravi Kumar, who was associated with part of this work in its early stages. We had the good fortune of having Professor Rutherford Aris at the National Chemical Laboratory early in 1984 when we had just commenced our work in the area of stochastic analysis. He has made many scholarly suggestions and also commented critically on some of our papers that were under preparation at that time. We would like to express our deep sense of gratitude to him.

The reader will notice some repetitions in the book. For instance, at the start of every chapter, indeed of almost every major section, there is a brief recapitulation of what has been said in the preceeding chapter or section. As chemical engineers are not generally exposed to stochastic methods of analysis, it is felt that such repetitions will be greatly to the advantage of the reader.

It would be surprising if a book of this kind were entirely free of inconsistencies and omissions. We would be glad to have our attention drawn to any such errors to enable us to make suitable corrections when a revision becomes possible.

In an attempt to focus on one major aspect of the vast untapped potential of stochastic methods in chemical engineering, we have presented in this book a critical evaluation of what is known in the limited area of chemically reacting systems, not unmixed with our own perception of this subject as a whole. We hope that this monograph may be of interest not just to chemical engineers but to those working on chemical reactions in general. Even so, if this book helps to stimulate interest among researchers in chemical engineering in an area that has not so far caught their fancy, we would be satisfied that it has served a useful purpose.

L. K. DORAISWAMY
B. D. KULKARNI

Notation

In a mathematical book of this kind it would be practically impossible to list *all* the symbols used. Thus only the main symbols common to all the chapters are listed here. The other symbols are clearly explained where they appear first, and sometimes repeated in the interest of clarity.

a	general constant in rate equation; activity of catalyst
a_s	external specific surface area of particle
A, A_1, A_2	reactant species; also phenomenological constant in Equation (4.45)
AZ	number of active sites occupied by species A
b	general constant in rate equation
BZ	number of active sites occupied by species B
d	spatial dimensionality
D	intensity of noise
$\mathscr{D}$	rate of exchange between adjacent cells
$\tilde{D}$	stochastic diffusivity
D_{ij}	diffusion coefficient
D_{PQ}, D_{PR}	intensities of Poisson noise entering in functions Q and R
$Da_{(i)}$	Damkhler number
E, E^{-1}	difference operator for the species denoted by subscript
$E(\cdot)$	mean or expected value of $(\cdot)$
F	general function as defined in text; also denotes flow rate to and from the reactor
$F(x)$	distribution function
G	general function defined as in the text
$h(x)$	function defined by Equation (6.67)
H	Hamiltonian defined by Equation (3.51)
k_i, k_{-i}	forward and backward reaction rate constants
k_d	deactivation rate constant
K	mass transfer coefficient; also denotes equilibrium constant

K_r	central moments
l	length of a cell
L	total number of sites
m_0	concentration of species A at the inlet of the reactor
m_{ij}	moments
n_0	concentration of species A_1 at the inlet of the reactor
N	normalisation constant; also number of species in the system
p_{eq}	equilibrium probability distribution
p_{st}	stationary probability
$P(X)$	probability of occurrence of event X
Q	number of particles at the boundary; also function defined by Equation (8.3)
$Q(N)$	fluctuating parameter defined by Equation (8.8)
$Q(x)$	generation rate of x
R	reaction rate operator as defined by Equation (5.8); also function defined by Equation (8.3)
$R(x)$	recombination rate of x
S_r	parameter defined by Equation (5.27)
t, t'	time variable
$T_\infty^\pm$	mean time spent in a state before the system moves to the next (+) or previous (−) state
$T(x)$	mean passage time for state variable x
U	function defined by Equation (6.11)
U_k	eigenfunction
V	volume; also denotes potential [Equation (4.128)]
w	Wiener process
$W, W_\pm$	transition probability per unit time
W_{nm}	probability of transition from state n to m per unit time
x	general macroscopic variable
x'	new state occupied by x
x_{10}, x_{20}	input concentrations for the species X_1, X_2
x_{1s}, x_{2s}	steady state values of x_1, x_2
X	random variable
y_m	most probable path
Y	product species
Z	vacant sites on catalyst surface
Z_0	total number of available sites on catalyst surface
α	coefficient in rate function; parameter defined by Equation (6.45); magnitude of the activation or inhibition [example (4.2)]; also denotes jump moments

$\boldsymbol{\alpha}_i$	matrices as defined in Equation (4.40)
β	coefficient in rate function
β_1, β_2	eigenvalues in the example (4.2)
δ	Dirac-delta function
ε	scaling parameter; measure of correlation time; strength of fluctuations; also denotes a smallness parameter
γ	coefficient in rate function; also coefficient of skewness
η	external coeffectiveness factor $[= 1/(1 + Daa)]$
η_{ij}	relaxation matrix defined by Equation (5.54)
λ	mean free path; also mean density of times t_i in the Poisson distribution
λ_1, λ_2	eigenvalues as defined in Equation (5.8)
$\lambda(x)$	rate for the generation process $(X \rightarrow X + 1)$
$\mathscr{L}$	projection operator
μ	coefficient of heterogeneity
μ_i	noncentral moments
$\mu(x)$	rate of the loss process $(X \rightarrow X - 1)$
ν	coefficient of variation
π	probability function
ϕ	characteristic function; function defined in Equations (6.10) and (6.88); also generally used to denote macroscopic part of the system variable
ϕ_{st}	steady state value of ϕ
$\rho(\cdot)$	probability density function defined by Equation (6.48)
σ	standard deviation
σ^2	variance of the random variable
ψ	moment generating function; also represents macroscopic part of the system variable
ψ_{st}	steady state value of ψ
τ	correlation time
τ_c	time scale governing the evolution of chemical reaction
τ_D	time scale for the diffusing species
Ω	system size parameters
$\xi(t)$	stochastic force; also external noise

Introduction

This book is concerned with the stochastic modelling of chemically reacting systems. If the need for such an analysis is relevant for gentlemanly reactions, it is even more so for systems that play truant. The macroscopic analysis of chemically reacting systems has been known for several decades, but the need to analyse such systems from a probabilistic point of view becomes evident when we visualise chemical reactions in their fundamental aspect. Thus, due to the discrete nature of events comprising chemical transformations, it is clear that the molecular population changes discretely with time. This observation seems to go against the conventional macroscopic description where changes in state of the system are assumed to occur continuously. In many situations the difference in the results of the two approaches is of no great pragmatic importance. On the other hand, there are situations where basic differences in approach do lead to substantial differences in end results. It is therefore essential that the more rigorous approach is followed.

The microscopic description, however, is far too rigorous, and contains all the information regarding the evolution of fast as well as slowly varying parameters of the system. The repeated averaging of the system leads to the elimination of the fast variables, and only the relevant macroscopic variables that are slowly varying remain. The eliminated fast variables constitute what is termed "internal noise"—which is generally negligible in the context of the macroscopic description. It must be noted that this noise, however small, has its origin within the system, and its incorporation in the analysis would naturally give more information about the actual behaviour of the system. The complexity of the attendant mathematics can be reduced if the system is treated as Markovian, leading finally to the so-called master equation that expresses the variation of the transition probability of the process. We thus end up with a stochastic approach that adequately takes account of internal fluctuations and at the same time avoids inclusion of any irrelevant information, such as that found in the microscopic approach.

While the formulation of the phenomenological master equation for a given situation is not a formidable task, its solution poses severe problems. It is of interest then to identify *a priori* situations for which the stochastic approach would lead to results significantly different from those obtained using the macroscopic method. It is in this context that the following fundamental law of statistical mechanics can be usefully invoked: for systems of nearly finite size (N) operating near the equilibrium points, the variation in the mean value of N is proportional to the square root of the size of the system. Clearly, the internal fluctuations would be unimportant in large volume systems, leading to the conclusion that the conventional macroscopic approach would be adequate for such systems. On the other hand, for finite size systems, the effects would be important and warrant a stochastic approach.

The same fundamental law also suggests that for systems operating near equilibrium points, fluctuations would be negligible. In other words, for systems operating far from equilibrium, fluctuations can manifest themselves strongly and alter the stationary points. This is especially so when non-idealities are present in the system. From a practical point of view, the first type of system is somewhat uncommon, but the second is rather frequent and merits stochastic analysis.

In practical systems, in addition to internal fluctuations, one can also visualise a stochastic force with its origin outside the system, leading to so-called external fluctuations. Typically, one can think of variations in the external parameters, such as flow rate, concentration, temperature etc., to exemplify this situation. These disturbances occur on a macroscopic scale, and a logical way to account for them would be to attempt a macroscopic description on which these external fluctuations are superimposed. This so-called Langevin approach should be distinguished from the microscopic analysis in which the internal fluctuations are allowed to remain. Since these fluctuations originate outside the system, they would have no relevance to the size or volume of the system as in the case of internal fluctuations. These can therefore be important even for large volume, globally stable systems and cannot, in general, be ignored. The consequences of taking these fluctuations into account in a variety of situations, such as bistable and unstable systems, have been analysed in recent years.

Finally, the question of sumultaneous accounting of both internal and external fluctuations has to be considered. This is a formidable task and no firm analysis is yet available. The important point to note at present is that these fluctuations, internal or external, even when accounted for individually, show drastic effects on system behaviour with far-reaching implications on the design and modelling of chemically reacting systems.

Analyses of chemically reacting systems using probabilistic methods have been carried out in the past by some investigators, notably L. T. Fan and D. Ramakrishna. The approach followed in this book uses the master equation for analysing internal fluctuations and the Langevin formalism for external fluctuations.

PART I

Basic Mathematical Framework

1 Basic Notions of Probability Theory

1.1 Introduction

THE PRESENT chapter summarises the mathematical framework for the modelling of the probabilistic processes considered in the subsequent chapters. The need for a probabilistic (or stochastic) description of physical and chemical processes arises naturally as a consequence of the inherent random nature of these systems such as, for example, when a molecular phenomenon is observed on a macroscopic scale or when one tracks the position of a microscopic particle in a fluid. It is also possible for a process to assume a stochastic character as a result of disturbances from external fluctuating forces. The equations of conservation for such systems would involve differential equations with random coefficients that are functions of the independent variable, time. A typical example would be the operation of a reactor with fluctuations in the values of variables (coefficients) such as volumetric flow rate, inlet composition, etc.

The two situations mentioned above clearly reveal the basic difference in the nature of stochasticity in the two cases. In the first case, one usually starts with a finer level of description. The macroscopic equations of the system are then obtained as a coarse grained approximation from this rigorous description. Here the information regarding the fluctuating behaviour of the system is inherent in the rigorous formulation. In situations of the second type one usually starts with the macroscopic description of the system and superimposes on it the fluctuations—the stochastic properties of which are supposedly known. In the subsequent chapters of this book, we will have occasion to consider both types of situations. We shall present in this chapter the basic notions of the probability theory required to handle these situations. The probability theory is concerned with the study of mathematical models of random phenomena. Conversely, random phenomena are described in terms of probability laws.

In this general introduction we have made use of terms such as *random phenomena*, *stochasticity*, etc. While the meaning of these terms is intuitively clear within the general context of this chapter, it is necessary that they are more explicitly defined before any rational mathematical development is undertaken. This is attempted in the following section. For more detailed treatment, reference may be made to the excellent books on the subject by Dixon and Masseyl (1957), Johnson and Leone (1964), Kendall and Stuart (1963, 1967, 1966), Feller (1968), Parzen (1960), Lindley (1965), Fisz (1963), Seinfeld and Lapidus (1975), Papoulis (1965), McShane (1974), Bury (1975).

1.2 Random Variable, its Distribution and Properties

To formalise the concept of the random variable, let us consider the example of a die set to roll N number of times. Each time the die is rolled the outcome is one of the integers 1 to 6. Let us represent the outcome of the tth trial by $X(t)$ which denotes the random variable and can assume any value between 1 and 6. As a second example, we consider the case of molecules impinging upon a catalyst surface—they arrive at the catalyst surface randomly. Let $X(t)$ denote the number of molecules present on the surface at any time t. Clearly $X(t)$ again represents a random variable. It will be noticed that in the examples considered X represents the state of the system, while the order variable (t) measures the number of trials or the time unit. In most processes that we come across the order variable actually represents the time, and $X(t)$ may therefore be regarded as a sequence of values of the state the system would attain at different times. The process is then referred to as a stochastic process.

In the first example both the state of the system and the number of trials take integer values. Such processes are called *discrete state–discrete time* processes. On the other hand, in the second example, the state of the system assumes discrete values but time runs continuously. Such processes are referred to as *discrete state–continuous time* processes. Similarly, we may think of stochastic processes where the state of the system is continuous and its monitoring is done at discrete time intervals or continuously: for example, the position of a particle in a turbulent flow stream. When such continuous state stochastic processes are monitored at discrete intervals of time we have a *continuous state–discrete time* process, while continuous monitoring of this system leads to a *continuous state–continuous time* process.

The nature of the random variable thus decides the nature of the stochastic process depending on whether it has a discrete or continuous character. Additionally, specification of the monitoring technique of the state of the system can completely identify the nature of the stochastic process, and the examples considered cover the possible types of processes that one may encounter.

Let us now consider an arbitrary stochastic process belonging to any of the sub-classifications noted above and call the outcome of such a process as an event. Clearly, depending on the nature of the stochastic process, the outcome could be a discrete number or a continuous spectrum of possible events. What one is most interested in knowing is which possible event would result from a particular trial or at a particular instant of time and is called the probability of occurrence of that event. Thus $\text{Prob}\{X\} = P(X)$ specifies the probability that event X will occur and represents a number as against X which denotes the state. The formal definition of probability requires some physical consideration of the process. Thus an event that is certain to occur is assigned a value of unity, while one that cannot occur is assigned a value of zero. Further, for two mutually exclusive events, X_1 and X_2, the probability of observing either X_1 or X_2 can be written as the sum of the individual probabilities of the two events.

To explain this point further let us again consider the example of a die rolled N times, during which the value $X = j$ occurred n_j times. The probability that the value $X = j$ occurs can then be defined as

$$\text{Prob}\{X = j\} = P(X = j) = \lim_{N \to \infty} \frac{n_j}{N} \tag{1.1}$$

Equation (1.1) represents one of the possible values of the probability of an outcome. Latent in this definition is the assumption of equal *a priori* probability—that X assumes a value j where j could be any integer between 1 and 6. This consideration of course arises from the physical situation and is clear at least for the example considered, even without an explicit statement. In general, however, it is necessary to specify the initial *a priori* probability, and an equation such as (1.1) represents the transformation of this initial *a priori* probability into the probability of outcome. Clearly, different probabilities of outcome would result for different initial *a priori* probabilities.

The initial *a priori* probability is generally fixed intuitively or from some kind of physical consideration, but is often not stated explicitly. Thus, in the use of the least squares method to obtain more accurate information from infected data, it is generally assumed that the errors have an *a priori* initial distribution of probability that is Gaussian. This implies selection of a proper variable that measures the observed quantity and where the condition is satisfied. Choice of any other variable may not satisfy this assumption and use of the least squares method would lead to different results for the same situation.

Let us now return to the pedagogical problem of defining the terms commonly used. It is customary in probability theory to denote the random variables by capital letters, while the values they assume are indicated by the corresponding lower case letters. Thus the probability that a discrete random variable X assumes a value x_j can be written as $P(X = x_j) = p(x_j)$, and $p(x_j)$ is referred to as the *probability mass function.*

For a continuous random variable we need to define the probability over a range of values. Thus the probability that X lies within the range x to $x + dx$ can be defined as

$$P(x < X < x + dx) = p(x)\,dx \tag{1.2}$$

where $p(x)$ is referred to as the *probability density function* (since it is a measure of the density of X in the region dx) and has the units of inverse x.

Both the probability mass and density functions obviously satisfy the condition of being always positive and lying in the range

$$0 \leqslant p(x_j) \leqslant 1, \qquad 0 \leqslant p(x) \leqslant 1 \tag{1.3}$$

Additionally, the requirement that their total value over all possible states should add up to unity leads to

$$\sum_j p(x_j) = 1 \quad \text{and} \quad \int_{-\infty}^{\infty} p(x)\,dx = 1 \tag{1.4}$$

for the discrete and continuous random variable cases, respectively.

The probability of a random variable can also be stated in terms of a distribution function of X which is defined as

$$F(x) = P(X \leqslant x) \tag{1.5}$$

Clearly this represents the probability that X (discrete or continuous) will not exceed a certain threshold value x and is dependent on the specification of this value. For the discrete variable case this can be expressed in terms of the probability mass function as

$$F(x) = \sum_j p(x_j) \tag{1.6}$$

and for the case of a continuous variable in terms of the probability density function as

$$F(x) = \int_{-\infty}^{\infty} p(x)\, dx \tag{1.7}$$

It of course follows that

$$p(x) = \frac{d}{dx} F(x) \tag{1.8}$$

for all real numbers x for which the derivative exists.

It is clear from the definitions of the distribution function that it is always positive, lying in the range $0 \leqslant F(x) \leqslant 1$ and always increasing with increase in x. Also the probability that X lies between any two values x_1 and x_2 can be written as

$$P(x_1 < X \leqslant x_2) = F(x_2) - F(x_1) = \int_{x_1}^{x_2} p(x)\, dx \tag{1.9}$$

It is possible in a given situation to have a combined discrete and continuous probability density function. In fact, when the probability density function is concentrated as a delta function at several discrete points, we have the situation corresponding to a discrete variable and the probability density function becomes

$$p(x) = p(1)\,\delta(x - x_1) + p(2)\,\delta(x - x_2) + \cdots p(N)\,\delta(x - x_n) + \cdots \tag{1.10}$$

where $p(1) \ldots p(n) \ldots$ denote the probabilities and $x_1 \ldots x_n \ldots$ the values which the random variable assumes at the discrete points $1 \ldots n \ldots$.

The specification of the random variable X in terms of the distribution function or probability mass or density function helps us to calculate certain other quantities of eventual interest. Thus the mean or the *expected value* $E(X)$ of a random variable X can be defined in terms of the distribution function as

$$E(X) = \int_{-\infty}^{\infty} x\,dF(x) \tag{1.12}$$

or in terms of the probability density function as

$$E(X) = \int_{-\infty}^{\infty} xp(x)\,dx \tag{1.13}$$

For the case of a discrete random variable X, the mean or the expected value can be defined as

$$E(X) = \sum_j xp(x_j) \tag{1.14}$$

Two functions most commonly used in characterising a distribution function are the *variance* and *standard deviation*. Both are based on derivations from the mean value of a distribution. The variance of X is defined as

$$\mathrm{Var}[X] = E[(X - E(X))^2] = E(X^2) - E^2(X) \tag{1.15}$$

that is, as the difference between the mean of the square and square of the mean. The standard deviation is defined as

$$\sigma[X] = \sqrt{\mathrm{Var}(X)} \tag{1.16}$$

Two other functions that are commonly used in characterising the distribution function are the *moment generating function* and the *characteristic function*. These are defined, respectively, as the Laplace transform and Fourier transform of the probability density function. Mathematically these may be represented as

$$\psi = E(e^{-tX}) \tag{1.17a}$$

$$\phi = E(e^{itX}) \tag{1.17b}$$

Note that Equations (1.15) and (1.17a) have been written in terms of the expected value $E(.)$. They can be expanded using the appropriate definition of $E(.)$ to obtain alternative forms.

An important property evident in the definitions [Equations (1.12)–

(1.17b)] is that the random variable X may or may not possess a finite mean, variance or generating function. It will, however, always have a characteristic function. Specification of the characteristic function therefore suffices to specify the probability law of a random variable.

The mean and the variance of a random variable as defined by Equations (1.12)–(1.15) are special cases of the more general properties of the probability distribution called its *moments*. Two types of moments of the probability distribution can be defined. These are referred to as the *noncentral moments* (μ_r) and the *central moments* (K_r) and are defined, respectively, as

$$\mu_r = E[X^r] \tag{1.18}$$

$$K_r = E[(X - \mu_1)^r] \tag{1.19}$$

where r refers to a general index specifying the rth moment and takes integer values of $0, 1, \ldots N$. Clearly, for $r = 0$, both μ_0 and K_0 take the value of unity; while for $r = 1$ and 2, respectively, μ_1 and K_2 represent the mean and variance of the random variable.

Higher order moments can similarly be obtained from the definitions given by Equations (1.18) and (1.19). Expanding the term $(X - \mu_1)^r$ in the definition of the central moment in the binomial series and making use of Equation (1.18) we obtain various central moments in relation to the noncentral moments:

$$K_0 = 1, \qquad K_1 = 0, \qquad K_2 = \mu_2 - \mu_1^2$$

$$K_3 = \mu_3 - 3\mu_2\mu_1 + 2\mu_1^3 \tag{1.20}$$

$$K_4 = \mu_4 - 4\mu_1\mu_3 + 6\mu_2\mu_1^2 - 3\mu_1^4, \quad \text{etc.}$$

Likewise the noncentral moments can be expressed in terms of central moments:

$$\mu_2 = K_2 + \mu_1^2; \qquad \mu_3 = K_3 + 3K_2\mu_1 + \mu_1^3$$
$$\mu_4 = K_4 + 4\mu_1 K_3 + 6K_2\mu_1 + \mu_1^4 \tag{1.21}$$

In addition to these moments, two other functions defined in terms of the ratio of two moments are often used in characterising a process. Thus a function defined as the ratio of the standard deviation to the mean is referred to as the *coefficient of variation* (v) and can be understood since standard deviation measures the dispersion about the mean value. The second function is defined as the ratio of the third central moment to the standard deviation $[r = K_3/\sigma^3]$ and is referred to as the *coefficient of skewness*. This follows from the fact that for symmetric distribution all odd order central moments are zero. The first central moment K_1 is always zero by definition. To define any deviation from symmetry we therefore need to take the next higher moment and the definition of coefficient of skewness follows.

The moment generating function defined earlier by Equation (1.17a) is most convenient for obtaining the several moments defined above. Thus, according to Equation (1.13), for a continuous random variable we have

$$\psi(t) = E(e^{-tX}) = \int_0^\infty e^{-xt} p(x)\,dx \tag{1.22}$$

Expanding the exponential term in series, performing the indicated integration, and using the definition of noncentral moments, Equation (1.18), we get

$$\begin{aligned} \psi(t) &= 1 - \mu_1 t + \frac{t^2}{2!}\mu_2 - \frac{t^3}{3!}\mu_3 + \cdots \\ &= 1 + \sum \frac{(it)^r}{r!}\left(\frac{d^r\psi(t)}{dt^r}\right)_{t=0} \end{aligned} \tag{1.23}$$

If the generating function of X is known, the noncentral moments of X can be readily obtained as coefficients of its expansion in series. The equation also defines the several moments more compactly as

$$\mu_r = (-1)^r\left(\frac{d^r\psi(t)}{dt^r}\right)_{t=0} \tag{1.24}$$

The central moments of the distribution can likewise be obtained from the generating function by multiplying it by e^{tK_X}

$$\begin{aligned} e^{tK_X}\psi(t) &= 1 - tE(X - K_X) + \frac{t^2}{2!}E[(X - K_X)^2] \\ &\quad - \frac{t^3}{3!}E[(X - K_X)^3] + \cdots \\ &= 1 - K_1 t + \frac{t^2}{2!}K_2 - \frac{t^3}{3!}K_3 + \cdots \end{aligned} \tag{1.25}$$

Alternatively, in terms of differentials, we obtain

$$K_i = (-1)^i \frac{d^i}{dt^i}[e^{tK_X}\psi(t)]_{t=0} \tag{1.26}$$

For a discrete random variable the generating function of X is defined as

$$\psi(Z) = \sum_j p(x_j)Z^j \tag{1.27}$$

where $(-1 \leqslant Z \leqslant 1)$. The moments of distribution in this case can likewise be obtained by differentiating the generating function successively and evaluating the derivatives at $Z = 1$.

Besides the generating function the characteristic function defined by Equation (1.17) can also be used to obtain the moments of distribution. Thus, for the discrete variable case, we can write Equation (1.17) as

$$\phi(t) = E(e^{itX}) = \sum_{j=0}^{N} e^{itx_j}p(x_j) \tag{1.28}$$

As before, expanding the exponential term and utilising the definition given by Equation (1.18), we obtain

$$\phi(t) = 1 + \sum_{j=1}^{N} \frac{\mu_i}{j!}(it)^j \tag{1.29}$$

Differentiation of Equation (1.28) r times gives

$$\frac{d^r\phi}{dt^r} = \sum_{j=0}^{N} (ix_j)^r e^{itx_j}p(x_j) \tag{1.30}$$

which is nothing but the expected value $E(i^rX^r e^{itX})$. Similarly, for a continuous random variable, we obtain a relation analogous to Equation (1.24) as

$$\mu_r = (i)^{-r}\left(\frac{d^r\phi}{dt^r}\right)_{t=0} \tag{1.31}$$

Finally, before we close the discussion on the properties of random variables, it is worthwhile to introduce additional parameters called cumulants that are similar to moments and are sometimes convenient to use. To give a formal definition of these parameters, let us consider the function

$$\psi(t) = \ln \phi(t) \tag{1.32}$$

Employing the definition of ϕ given by Equation (1.29) in (1.32) and expanding the logarithmic function we obtain

$$\psi(t) = \sum_{j=1}^{\infty} \frac{\mu_j}{j!} (it)^j - \frac{1}{2}\left\{\sum_{j=1}^{\infty} \frac{\mu_j}{j!} (it)^j\right\}^2 + \frac{1}{3}\left\{\sum_{j=1}^{\infty} \frac{\mu_j}{j!} (it)^j\right\}^3 - \cdots \tag{1.33}$$

which can be more compactly written as

$$\psi(t) = \sum_{j=1}^{\infty} \frac{K_j}{j!} (it)^j \tag{1.34}$$

The coefficients K_j are referred to as *cumulants* and can be identified in terms of noncentral moments by comparing coefficients of like order terms in t:

$$K_1 = \mu_1, \qquad K_2 = \mu_2 - \mu_1^2, \qquad K_3 = \mu_3 - 3\mu_1\mu_2 + 2\mu_1^3$$
$$K_4 = \mu_4 - 3\mu_2^2 - 4\mu_1\mu_3 + 14\mu_1^2\mu_2 - 8\mu_1^4 \tag{1.35}$$

Equation (1.34) expresses the generating function in terms of the cumulants. Employing the definition given by Equation (1.29) in Equation (1.32) the characteristic function can also be expressed in terms of the cumulants as

$$\phi(t) = 1 + \sum_{j=1}^{\infty} \frac{\mu_j}{j!} (it)^j = \exp\left[\sum_{j=1}^{\infty} \frac{K_j}{j!} (it)^j\right]$$
$$= 1 + \sum_{j=1}^{\infty} \frac{K_j}{j!} (\mathrm{it})^j + \frac{1}{2!}\left[\sum_{j=1}^{\infty} \frac{K_j}{j} (it)^j\right]^2 + \cdots \tag{1.36}$$

1.3 Multivariate Distributions: Joint, Marginal and Conditional Distributions

In the previous section we discussed the probability functions, distribution functions and general properties of the single random variable case. The present section extends this discussion to include cases where more than one random variable is involved. Thus (Seinfeld and

Lapidus, 1975) consider the case of two random variables X and Y both of which are discrete and can assume N_X and N_Y possible values. When X and Y are measured, there are $N_X N_Y$ possible values, and to characterise the random variables X and Y it would be necessary to know the probability that X assumes a value x_{j1} and Y a value y_{j2}. Thus, following the single variable case, we can define the joint probability mass function as

$$\text{Prob}\{X = x_{j1}, Y = y_{j2}\} = P_{XY}(j_1, j_2) \tag{1.37}$$

which is always positive. Further, the total value of the probability mass function must add up to unity, and we have the following relation similar to Equation (1.4)

$$\sum_{j_1=1}^{N_x} \sum_{j_2=1}^{N_y} P_{XY}(j_1, j_2) = 1 \tag{1.38}$$

The *joint distribution function* for this case can be written as

$$F_{XY}(j_1, j_2) = \text{Prob}\{X \leqslant x_{j1}, Y \leqslant y_{j2}\} = \sum_{l_1=1}^{j_1} \sum_{l_1=1}^{j_2} P_{XY}(l_1, l_2) \tag{1.39}$$

The distribution function $F_{XY}(j_1, j_2)$ begins with zero $[F_{XY}(l, j_1) = F(j_2, l) = 0]$, ends with $F_{XY}(N_X N_Y) = 1$, and is always positive.

The corresponding probability functions for the case of continuous random variables X and Y can be defined as

$$\text{Prob}\{x \leqslant X \leqslant x + dx, y \leqslant Y \leqslant y + dy\} = P_{XY}(x, y)\, dx\, dy \tag{1.40}$$

where $P_{XY}(x, y)$ is referred to as the *joint probability density function* and obviously satisfies the relation

$$\int_{-\infty}^{\infty} \int_{-\infty}^{\infty} p_{XY}(x, y)\, dx\, dy = 1 \tag{1.41}$$

The distribution function for this case can be written as

$$F_{XY}(x, y) = \text{Prob}\{X \leqslant x, Y \leqslant y\} = \int_{-\infty}^{x} \int_{-\infty}^{y} p_{XY}(l, k)\, dl\, dk \tag{1.42}$$

and possesses the following boundary values:

$$\lim_{x \to -\infty} F_{XY}(x, y) = \lim_{y \to -\infty} F_{XY}(x, y) = 0 \tag{1.43}$$

and

$$F_{XY}(\infty, \infty) = 1 \tag{1.44}$$

The *joint probability density function* can be obtained from the distribution function by differentiating it at the point (x, y):

$$p_{XY}(x, y) = \frac{\partial^2 F_{XY}(x, y)}{\partial x \, \partial y} \tag{1.45}$$

In many instances in the multirandom variable case, we are interested in knowing the probability that a variable X assumes a value x_j regardless of the value the other random variable assumes. To illustrate this point let us consider a case of N random variables $X_1, X_2, \ldots X_N$ with each random variable assuming $N_1, N_2, \ldots N_N$ possible values. Let us further assume that we are interested in knowing the probability that the M variables $X_1, X_2, \ldots X_M$ take the values of $x_1, x_2, \ldots x_M$ regardless of the values of the remaining variable $(N - M)$. This can be written as

$$\text{Prob}\{X_1 = x_1, X_2 = x_2, \ldots X_M = x_M\} = p_{X_1X_2\ldots X_M}(x_1, x_2, \ldots x_m) \tag{1.46a}$$

$$= \sum_{N_1=1}^{N_1} \cdots \sum_{N_{H+1}=1}^{N_{H+1}} \cdots \sum_{N_N=1}^{N_N} p_{X_1X_2\ldots X_M\ldots X_N}(x_1, x_2, \ldots x_M \ldots x_N) \tag{1.46b}$$

Equation (1.46a) represents the *marginal probability mass function*. The general definition for the case of N variables assumes the following simpler form when only two random variables, say X and Y with N_X and N_Y possible values, are involved:

$$\text{Prob}\{X = x_{j1}\} = p_X(x_{j1}) = \sum_{j_2=1}^{N_Y} p_{XY}(x_{j1}, y_{j2}) \tag{1.47}$$

The corresponding *marginal distribution functions* can then be defined as

$$\begin{aligned} F_X(x_{j1}) &= \text{Prob}\{X \leqslant x_{j1}\} \\ F_Y(y_{j2}) &= \text{Prob}\{Y \leqslant y_{j2}\} \end{aligned} \tag{1.48}$$

In the case of continuous random variables (X, Y), if we are interested in the probability that X assumes a value in the range x to $x + dx$ regardless of the value of y, we can similarly define the *marginal probability density function* and *marginal distribution function*:

$$P_X(x)\,dx = \int_{-\infty}^{\infty} p_{XY}(x, y)\,dy \tag{1.49}$$

$$F_X(x) = \text{Prob}\{X \leqslant x\}, \qquad F_Y(y) = \text{Prob}\{Y \leqslant y\} \tag{1.50}$$

The marginal probability density function can be obtained from the marginal probability distribution function by differentiation:

$$p_X(x) = \frac{dF_X(x)}{dx}, \qquad p_Y(y) = \frac{dF_Y(y)}{dy} \tag{1.51}$$

The usual condition of the marginal distribution functions being non-negative and attaining the values of 0 and 1 at the boundaries is of course implied in the definitions.

In addition to the joint and marginal probabilities, depicting two different situations, we may be interested in defining another type of probability called the conditional probability. Thus in a two variable case we may be interested in knowing the probability that X assumes a value x_{j1} given that $Y = y_{j2}$. This probability is called the *conditional probability* and is defined as

$$p(x_{j1}|y_{j2}) = \text{Prob}\{X = x_{j1}, \text{ if } Y = y_{j2}\} \tag{1.52}$$

where $p(x_{j1}|y_{j2})$ is referred to as *conditional probability mass function.* Employing the definitions of joint and marginal probability mass functions in Equation (1.52) we can obtain

$$p(x_{j1}|y_{j2}) = \frac{p_{XY}(x_{j1}, y_{j2})}{p_Y(y_{j2})}$$

and (1.53)

$$p(y_{j2}|x_{j1}) = \frac{p_{XY}(x_{j1}, y_{j2})}{p_X(x_{j1})}$$

Equations (1.53) can be combined to give the following expression known as Baye's rule:

$$p(x_{j1}|y_{j2}) = \frac{p(y_{j2}|x_{j1})p_X(x_{j1})}{p_Y(y_{j2})} \qquad \text{(Baye's rule)} \tag{1.54}$$

Stated in words, the rule says that the probability that $X = x_{j1}$ given that $Y = y$ is equal to the probability that $Y = y$ given that $X = x$ multiplied by the ratio of the probabilities that $X = x_{j1}$ and $Y = y_{j2}$.

We can similarly define the conditional probability density function for a two random variable situation as

$$p(x|y)\,dx = \text{Prob}\{x \leqslant X \leqslant x + dx \text{ given } y \leqslant Y \leqslant y + dy\} \quad (1.55)$$

$p(x|y)$ denotes the conditional probability of finding X in the range x to $x + dx$ given that Y lies in the range y to $y + dy$. As before, employing the definition of joint and marginal probability density functions, the conditional probability density function can be written as

$$p(x|y) = \frac{p_{XY}(x, y)}{p_Y(y)}, \qquad p(y|x) = \frac{p_{XY}(x, y)}{p_X(x)} \quad (1.56)$$

or, rearranging the equations, as

$$p(x|y) = \frac{p(y|x)p_X(x)}{p_Y(y)} \quad \text{or} \quad p(y|x) = \frac{p(x|y)p_Y(y)}{p_X(x)} \quad (1.57)$$

which is Baye's rule for continuous variables.

The conditional distribution function can be written for the continuous random variable as

$$\begin{aligned} F(x|y) &= \text{Prob}\{X \leqslant x \text{ if } y < Y \leqslant y + dy\} \\ &= \frac{\text{Prob}\{X \leqslant x, y < Y \leqslant y + dy\}}{\text{Prob}\{y < Y \leqslant y + dy\}} \end{aligned} \quad (1.58)$$

or, equivalently as

$$F(x|y) = \frac{\int_{-\infty}^{\infty} p_{XY}(l, y)\,dl}{p_Y(y)} \quad (1.59)$$

The equations presented for the two variable case can be easily extended to an N variable system, and the probability that M components assume the values $(x_1, x_2, \ldots x_M)$ on condition that the remaining variables $(M + 1, \ldots N)$ assume the prescribed values $(x_{M+1}, \ldots x_N)$ can be readily written:

$$p(x_1, x_2, \ldots x_M \mid x_{M+1}, \ldots x_N)\,dx_1 \ldots dx_M = \frac{p_{X_1 \cdots X_N}(x_1, x_2, \ldots x_N)}{p_{X_{M+1} \cdots X_N}(x_{M+1}, \ldots x_N)} \quad (1.60)$$

Equations (1.54), (1.56) and (1.60) suggest that the total joint probability is equal to the marginal probability of the remaining variables $(X_{M+1}, \dots X_N)$ having the values $(x_{M+1}, \dots x_N)$ times the conditional probability that if this is so then the remaining variables $(X_1, X_2, \dots X_M)$ have the values $(x_1, x_2, \dots x_M)$. An especially interesting case arises when the total number of random variables (N) can be divided into two sets $(1, 2, \dots M)$ and $(M+1, M+2, \dots N)$ such that the total joint probability function factorises. Thus we have

$$p_{1,2,\dots N}(x_1, x_2, \dots x_N = p_{1,2,\dots M}(x_1, x_2, \dots x_M) p_{M+1,\dots N}(x_{M+1}, \dots x_N) \tag{1.61}$$

$p_{1,2,\dots M}(x_1, x_2, \dots x_M)$ in this equation can be recognised as the marginal probability density function of the variables $(X_1, \dots X_M)$. At the same time we can recognise it as the conditional probability density

$$p(x_1, x_2, \dots x_M \mid x_{M+1}, x_{M+2}, \dots x_N) = p_{1,2,\dots M}(x_1, x_2, \dots x_M) \tag{1.62}$$

The set of variables $(1, 2, \dots M)$ is then said to be statistically independent of the other set $(M+1, M+2, \dots N)$.

Let us now return to define the properties of these multivariate distributions in terms of the generating and characteristic functions. The general definitions given earlier for the case of a single variable now need to be expanded to include other variables and can be written for the case of a jointly distributed function:

$$\psi_{X_1, X_2, \dots X_N}(v_1, v_2, \dots v_N) = E\left(\exp\left\{\sum_{r=1}^{N} v_r X_r\right\}\right) \tag{1.63}$$

$$\phi_{X_1, X_2, \dots X_N}(v_1, v_2, \dots v_N) = E\left(\exp\left\{i \sum_{r=1}^{N} v_r X_r\right\}\right) \tag{1.64}$$

where $E(.)$ as before refers to the expected value and can be defined as

$$E(.) = \int_{-\infty}^{\infty} \dots \int_{-\infty}^{\infty} (.)\, dF_{X_1, X_2, \dots X_N}(x_1, x_2, \dots x_N) \tag{1.65}$$

with the number of integrals corresponding to the number of variables. $E(.)$ can also be defined in terms of the joint probability density function for the case of continuous random variables and in terms of joint probability mass function for the case of discrete variables.

For a general function $f(X, Y)$ we can write Equation (1.65) alternatively as

$$E[f(X, Y)] = \int_{-\infty}^{\infty}\int_{-\infty}^{\infty} f(X, Y)p_{XY}(x, y)\,dx\,dy \qquad (1.66)$$

For the function $f(X, Y) = X + Y$ where the random variables are independent we can write this equation as

$$E(X + Y) = \int_{-\infty}^{\infty}\int_{-\infty}^{\infty} (x + y)p_{XY}(x, y)\,dx\,dy \qquad (1.67)$$

which leads, on separation, to

$$E(X + Y) = E(X) + E(Y) \qquad (1.68)$$

The expected value of a sum of independent random variables therefore equals the sum of expected values of the individual random variables.

Similarly, if the function $f(X, Y)$ is given as a product of the random variables ($= X^rY^s$), the expected value of $f(X, Y)$ can be calculated:

$$E[f(XY)] = E(X^rY^s) = \int_{-\infty}^{\infty}\int_{-\infty}^{\infty} (x^ry^s)p_{XY}(x, y)\,dx\,dy$$
$$= E(X^r)E(Y^s) = \mu_{rs} \qquad (1.69)$$

Thus the expected value of the product function is simply the product of the expected values of individual random variables. One can easily recognise that the lhs of Equation (1.69) is the conventional definition of the noncentral moment which for specific values of r and s can be identified. Hence $\mu_{10} = \mu_{01} = E(Y)$. These functions, of course, also represent the expected values of the marginal distributions. Similarly higher order noncentral moments can be calculated.

The central moments can be defined in analogy with Equation (1.19) for the single variable case:

$$K_{rs} = E[(X - \mu_{10})^r(Y - \mu_{01})^s] \qquad (1.70)$$

The central moment K_{11} is of special significance and is referred to as the *covariance* of variables X and Y. In explicit form it may be written as

$$K_{11} = E[(X - \mu_{10})(Y - \mu_{01})]$$
$$= E(XY) - \mu_{01}E(X) - \mu_{10}E(Y) + \mu_{10}\mu_{01} \qquad (1.71)$$

$E(XY)$ in this equation represents the expected value of the product function (XY) and can be written for independent random variables in terms of the product of the expected values of the individual random variables as $E(XY) = E(X)E(Y)$. Further, replacing $E(X) = \mu_{10}$ and $E(Y) = \mu_{01}$ in Equation (1.71) for independent random variables, we obtain $K_{11} = 0$. Thus two jointly distributed random variables are said to be uncorrelated if their covariance vanishes.

It is customary to express the covariance in relation to the variance in terms of a correlation coefficient:

$$\text{Correlation coefficient } (\beta) = \frac{\text{covariance } (X, Y)}{\sigma(X)\sigma(Y)} \tag{1.72}$$

Clearly, even the correlation coefficient vanishes when the jointly distributed random variables X and Y are independent of each other. The definition also clarifies that the second moments should be finite. The same condition may be translated in terms of the variances $[\text{Var}(X), \text{Var}(Y)]$. For this purpose we invoke the definition of the variance given by Equation (1.19) and write the variance of Z defined as $(X + Y)$:

$$\sigma^2[Z] = E\{[(X - \mu_{10}) + (Y - \mu_{01})]^2\} \tag{1.73}$$

which on expansion and proper substitution yields

$$\sigma^2[Z] = \sigma^2[X] + \sigma^2[Y] + 2K_{11} \tag{1.74}$$

In view of the fact that the covariance K_{11} is zero for independent parameters, this equation suggests that for independent random variables the variance of the sum of the variables equals the sum of the variances of the individual random variables.

The independence of the two random variables can also be noted from their generating and characteristic functions. Thus using $Z = X + Y$ in the definitions and performing the indicated operations we obtain

$$\psi_Z(k) = \psi_X(k)\psi_Y(k) \tag{1.75}$$

and

$$\phi_Z(k) = \phi_X(k)\phi_Y(k) \tag{1.76}$$

Hence the generating and characteristic functions of a sum of two independent random variables equal, respectively, the product of the corresponding marginal generating and characteristic functions of the two random variables.

The generating or the characteristic functions [Equations (1.63) and

(1.64)] can be expanded in series to obtain the moments of distribution. Taking the example of a characteristic function for a two random variable case, and expanding it in series, we obtain

$$\begin{aligned}\phi_{XY}(t_1 t_2) &= E[\exp i(t_1 X + t_2 Y)]\\ &= 1 + i(t_1 x + t_2 y) - \frac{1}{2!}(t_1 x + t_2 y)^2 + \frac{i}{3!}(t_1 x + t_2 y)^3 \ldots\\ &= 1 + it_1 E(X) + it_2 E(Y) - \frac{1}{2!} t_1^2 E(X^2)\\ &\quad - \frac{1}{2!} t_2^2 E(Y^2) - t_1 t_2 E(XY) + \cdots \end{aligned} \tag{1.77}$$

The coefficients in the expansion of the characteristic function variously represent the marginal moments, product moments, etc.

We shall now turn to the properties of conditional distributions. For the case of two jointly distributed discrete random variables X and Y, the *conditional expectation* of Y given that $X = x$ is given by

$$E(Y \mid X = x) = \sum_{\text{Over all } y} y p_Y|_X(y|x) \tag{1.78}$$

The conditional expectation possesses three important properties. First, given the conditional expectation of Y for given X, we can obtain the *unconditional expectation* of Y as

$$E(Y) = \sum_{\text{Over all } y} E(Y \mid X = x) p_X(x) \tag{1.79}$$

Second, if X and Y are independent then the conditional and unconditional expectations of Y are identical. In other words, the conditional expectation of Y now does not depend on X. Finally, we state without proof the third property, viz.

$$E[f(X, Y) \mid X = x] = E[f(x, Y) \mid X = x] \tag{1.80}$$

which implies that the conditional mean of $f(X, Y)$ given $X = x$ is the same as the conditional mean of $f(x, Y)$ given $X = x$.

For the case of continuous random variables X and Y the conditional expectation of Y given that $X = x$ can be written as follows:

$$E(Y \mid X = x) = \int_{-\infty}^{\infty} y p_Y|_X(y|x)\, dy \tag{1.81}$$

or, more generally, for an arbitrary function $f(x)Y$ as

$$E[f(X)Y \mid X = x] = E[f(X)E(Y \mid X = x)]$$
$$= \int_{-\infty}^{\infty}\int_{-\infty}^{\infty} f(x)yp_{XY}(x, y)\,dx\,dy \qquad (1.82)$$

The conditional variance $\mathrm{Var}(Y \mid X = x)$ can be similarly defined:

$$\mathrm{Var}(Y \mid X = x) = E[Y - E(Y \mid X = x)]^2$$
$$= \int_{-\infty}^{\infty} [y - E(Y \mid X = x)]^2 p_Y|_X(y|x)\,dy \quad (1.83)$$

The equation on expansion yields the unconditional variance of Y in terms of the conditional variance:

$$\mathrm{Var}[Y] = E[\mathrm{Var}(Y \mid X = x)] + \mathrm{Var}[E(Y \mid X = x)] \qquad (1.84)$$

or, in words, the unconditional variance of Y equals the sum of the mean of the conditional variance and the variance of the conditional mean.

The *conditional noncentral and central moments* can be expressed in terms of the conditional characteristic or generating functions. Thus we can define the conditional characteristic function as

$$\phi_{Y|X}(t|x) = E(e^{itY} \mid X = x)$$
$$= \int_{-\infty}^{\infty} e^{ity} p_Y|_X(y|x)\,dy \qquad (1.85)$$

Also from a knowledge of the conditional characteristic function one can obtain the unconditional characteristic function

$$\phi_Y(t) = E[E(e^{itY} \mid X = x)] = \int_{-\infty}^{\infty} \phi_Y|_X(t|x)\,dF_X(x) \qquad (1.86)$$

1.4 Operations with Random Variables

The previous sections were concerned with a general description regarding the probability laws and properties of single and multirandom variable cases. Practical necessity would often require certain operations such as transformations, integrations, differentiations, etc. to be performed on these random variables. The present section outlines the basic concepts pertaining to these operations—again with the limited objective of serving the needs of the subsequent chapters.

1.4.1 Transformations of Random Variables

It is often necessary to obtain the probability density of a random variable that is related to another random variable (through a functional relation) the probability density for which is known. Thus consider a functional relation of the form

$$Y = g(X) \tag{1.87}$$

where X and Y are random variables with X possessing a known probability density function $p(x)$. We are interested in obtaining the corresponding probability density function for Y. An equation such as (1.87) may arise as a result of physical consideration of the system or may be artificially designed to transform the known distribution into one of standard form with zero mean and unit variance.

The functional relation suggests that for $X = x$, the random variable Y possesses some value, say, $Y = y$. The range of values of x and y would of course be different. Also, for some small deviation $|dx|$ in x, we would have a corresponding deviation $|dy|$ in y. The probability that X lies between x and $x + dx$ is therefore the same as the probability of finding Y in the range y to $y + dy$. We can therefore write

$$p(x)|dx| = p(y)|dy| \tag{1.88}$$

or

$$p(y) = \frac{p(x)}{|dy/dx|} = \frac{p(x)}{|dg/dx|} \tag{1.89}$$

where $|dy/dx|$ represents the absolute value of the Jacobi determinent. The equation assumes that there is one-to-one correspondence between the values of x and y. It is possible, especially if the relation $y = f(x)$ is nonlinear, that for a given value of y, more than one value of x exists. In situations of this type Equation (1.89) gets modified to

$$p(y) = \frac{p(x_1)}{|dg/dx_1|} + \frac{p(x_2)}{|dg/dx_2|} + \cdots \frac{d(x_n)}{|dg/dx_n|} \tag{1.90}$$

where $(x_1, \ldots x_n)$ represent different values of x satisfying the relation $y = g(x)$.

In the case of jointly distributed variables with a known joint probability density function $p_{XY}(x, y)$, we can compute the joint probability density function for the random variables U and V that are related to the random variables X and Y through some functional relationship. Thus consider the functions

$$U = g(X, Y), \qquad V = h(X, Y) \tag{1.91}$$

An equation similar to (1.88) for this case can be written:

$$p_{UV}(u, v)|du\, dv| = p_{XY}(x, y)|dx\, dy|$$

or

$$p_{UV}(uv) = \frac{p_{XY}(x, y)}{|du\, dv/dx\, dy|} = \frac{p_{XY}(x, y)}{J\left[\dfrac{u, v}{x, y}\right]} \tag{1.92}$$

where

$$J\left[\frac{u, v}{x, y}\right] = \det \begin{vmatrix} \dfrac{\partial g}{\partial x} & \dfrac{\partial g}{\partial y} \\ \dfrac{\partial h}{\partial x} & \dfrac{\partial h}{\partial y} \end{vmatrix} \tag{1.93}$$

Again, when x and y are multivalued functions, Equation (1.92) gets modified to

$$p_{UV}(u, v) = \frac{p_{XY}(x, y)}{|J(u, v/x_1 y_1)|} + \cdots + \frac{p_{XY}(x_n, y_n)}{|J(u, v/x_n, y_n)|} \tag{1.94}$$

To illustrate, let us consider the commonly used example of the sum of two random variables denoted as $Z = X + Y$. Given the joint probability density function of (x, y) we are interested in knowing the joint probability density function $p_Z(z)$. If $f_Z(z)$ represents the distribution function for Z, then

$$f_Z(z) = \text{Prob}\{Z \leqslant z\} = \int_{-\infty}^{\infty}\int_{-\infty}^{z-x} p_{XY}(x, y)\, dx\, dy \tag{1.95}$$

Noting that $\xi = x + y$ we can rewrite this equation as

$$f_Z(z) = \int_{-\infty}^{\infty}\int_{-\infty}^{\xi-x} p_{XY}(x, \xi - x)\, dx\, dy \tag{1.96}$$

Changing the order of integration and noting that $p_Z(z) = (df_Z(z)/dz)$, we obtain

$$p_Z(z) = \int_{-\infty}^{\infty} p_{XY}(x, z-x)\,dx = \int_0^{\infty} p_X(x)p_Y(z-x)\,dx \tag{1.97}$$

which is the convolution integral. For a general function such as Equation (1.87) we can write equivalently

$$p_Y(y) = \int \delta[g(x) - y]p_X(x)\,dx$$

1.4.2 Integration and Differentiation of Stochastic Variables

Before we begin with the formal definition of the integral or the differential of a stochastic variable, it is necessary to understand the notion of *mean value function* and the *correlation kernel* of a stochastic process. These properties play the same role in a stochastic *process* as their counterparts—the mean and the variance—used in defining a stochastic *variable*. Of course, the mean and the variance do not, in general, define the probability law completely and several higher order parameters are necessary. Nevertheless these two properties are often sufficient as, for example, when the random variable X obeys a normal distribution. In any event, when the nature of the distribution function is unknown, these properties along with certain other inequalities such as the Chebyshev inequality yield useful estimates regarding the nature of distribution.

In analogy with the mean of a stochastic variable, the *mean value function* of a stochastic process is simply defined as the expectation value of $X(t)$. Thus

$$m(t) = E[X(t)] \tag{1.98}$$

The *covariance kernel* is defined as

$$K(x, t) = \text{Cov}[X(s), X(t)] \tag{1.99}$$

and can be easily calculated. For a typical case of $X(t) = X_0 + Vt$ where X_0 and V are random variables, we can easily calculate these properties:

$$m(t) = E[X_0 + Vt] = E[X_0] + tE(V) \tag{1.100}$$

$$\begin{aligned} K(s,t) &= \text{Cov}[X(s), X_0 + Vt] \\ &= \text{Var}[X_0] + st\,\text{Var}[V] + (s+t)\,\text{Cov}[X_0, V] \end{aligned} \tag{1.101}$$

Note that knowledge of the joint probability density or distribution function is not necessary. Information on their mean, variance and covariance is sufficient to determine these functions. While the ease of their calculation is clearly an advantage, their chief usefulness is evident in studies of the behaviour of sample averages over a stochastic process, of integrals and differentials of stochastic processes, and in the identification of so-called ergodic and nonergodic processes. We shall return to the discussion of these processes in a later section on the stationarity of stochastic processes.

For the present purpose, we start with the sample mean defined by

$$M = \frac{1}{T}\int_0^T X(t)\,dt \tag{1.102}$$

Clearly the definition of sample mean or average involves the integration of a stochastic quantity and knowledge concerning the conditions for the integral to exist is essential.

A natural way to decompose the integral is to employ summation over finite time elements and write the integral as

$$\int_{\tau_1}^{\tau_2} X(t)\,dt = \lim \sum_{k=1}^{n} X(t_k)(t_k - t_{k-1}) \tag{1.103}$$

The summation is taken over n points lying within the interval τ_1 to τ_2. The limiting operation in Equation (1.103) can be defined for various modes of convergence. A sequence of random variables $X_1, X_2, X_3, \ldots X_n$ converges to X in mean square if each random variable X_N has a finite mean square and if

$$\lim_{n\to\infty} E[|X_n - X|^2] = 0 \tag{1.104}$$

For a continuous random variable $X(t)$ possessing a finite second moment with mean value function and covariance kernel defined continuously over (s,t) as in Equations (1.98)–(1.99), the integral $\int_{\tau_1}^{\tau_2} X(t)\,dt$ is said to be well defined as a limit in mean square.

The various other properties of the integral may then be defined as follows:

$$\text{Mean} = E\left(\int_{\tau_1}^{\tau_2} X(t)\,dt\right) = \int_{\tau_1}^{\tau_2} E[X(t)]\,dt = \int_{\tau_1}^{\tau_2} m(t)\,dt \quad (1.105)$$

$$\text{Mean square} = E\left[\left|\int_{\tau_1}^{\tau_2} X(t)\,dt\right|^2\right]$$

$$= E\int_{\tau_1}^{\tau_2}\int_{\tau_1}^{\tau_2} X(s)X(t)\,ds\,dt$$

$$= \int_{\tau_1}^{\tau_2}\int_{\tau_1}^{\tau_2} E[X(s)X(t)]\,ds\,dt \quad (1.106)$$

$$\text{Variance} = \text{Var}\left[\int_{\tau_1}^{\tau_2} X(t)\,dt\right] = \int_{\tau_1}^{\tau_2}\int_{\tau_1}^{\tau_2} \text{Cov}[X(s), X(t)]\,ds\,dt$$

$$= \int_{\tau_1}^{\tau_2}\int_{\tau_1}^{\tau_2} K(s,t)\,ds\,dt \quad (1.107)$$

Also, the following relations (for various τ's > 0) can be easily derived from these equations:

$$E\left[\int_{\tau_1}^{\tau_2} X(s)\,ds \int_{\tau_3}^{\tau_4} X(t)\,dt\right] = \int_{\tau_1}^{\tau_2} ds \int_{\tau_3}^{\tau_4} dt\, E[X(s)X(t)] \quad (1.108)$$

$$\text{Cov}\left[\int_{\tau_1}^{\tau_2} X(s)\,ds, \int_{\tau_3}^{\tau_4} X(t)\,dt\right] = \int_{\tau_1}^{\tau_2} ds \int_{\tau_3}^{\tau_4} dt\, K(s,t) \quad (1.109)$$

We see in Equations (1.105)–(1.107) that the operations of integration and forming the expectation values commute.

In similar fashion one can write the derivative of a continuous stochastic process $[X(t)]$ with finite second moment as

$$X'(t) = \lim_{h\to 0} \frac{X(t+h) - X(t)}{h} \quad (1.110)$$

The limiting operation in Equation (1.110) is evaluated in the sense of convergence in the mean square and exists provided the corresponding limiting operations for the expectation and covariance kernel exist, i.e. if the following limiting operations exist:

$$\lim_{h\to 0} E\left[\frac{X(t+h) - X(t)}{h}\right] \quad (1.111)$$

$$\lim_{h\to 0,k\to 0} \operatorname{Cov}\left[\frac{X(t+h)-X(t)}{h}, \frac{X(t+k)-X(k)}{k}\right] \tag{1.112}$$

Alternatively, the conditions can be stated in terms of mean value function and mixed derivative as

(i) mean value function should be differentiable, and
(ii) the mixed derivative $[\partial^2/(\partial s\,\partial t)]K(s,t)$ should exist and be continuous.

With these conditions met, the derivative of $X(t)$ is well defined and the following relations can be established:

$$\text{Mean} = E[X'(t)] = E\left[\frac{d}{dt}X(t)\right] = \frac{d}{dt}[E(X(t))] = m'(t) \tag{1.113}$$

$$\begin{aligned}\text{Covariance} = \operatorname{Cov}[X'(s), X'(t)] &= \operatorname{Cov}\left[\frac{d}{ds}X(s), \frac{d}{dt}X(t)\right] \\ &= \frac{d}{ds}\frac{d}{dt}\operatorname{Cov}[X(s), X(t)] \\ &= \frac{\partial^2}{\partial s\,\partial t}K(s,t)\end{aligned} \tag{1.114}$$

$$\begin{aligned}\text{Covariance} = \operatorname{Cov}[X'(s), X(t)] &= \operatorname{Cov}\left[\frac{d}{ds}X(s), X(t)\right] \\ &= \frac{d}{ds}\operatorname{Cov}[X(s), X(t)] = \frac{\partial}{\partial s}K(s,t)\end{aligned} \tag{1.115}$$

Again, it is clear from the relations above that the operations of differentiation and forming expectation values commute.

1.5 Difference (or Differential-Difference) Equations in Probability Theory

The logical application of the probability concepts, defined and discussed in the earlier sections, to the physical processes under consideration often leads to a set of equations which can be easily recognised as the difference equations. Frequently, the set of equations formulated also represent the differential-difference equations. The general theory of this type of equations is well established and in the present section we shall briefly review the basic preliminaries.

Let $p(x)$ represent some function of argument x that takes nonnegative integer values and $p(x+1)$ the next higher value. The first difference of $p(x)$ is then defined as

$$\Delta p(x) = p(x+1) - p(x) \tag{1.116}$$

Following this procedure the higher order differences can be identified and we can write the general k^{th} order difference equation

$$\Delta^{k+1} p(x) = \Delta^k p(x+1) - \Delta^k p(x), \qquad k > 0 \tag{1.117}$$

It is customary to refer to $p(x+1)$ by $Ep(x)$ where E is referred to as a displacement operator. The general Equation (1.117) can then be written in terms of the operator

$$E^k p(n) = p(n+k) \tag{1.118}$$

in which Δ can be identified as equal to $(E-1)$. It should be noted that E in these equations is different from the E operator used in defining the mean or the expectation value of the function. However, in view of the common usage of this terminology, we continue to use the same symbol for both the operators, and identification of the proper operation implied is usually a trivial task.

It may be noted from the above discussion that a difference equation represents an equation that involves the functions evaluated at arguments which differ by some fixed number of values. A general k^{th} order homogeneous difference equation is written as

$$p(x) + a_1 p(x+1) + a_2 p(x+2) + \cdots a_k p(x+k) = 0 \tag{1.119}$$

or simply in terms of the E operator as

$$q(E)p(x) = 0 \tag{1.120}$$

where $q(E)$ is given by

$$q(E) = 1 + a_1 E + \cdots a_k E^k \tag{1.121}$$

and is referred to as the characteristic equation. Suppose that $m_1, m_2, \ldots m_k$ represent the real distinct roots of the characteristic

Equation (1.121); the general solution to the equation is given by

$$p(x) = C_1 m_1^x + C_2 m_2^x + \cdots C_k m_k^x \tag{1.122}$$

where the various C's can be evaluated subject to appropriate initial conditions. In the event some of the roots are equal (say $m_1 = m_2$), the solution becomes

$$p(x) = (C_1 + C_2 x) m_1^x + C_3 m_3^x + \cdots C_k m_k^x \tag{1.123}$$

Also, when the roots are complex, the general solution could be reduced to the corresponding form involving trigonometric arguments.

It is apparent from the above that a close similarity between the theories of difference and ordinary differential equations (ODE) exists. Thus the solution of a nonhomogeneous difference equation, following a procedure analogous to that for ordinary differential equations, can be written as the sum of the solution for the corresponding homogeneous equation and the particular solution of the complete equation. The method of obtaining the particular solution, as in the case of ODE, depends largely upon the form of the perturbing function. We shall indicate in the following examples some of the different methods that can be used to solve these equations.

Example 1.1

Find the probability mass function p_x of a random variable X satisfying the following difference equations:

$$p_{x+2} - (1 + a)p_{x+1} + ap_x = 0, \qquad x = 0, 1, 2, \ldots \tag{1}$$

$$-p_1 + ap_0 = 0, \qquad 0 < a < 1 \tag{2}$$

Method of characteristic equation:

The second order difference equation can be written in terms of the difference operator as

$$q(E)p_x = 0 \tag{3}$$

where the characteristic equation $q(E)$ can be identified as

$$m^2 - (1 + a)m + a = 0 \tag{4}$$

and possesses the following two roots:

$$m_1 = a, \qquad m_2 = 1 \tag{5}$$

The general solution to the equation can then be obtained as

$$p_x = C_1 + C_2 a^x \tag{6}$$

The constants of the equation can be evaluated from the given condition (2):

$$p_1 = C_1 + C_2 a = a p_0 \tag{7}$$

Also Equation (6) gives

$$p_0 = C_1 + C_2 \tag{8}$$

Solving (7) and (8) simultaneously the constants can be evaluated:

$$C_1 = 0, \qquad C_2 = p_0 \tag{9}$$

The general solution can then be written:

$$p_x = p_0 a^x \tag{10}$$

The unknown probability p_0 can be evaluated from the consideration that $\sum p_x = 1$ and obtained as $p_0 = (1 - a)$. We have therefore

$$p_x = (1 - a)a^n \tag{11}$$

Method of induction:

Equation (1) can be alternatively written as

$$p_{x+1} - (1 + a)p_x + a p_{x-1} = 0 \tag{12}$$

where $p(x + 1)$ has been replaced by p_x; further replacing x by $(x - 1)$ we obtain

$$p_x - (1 + a)p_{x-1} + a p_{x-2} = 0 \tag{13}$$

and the procedure can be continued.

Equation (12) can be rearranged as

$$p_{x+1} - a p_x = p_x - a p_{x-1} \tag{14}$$

where the rhs can be redefined from Equation (12) in terms of lower order probability difference. The procedure is continued to finally obtain

$$p_{x+1} - a p_x = p_x - a p_{x-1} = p_{x-1} - a p_{x-2} = \cdots = p_1 - a p_0 \tag{15}$$

Utilising the imposed condition (2), we have

$$p_x = a p_{x-1} = a(a p_{x-2}) = \cdots = a^n p_0 \tag{16}$$

Since the probability mass function should satisfy $\sum p_x = 1$, we can obtain $p_0 = 1 - a$, so that

$$p_x = (1 - a)a^n, \qquad n \geqslant 0 \tag{17}$$

which is identical to that obtained using the method of characteristics.

Method of generating function:

For the case of a discrete random variable, the generating function defined by Equation (1.16) can be obtained as

$$\psi(s) = \sum_{x=0}^{\infty} p_x s^x \tag{18}$$

Multiplying Equation (1) by s^x $(x > 1)$ and adding over $x = 1, 2, \ldots,$ we get

$$\sum_{x=1}^{\infty} p_{x+1} s^x - (1 + a) \sum_{x=1}^{\infty} p_x s^x + a \sum_{x=1}^{\infty} p_{x-1} s^x = 0 \tag{19}$$

which can be rearranged using definition (18) as

$$\frac{1}{s}[\psi(s) - p_0 - p_1 s] - (1 + a)[\psi(s) - p_0] + as\psi(s) = 0 \tag{20}$$

or as

$$\psi(s) = \frac{p_0}{1 - as} = p_0(1 + as + a^2 s^2 + \cdots) \tag{21}$$

The coefficient of s^x, according to Equations (16) and (21), can now be identified:

$$p_x = p_0 a^x \tag{22}$$

Finding p_0 as before from the requirement that $p_x = 1$ we have finally

$$p_x = (1 - a)a^x \qquad (x \geqslant 0)$$

The method of generating function is a powerful tool that can also be used to solve inhomogeneous difference equations or equations where the coefficients are not constant but functions of the argument. The general technique can also be used to obtain the solution to certain differential-difference equations. We shall indicate this in the next example.

Example 1.2

Find the probability function satisfying the following equations:

$$p'_x(t) = -\lambda[p_x(t) - p_{x-1}(t)], \qquad x \geqslant 1, \quad \lambda > 0 \tag{1}$$

$$p'_0(t) = -\lambda p_0(t) \tag{2}$$

subject to initial conditions

$$p_x(0) = 0, \qquad x > 0 \qquad \text{and} \qquad p_0(0) = 1 \tag{3}$$

The two commonly used methods for obtaining the solution to differential-difference equations are (1) the method of Laplace transform, and (2) the method of generating function. In the present example we shall obtain the solution using these two methods.

Method of Laplace transform:

Defining the Laplace transform of probability function as

$$\mathscr{L}p_x(t) = \int_0^\infty e^{-st}p_x(t)\,dt = f_x(s) \tag{4}$$

and employing the definition in Equation (1) we get

$$sf_x(x) = \lambda[f_{x-1}(s) - f_x(s)] \tag{5}$$

Note that the use of Laplace transform has reduced the differential-difference equation to an ODE which can be rearranged as

$$f_x(s) = \frac{\lambda}{\lambda + s} f_{x-1}(s), \qquad x \geqslant 1 \tag{6}$$

Similarly, taking the Laplace transform of Equation (2) and using the appropriate initial condition, we get

$$f_0(s) = 1/(s + \lambda) \tag{7}$$

The solution to the difference Equation (6) can be obtained as

$$f_{x-1}(s) = C_1\left(\frac{\lambda}{\lambda + s}\right)^{x-1}, \qquad x \geqslant 1 \tag{8}$$

where C_1 can be evaluated using Equation (7) in Equation (8) as equal to $(1/s + \lambda)$. We have therefore the final solution:

$$f_{x-1}(s) = \frac{\lambda^{x-1}}{(\lambda + s)^x}, \qquad x \geqslant 1$$

or

$$f_x(s) = \frac{\lambda^x}{(\lambda + s)^{x+1}}, \qquad x \geqslant 0 \tag{9}$$

Inverting Equation (9) we obtain the probability function:

$$p_x(t) = \frac{\exp(-\lambda t)(\lambda t)^x}{x!}, \qquad x = 0, 1, 2, \ldots \tag{10}$$

Method of generating function:

Following the general definition of generating function we can write

$$\begin{aligned} \psi(s, t) = E(s^{X(t)}) &= \sum_{x=0}^{\infty} \text{Prob}\{X(t) = X\} s^x \\ &= \sum_{x=0}^{\infty} p_x(t) s^x \end{aligned} \tag{11}$$

The generating function can be differentiated as per the instructions set down in the previous section to obtain

$$\frac{\partial}{\partial t} \psi(s, t) = \sum_{x=0}^{\infty} \frac{\partial}{\partial t} p_x(t) s^x = \sum_{x=0}^{\infty} p'_x(t) s^x \tag{12}$$

where

$$\sum_{x=0}^{\infty} p'_x(t) s^x = \sum_{x=1}^{\infty} p'_x(t) s^x + p'_0(t) \tag{13}$$

Similarly, we can write

$$\sum_{x=0}^{\infty} p_x(t) s^x = \sum_{x=1}^{\infty} p_x(t) s^x + p_0(t) \tag{14}$$

where, due to Equation (11), we have

$$\sum_{x=1}^{\infty} p_x(t) s^x = \psi(s, t) - \psi_0(s, t) \tag{15}$$

Also,

$$\sum_{x=1}^{\infty} p_{x-1}(t) s^x = s\psi(s, t)$$

Let us now multiply Equation (1) by s^x and add over for $x = 1, 2, \ldots$. Further, making use of the several relations derived above, the equation can be rewritten as

$$\frac{\partial}{\partial t}\psi(s,t) = \psi(s,t)\{\lambda(s-1)\} \tag{16}$$

the solution to which is

$$\psi(s,t) = e^{-\lambda t}\left\{\sum_{x=0}^{\infty}\frac{(\lambda st)^x}{x!}\right\} \tag{17}$$

so that the coefficient of s^x in Equations (11) and (18) can be identified as

$$p_x(t) = \frac{e^{-\lambda t}(\lambda t)^x}{x!}, \qquad x \geqslant 0 \tag{18}$$

As would be noted, the use of generating function reduces the differential-difference equation to an ordinary differential equation which can then be readily solved.

1.6 Common Probability Distributions

In this section we shall present the probability distributions that are frequently invoked to treat the physical processes. Many of these distributions have already been discussed in great detail (see, for example, Feller, 1968; Populis, 1965).

The properties of these distributions along with the values for different arguments have also been variedly tabulated (see, for example, Abramovitz and Stegun, 1964). For the sake of completeness, however, we present in Table 1.1, the basic probability mass or density functions for these distributions, along with the important moments.

1.7 Illustrative Examples

The previous sections were concerned with a brief discussion of the basic concepts in probability theory formulated in terms of mathematical

TABLE 1.1
Some common probability density and distribution functions

Type of distributions	Density function	Distribution function	Moments
Normal Gaussian	$\frac{1}{\sigma\sqrt{2\pi}}\exp\left\{-\frac{1}{2}\left(\frac{x-\mu}{\sigma}\right)^2\right\}$ $\infty<x<\infty$	$\frac{1}{\sqrt{2\pi}}\int_{-\infty}^{\infty}\exp\left\{-\frac{1}{2}\left(\frac{x-\mu}{\sigma}\right)^2\right\}\frac{\mathrm{d}x}{\sigma}$	$\mu_1=\mu$ $\mu_2=\sigma^2+\mu^2$ $\mu_3=3\sigma^2\mu+\mu^3$
Gamma	$\frac{1}{\sigma\Gamma(\lambda)}\left(\frac{x}{\sigma}\right)^{\lambda-1}\exp-\left(\frac{x}{\sigma}\right)$ $0<x<\infty$	$\frac{1}{\Gamma(\lambda)}\int_0^x\left(\frac{x}{\sigma}\right)^{\lambda-1}\exp\left\{-\frac{x}{\sigma}\right\}\frac{\mathrm{d}x}{\sigma}$	$\mu_1=\frac{n}{\sigma}$
Exponential	$\frac{1}{\sigma}\exp\left\{-\frac{x}{\sigma}\right\}$ $0\leqslant x<\infty$	$1-\exp\left\{-\frac{x}{\sigma}\right\}$	$\mu_r=r+\sigma^r$
Chi squared	$\frac{1}{2\Gamma(\nu/2)}\left(\frac{x}{2}\right)^{\nu/2-1}\exp\left\{-\frac{x}{2}\right\}$ $0<x<\infty$	$\frac{1}{\Gamma(\nu/2)}\int_0^x\left(\frac{x}{2}\right)^{\nu/2-1}\exp\left\{-\frac{x}{2}\right\}\frac{\mathrm{d}x}{2}$	$\mu_1=\nu$ $\mu_2=2\nu$
Poisson	$\frac{(ax)^n}{n!}e^{-ax}$ $x\geqslant 0$	$\int^x p(x)\,\mathrm{d}x$	$\mu_1=K_2=ax$ $\mu_2=\mu_1^2+\mu_1$
	$\frac{1}{\mu_2-\mu_1}$ $\mu_1\leqslant x\leqslant\mu_2$	$\frac{x-\mu_1}{\mu_2-\mu_1}$	$\mu_r'=\frac{\mu_2^{r+1}-\mu_1^{r+1}}{(r+1)(\mu_2-\mu_1)}$ $\mu_2=(\mu_2-\mu_1)^2/12$ $\mu_3=0$

equations. While more elaborate discussions of these concepts can be found in the references cited at the beginning of this chapter, it is considered desirable to illustrate some of these concepts by solving specific examples.

Example 1.3

The probability density function of a continuous random variable X is given by

$$p(x) = 1 \qquad 1 \leqslant x \leqslant 0$$

$$p_{(x)} = 0 \qquad \text{Otherwise}$$

Calculate the moments using the method of Laplace transform and generating function.

Laplace transform:

For a continuous random variable X possessing a probability density function $p(x)$, the distribution function can be written as

$$\text{Prob}\{X \leqslant x\} = F(x) = \int_0^\infty p(x)\,dx \tag{1}$$

Taking the Laplace transform of $F(x)$ and $p(x)$ we have

$$\mathscr{L}[F(x)] = \int_0^\infty e^{-st} F(x)\,dx \tag{2}$$

and

$$\mathscr{L}[p(x)] = \int_0^\infty e^{-st} p(x)\,dx = \int_0^\infty e^{-st}\,dF(x) = p^*(s) \tag{3}$$

where we have denoted Equation (3) by $p^*(s)$. Equation (2) can now be integrated by parts and rearranged to obtain

$$\mathscr{L}[F(x)] = p^*(s)/s \tag{4}$$

Equation (3) can be differentiated with respect to s and the general n^{th} order differential can be written as

$$\frac{d^n}{ds^n} p^*(s) = (-1)^n \int_0^\infty (x)^n e^{-st}\,dF(x) \tag{5}$$

Note from Equation (1.12) in the text that the rhs of Equation (5) represents $E(x^n)$ provided $s = 0$. We can thus relate the moments to the

Laplace transform of the density function as

$$E(X^n) = \left[(-1)^n \frac{d^n}{ds^n} p^*(s) \right]_{s=0} \tag{6}$$

For the given probability distribution $p(x)$, using Equation (3) above we can calculate $p^*(s)$ thus:

$$p^*(s) = \int_0^\infty e^{-sK} p(x)\, dx = \int_0^1 e^{-sK} 1\, dx + \int_1^\infty e^{-sK}(0)\, dx$$
$$= (1 - e^{-s})/s \tag{7}$$

The first and second noncentral moments can then be calculated from

$$E(X) = -\frac{d}{ds}[p^*(s)] = -\left\{ \frac{d}{ds} \frac{1 - e^{-s}}{s} \right\}_{s=0} = \tfrac{1}{2} \tag{8}$$

$$E(X^2) = \frac{d^2}{ds^2} p^*(s) = \frac{d}{ds^2} \left[\frac{1 - e^{-s}}{s} \right]_{s=0} = \tfrac{1}{3} \tag{9}$$

and the variance from

$$\text{Var}[X] = E(X^2) - [E(X)]^2 = \tfrac{1}{3} - \tfrac{1}{4} = \tfrac{1}{12} \tag{10}$$

Generating function method:

The generating function of the distribution is given by

$$\psi(s) = E(e^{sX}) = \int_0^\infty e^{-sX} p(x)\, dx$$
$$= \frac{(1 - e^{-s})}{s} \tag{11}$$

which can be expanded in series as

$$\psi(s) = 1 + \frac{s}{2!} - \frac{s^2}{3!} + \cdots \tag{12}$$

The coefficients of s^n $(n = 1, 2, \ldots)$ etc. define the appropriate moments.

One can easily see from the probability density function that this example provides an illustration of uniform distribution.

Example 1.4

A continuous random variable X obeys the normal distribution

$$p(x) = \frac{1}{\sqrt{2\pi\sigma}} \exp\left[-\frac{1}{2}\left(\frac{x-\mu}{\sigma}\right)^2\right] \tag{1}$$

Find its characteristic function and significant moments.

The characteristic function $\phi(t)$ is given by

$$\phi(t) = E(e^{itX}) = \frac{1}{\sqrt{2\pi\sigma}} \int_0^\infty e^{itx} \exp\left[-\frac{1}{2}\left(\frac{x-\mu}{\sigma}\right)^2\right] dx \tag{2}$$

where the indicated integration can be performed to obtain

$$\begin{aligned} \phi(t) &= e^{-\mu^2/2\sigma^2}\left[\exp\left\{\frac{\sigma^2(it+\mu/\sigma^2)^2}{2}\right\}\right] \\ &= e^{it\mu}\, e^{-t^2\sigma^2/2} \end{aligned} \tag{3}$$

Equation (3) can be expanded in series to obtain

$$\phi(t) = 1 + it\mu - \frac{t^2\sigma^2}{2} - t^2\mu^2 \tag{4}$$

where the noncentral moments ($\mu_1, \mu_2, \mu_3, \mu_4$, etc.) can be easily identified as

$$\begin{aligned} &\mu_1 = \mu, \qquad \mu_2 = \sigma^2 + \mu^2, \qquad \mu_3 = 3\sigma^2\mu + \mu^3, \\ &\mu_4 = 3 + 6\mu^2\sigma^2 + \mu^4 \ldots \quad \text{etc.} \end{aligned} \tag{5}$$

Utilising the relationship between the central and noncentral moments we can obtain the central moments:

$$\begin{aligned} K_1 &= 0 \\ K_2 &= \mu_2 - \mu_1^2 = \sigma^2 \\ K_3 &= \mu_3 - 3\mu_2\mu_1 + 2\mu_1^3 = 0 \end{aligned} \tag{6}$$

In fact one can easily observe that all odd central moments become zero for this distribution implying a symmetrical distribution. Likewise the even central moments can be written in terms of the following recursive relation:

$$K_{2k} = 1.3.5\ldots(2k-1)\sigma^2 \tag{7}$$

and for odd moments we have

$$K_{2k+1} = 0 \tag{8}$$

The use of cumulants is especially suitable for this distribution. Thus

$$\alpha = \mu, \qquad \alpha_2 = \sigma^2, \qquad \alpha_3 = \alpha_4 = \cdots = 0 \tag{9}$$

All higher order cumulants ($>\alpha_3$) for this distribution vanish.

The results obtained here could be generalised to a sum of n independent random variables $Y = X_1 + X_2 \ldots X_{n1} \ldots$ each of which is a Gaussian variable. Following the observation that the characteristic function of Y can be written as the product of the characteristic functions of the individual random variables, we have

$$\phi(Y) = \phi(X_1)\phi(X_2)\ldots\phi(X_n)$$

The mean and variance of Y can thus be written as the sum of the mean and the variances of the individual variables. The random variable Y also then follows the Gaussian distribution.

Example 1.5

The joint probability mass function of two random variables X and Y is given: $p_{XY}(j, k) = (1 - m)^{j+k}m^2$. Find the joint probability and marginal probability generating functions.

Given the joint probability mass function $p_{XY}(j, k)$ we can write the joint probability generating function as

$$\psi(j, k) = \sum_{j=0}^{\infty} \sum_{k=0}^{\infty} S_1^j S_2^k p_{XY}(j, k) \tag{1}$$

$$= \sum_{j=0}^{\infty} \sum_{k=0}^{\infty} S_1^j S_2^k (1 - m)^{j+k} m^2 \tag{2}$$

$$= m^2 \sum_{j=0}^{\infty} [S_1(1 - m)]^j \sum_{k=0}^{\infty} [S_2(1 - m)]^k \tag{3}$$

$$= m^2\{1 + S_1(1 - m) + [S_1(1 - m)]^2 + \cdots\} \times \{1 + S_2(1 - m) + [S_2(1 - m)]^2 + \cdots\} \tag{4}$$

$$= \frac{m^2}{[1 - S_1(1 - m)][1 - S_2(1 - m)]} \tag{5}$$

The marginal probability generating function for X and Y is likewise given by

$$\psi_x(j) = \sum_{k=0}^{\infty} p_{XY}(j)S_1^j, \qquad j = 0, 1, 2, \ldots \tag{6}$$

$$\psi_y(k) = \sum_{j=0}^{\infty} p_{XY}(k)S_2^k, \qquad k = 0, 1, 2, \ldots \tag{7}$$

It will be noticed that Equations (6) and (7) can be obtained from Equation (1) by setting S_2 and S_1 each equal to 1. We have therefore from Equation (15)

$$\psi_x(j) = \frac{m}{[1 - S_1(1 - m)]}, \qquad \psi_y(k) = \frac{m}{1 - S_2(1 - m)} \tag{8}$$

The corresponding marginal probability mass functions are given by

$$p_X(j) = \sum_{k=0}^{\infty} p_{XY}(j, k) = m^2(1 - m)^j \sum_{k=0}^{\infty} (1 - m)^k, \qquad k = 0, 1, 2, \ldots$$

$$p_Y(k) = \sum_{j=0}^{\infty} p_{XY}(j, k) = m^2(1 - m)^k \sum_{j=0}^{\infty} (1 - m)^j, \qquad j = 0, 1, 2, \ldots$$

The conditional distributions of X and Y are thus

$$p_{XY}(jk) = \frac{p_{XY}(j, k)}{p_Y(k)} = \frac{(1 - m)^{j+k}m^2}{m(1 - m)} = m(1 - m)^{j+k-1}$$

$$p_{YX}(kj) = \frac{p_{XY}(j, k)}{p_X(j)} = m(1 - m)^{j+k-1}$$

REFERENCES

Abramovitz, M. and Stegun, L. A., *Handbook of Mathematical Functions*, U.S. Department of Commerce, National Bureau of Standards, 1964.

Bury, K. V., *Statistical Models in Applied Science*, Wiley, New York, 1975.

Dixon, W. J. and Masseyl, F. J. Jr., *Introduction to Statistical Analysis*, 2nd Ed., McGraw-Hill, New York, 1957.

Feller, W., *An Introduction to Probability Theory and Its Applications*, 3rd Ed., Vol. I, Wiley, New York, 1968.

Fisz, N., *Probability Theory and Mathematical Statistics*, 3rd Ed., Wiley, New York, 1963.

Johnson, N. L. and Leone, F. C., *Statistics and Experimental Design*, Vols. I and II, Wiley, New York, 1964.

Kendall, M. G. and Stuart, A., *The Advanced Theory of Statistics*, 2nd Ed., Vol. I, Hafner, New York, 1963.

Kendall, M. G. and Stuart, A., *The Advanced Theory of Statistics*, Vol. III, Hafner, New York, 1967.

Kendall, M. G. and Stuart, A., *The Advanced Theory of Statistics*, Vol. II, Hafner, New York, 1966.

Lindley, D. V., *Introduction to Probability and Statistics*, Vols. I and II, Cambridge University Press, 1965.

McShane, E. J., *Stochastic Calculus and Stochastic Models*, Academic Press, New York, 1974.

Papoulis, A., *Probability, Random Variables and Stochastic Processes*, McGraw-Hill, New York, 1965.

Parzen, E., *Modern Probability Theory and Its Applications*, Wiley, New York, 1960.

Seinfeld, J. M. and Lapidus, L., *Mathematical Methods in Chemical Engineering, Vol. 3, Process Modeling, Estimation and Identification*, Prentice-Hall, Englewood Cliffs, New Jersey, 1975.

2 Markov Processes and Master Equation

THE PREVIOUS chapter was concerned with the formulation and general properties of stochastic processes. In the present chapter, we shall examine a special class of stochastic processes that are referred to as homogeneous Markov processes. A convenient mathematical description of the class of homogeneous Markov processes is the so-called master equation. Thus this chapter generally discusses these Markov processes and their representation in terms of the master equation. It may be mentioned that in applying the theory of homogeneous Markov processes to chemically reacting systems, a complete presentation of the Markov processes as a whole is unnecessary. What is required for this purpose is a description of these processes as they are relevant to a probabilistic analysis of reacting systems. Such a presentation follows in the sequel. For a more complete description of the Markov processes *per se* reference may be made to Bharucha-Reid (1960), Doob (1953), Feller (1968), Fisz (1963), Gardiner (1983), Howard (1971), Karlin (1966), King (1971), Mortensen (1969), Papoulis (1965), Stratonovich (1966), Seinfeld and Lapidus (1974), van Kampen (1981).

2.1 Markov Processes—General Introduction

Inherent in the modelling of system behaviour employing the deterministic formalism is the notion that the factors determining the evolution of the system to the next state are entirely contained in the prevailing state. One is logically tempted to ask whether a similar notion can be imbedded in the use of the probabilistic formalism in describing the system behaviour. The intention is clear. From a knowledge of the current state of the system we would like to deduce the probability that the system will move to a given state at some later time, the process or the time-history of how the current state was reached being immaterial. In other words, we would like to describe a random phenomenon, whenever possible, as a stochastic process which is visualised as a series of transitions between different states of the system and which possesses the important property of exclusive determinism as regards its evolution to the next stage purely from a knowledge of the current state. Processes that subscribe to this condition are referred to as Markov processes and form an important and practically useful class of general stochastic processes.

The Markov processes can be further subclassified depending on the nature of the state-space (X) and the ordering parameter (t). Thus discrete state–discrete time processes that satisfy the Markovian assumption are said to form a *discrete parameter Markov chain*, while the discrete state–continuous time processes are said to form a *continuous parameter Markov chain*. Similarly, continuous state–discrete time and continuous state–continuous time processes are correspondingly referred to as *discrete parameter Markov processes* and *continuous parameter Markov processes*. The present section discusses the properties and general formulation of these processes.

Before we begin with the formal description of these processes, a few points that have general validity may be mentioned. First, we must appreciate that the Markovian assumption is more of a mathematical convenience and real systems may be described only approximately as Markov processes. This necessitates the choice of a proper variable to denote the state-space. To illustrate the point we can think of a simple decomposition type of reaction where the reactant breaks up due to collisions. Thus, if we were to monitor the concentration of the reactant, the change in concentration between times t and $t + \Delta t$ has a certain probability distribution that would depend only on the concentration at time t. The mode by which this concentration was reached, or the previous concentration history, is immaterial. Thus a process described with concentration as a random variable can be treated as a Markov process. Let us now suppose that such a reaction occurs in presence of a catalyst that loses its activity in proportion to the concentration of reactant it processes. Clearly, the change in concentration now depends not only on the concentration at that instant, but also on the activity of the catalyst, which requires a knowledge of the concentration history. Concentration cannot any more then be used as a random variable to describe the system as a Markov process. It is possible, however, to treat the system as a two-component (concentration-activity) Markov process.

The discussion presented above brings to light an additional point—that the Markov process need not necessarily be restricted to a single variable case but can be generalised for an arbitrary n-variable case. As typical example, one could think of an n-component reacting mixture where concentrations of the n species could be chosen to represent the Markov process. It is necessary to choose all n components. In a random process with n components, if we were to monitor, say, any r components ($r < n$) the resultant r-component process is still random. However, for a Markovian process the corresponding r-component process, in general, would not be Markovian. It would not be possible to describe the evolution of an n-component process from current knowledge of only r components.

The choice of random variables is therefore critical, and it is meaningless to ask whether a given process is Markovian unless the variables are specified. Alternatively, when we describe a process as Markovian, it is also necessary to specify the variable(s). In reality any closed system can be described as a Markovian process by specifying all its microvariables. In view of the complexities of such description, our efforts are always to find a few variables of the system that can be used to describe it as a Markov process. Such a description, necessarily approximate, yields only coarse grained information—which is often sufficient for most practical systems.

With these general comments regarding Markov processes, we shall now turn to their formal description.

2.2 Discrete Parameter Markov Chain

To illustrate the basic concept, let us consider an experiment which has only two possible outcomes, such as tossing a coin. We further specify that the probability of one outcome, say head, is given by m. The requirement of total probability equal to one specifies the probability of achieving the second outcome, say tail, as $(1 - m)$. Let us assume that the experiment is repeated n number of times and the possible outcomes are listed as $X_1, X_2, \ldots X_n$, where X denotes the random variable specifying the state of the system and the subscripts $1, 2, \ldots n$ denote the trial number. The sequence of random variables thus formed does not form a Markov chain, for the outcome of the $(n + 1)^{\text{th}}$ trial does not depend on the nth trial.

Let us now form a partial sum of these independent random variables and denote it by

$$Y_n = X_1 + X_2 \cdots + X_n = j \tag{2.1}$$

In the $(n + 1)^{\text{th}}$ experiment, we know that the result could be either head or tail. If to each head and tail we assign values of 1 and 0, respectively, then we know that X_{n+1} is either 1 or 0. The partial sum given by Equation (2.1) after the $(n + 1)^{\text{th}}$ experiment would therefore take a value of either j or $(j + 1)$. The outcome of the $(n + 1)^{\text{th}}$ experiment for the sum thus depends on the value j of the nth stage, and the chain so formed constitutes a Markov chain.

The probability of outcome of head or tail at the $(n + 1)^{\text{th}}$ experiment, being independent of the previous trials, can be represented as (m) and $(1 - m)$, respectively. In mathematical terms, we can summarise the results as

$$\text{Prob}\{ Y_{n+1} = j + 1 \mid Y_n = j\} = m \tag{2.2}$$

$$\text{prob}\{ Y_{n+1} = j \mid Y_n = j\} = (1 - m) \tag{2.3}$$

It is seen that the conditional probability of Y_{n+1} given Y_n depends only on Y_n and how one arrives at Y_n is of no consequence. The simple example illustrates the case of a discrete parameter–discrete state Markov chain.

2.2.1 Transition Probability

For a more formal definition, let us consider the stochastic process $X(t)$ where both X and t assume discrete integer values. The random variable X observed at a discrete set of times and denoted by $X_0, X_1, \ldots X_n$ represents the state of the system at times indicated by the subscripts. Thus X_n represents the state of the system at time n. In general, description of the sequence of random variables thus formed would require specifying its joint probability mass function. If, however, the system satisfies the Markov property the situation gets considerably simplified. Thus for any set of k time points $1, 2, \ldots k$ the conditional distribution of X_k given the values $x_1, x_2, \ldots x_{k-1}$ of the remaining variables depends only on x_{k-1}, and we can write

$$[\text{Prob } X_k = x_k \mid X_0 = x_0, X_1 = x_1, \ldots X_{k-1} = x_{k-1}]$$
$$= \text{Prob}[X_k = x_k \mid X_{k-1} = x_{k-1}] \quad (2.4)$$

The single value observed in the last step contains as much information about the future evolution of the system as do the values for the entire previous steps. This conditional probability mass function is referred to as the transition probability and is written as $p^{(k)}_{x_{k-1}|x_k}$ where the superscript k denotes the value at the kth step.

As is evident from Equation (2.4) the outcome of a kth trial does not have a fixed probability Prob$\{X_k = x_k\}$, but is defined in terms of the conditional probability that involves a pair of states (x_{k-1}, x_k). It represents the probability of transition from the state x_{k-1} at the $(k-1)^{\text{th}}$ trial to the state x_k at the kth trial. The transition probabilities are thus basic to defining the chain.

The example cited at the beginning of this section possesses only two states (0 and 1) as the possible outcomes for any trial. Each one of the random variables $X_1, X_2, \ldots X_{n-1}, X_n$ can therefore possess either of these states. Our question is: if at the $(n-1)^{\text{th}}$ trial the state occupied is 0 or 1 what is the transition probability that at the nth trial the state occupied will be 0 or 1. The transition probability matrix for the two-state Markov chain can then be written as

$$P = \begin{bmatrix} p_{00} & p_{01} \\ p_{10} & p_{11} \end{bmatrix} \quad (2.5)$$

The simple arguments can be generalised for an M-state Markov chain wherein we would need to specify the $M \times M$ matrix of transition probabilities defined as

$$P = \begin{bmatrix} p_{11} & p_{12} & \cdots & p_{1M} \\ p_{21} & p_{22} & \cdots & p_{2M} \\ p_{M1} & p_{M2} & \cdots & p_{MM} \end{bmatrix} \quad (2.6)$$

In the examples of the two-state or M-state Markov chain, if the transition probabilities do not depend on the trial number, it is referred to as a *homogeneous Markov chain.* The Markov chain then has a stationary transition probability. The superscript n in the transition probability $p^{(n)}_{x_{k-1}|x_k}$ is now superfluous and we can write the transition probability merely as $p_{x_{k-1}|x_k}$ or even more simply as p_{ij} where i and j refer to two states. Clearly the transition probabilities must satisfy the relations

$$0 \leqslant p_{ij} \leqslant 1 \qquad 1 \leqslant i, \; j \leqslant M \tag{2.7}$$

$$\sum_{j=1}^{M} p_{ij} = 1 \qquad j = 1, 2, \ldots M \tag{2.8}$$

A square matrix such as the one defined by Equation (2.5) or (2.6), possessing nonnegative elements [in view of Equation (2.7)] and the elements in a row adding up to unity [Equation (2.8)], is referred to as a stochastic or Markov matrix. Additionally, when the elements in a column also add up to unity, the matrix is said to be doubly stochastic. The stochastic matrices possess the following useful properties:

(1) the eigenvalue is unity,
(2) the associated eigenvector is a column vector with every element of it being equal to unity, and
(3) unit spectral radius.

These properties are especially useful in analysing the Markov chain.

The transition probabilities as defined above afford a simple means of computing the joint probability distribution of a sequence of variables $X_0, X_1, \ldots X_n$. Thus the joint probability mass distribution function for the sequence can be written as

$$\text{Prob}\{X_0 = x_0, X_1 = x_1, \ldots X_n = x_n\} = p(x_0, x_1, x_2, \ldots x_n) \tag{2.9}$$

where the lhs could be further written in terms of conditional distribution following Equation (1.60) as

$$\text{Prob}\{X_n = x_n \mid X_{n-1} = x_{n-1}, \ldots X_0 = x_0\} \; \text{Prob}\{X_{n-1} = x_{n-1}, \ldots X_0 = x_0\} \tag{2.10}$$

The procedure could be continued to yield

$$\text{Prob}\{X_n = x_n \mid X_{n-1} = x_{n-1}\} \, \text{Prob}\{X_{n-1} = x_{n-1} \mid X_{n-2} = x_{n-2}\} \; \text{Prob}\{X_0 = x_0\} \tag{2.11}$$

One can easily recognise the several terms in Equation (2.11) as the transition probabilities and the last term as the initial probability distribution. We can thus express the joint probability distribution of random variables in terms of their transition probabilities and initial distributions as

$$p_{n-1|n}\, p_{n-2|n-1} \ldots \operatorname{Prob}\{X_0 = x_0\} \tag{2.12}$$

Specification of the transition probability matrix along with the initial distribution completely specifies the Markov chain $\{X_n, n = 0, 1, \ldots\}$.

2.2.2 Chapman–Kolmogorov Equation

In the discussions presented so far we were concerned with the outcome of an nth experiment or trial given the outcomes of the preceding, i.e. $(n-1)^{\text{th}}$, trial. The transition probabilities defined are thus one-step or unit-step transition probabilities. In many instances, we would be interested in knowing the higher transition probabilities, i.e. if j represents the outcome of an nth experiment, we would like to know the probability $p_{jk}(m)$ that the state k is reached from state j in the $(m+n)^{\text{th}}$ trial. This higher transition probability can also be written as

$$p_{jk}(m) = \operatorname{Prob}\{X_{n+m} = k \mid X_n = j\} \tag{2.13}$$

Note that nowhere does the trial number n appear on the lhs. The chain is therefore homogeneous and $p_{jk}(m)$ are *stationary or constant transition probabilities.* The unit-step or one-step transition probability is a special case of Equation (2.13) when m takes the value of unity.

To illustrate the higher transition probabilities, let us consider the case of $m = 2$. Equation (2.13) can then be written as

$$p_{jk}(2) = \operatorname{Prob}\{X_{n+2} = k \mid X_n = j\} \tag{2.14}$$

The equation implies that the state k is reached from state j in two steps through some intermediate step l. We can therefore write Equation (2.14) more explicitly as

$$\begin{aligned} p_{jk}(2) &= \operatorname{Prob}\{X_{n+2} = k, X_{n+1} = l \mid X_n = j\} \\ &= \operatorname{Prob}\{X_{n+2} = k \mid X_{n+1} = l, X_n = j\} \\ &\times \operatorname{Prob}\{X_{n+1} = l \mid X_n = j\} \\ &= p_{jl}\, p_{lk} \end{aligned} \tag{2.15}$$

The intermediate states l can take values $l = 1, 2, \ldots$ that are mutually exclusive. We need to sum, therefore, over all intermediate states and write the relation as

$$p_{jk}(2) = \sum_l p_{jl} p_{lk} \tag{2.16}$$

The equation for a two-step transition could be generalised for any arbitrary m-step transition as

$$p_{jk}(m + n) = \sum_l p_{jl}(n) p_{lk}(m) \tag{2.17}$$

The equation is commonly referred to as the *Chapman–Kolmogorov equation.*

The result can also be expressed in terms of transition matrices where p_{jk} denote the elements of the unit-step transition and $p_{jk}(m)$ the elements of the m-step transition. For an m-step transition ($m = 2$) we have the transition matrix $p(2)$ whose elements are identified in Equation (2.15). It can be seen from this equation that $P(2)$ can be written as

$$P(2) = P \,.\, P = P^2 \tag{2.18}$$

For the general arbitrary case we can thus write

$$P(m + 1) = P \,.\, P^m = P^m \,.\, P \tag{2.19}$$

$$P(m + n) = P^m \,.\, P^n = P^n \,.\, P^m \tag{2.20}$$

To illustrate this point further, let us revert to the example of the m-state Markov chain, the unit-step transition probability matrix for which is given by Equation (2.6). If the transition from one state to another occurs in n steps, we can write the transition probability matrix as

$$p(n) = \begin{bmatrix} p_{11}(n) & p_{12}(n) & \cdots & p_{1M}(n) \\ p_{21}(n) & p_{22}(n) & \cdots & p_{2M}(n) \\ p_{M1}(n) & p_{M2}(n) & \cdots & p_{MM}(n) \end{bmatrix} \tag{2.21}$$

where the individual $\{p_{ij}(n)\}$ can be identified from Equation (2.16). The rows of $P(n)$ specify the probabilities that starting from the states defined for the particular row, the system would evolve itself to any one of the M states after n transitions. The row vectors $\mathbf{R}(0)$ and $\mathbf{R}(n)$ specifying the initial and final probabilities of the system being in state $1, 2, \ldots M$ can be written as

$$\mathbf{R}(0)P(n) = \mathbf{R}(n) \tag{2.22}$$

where

$$\mathbf{R}(0) = [R_1(0), R_2(0), \ldots R_M(0)]$$

$$\mathbf{R}(n) = [R_1(n), R_2(n), \ldots R_M(n)]$$

and constitute the fundamental Markov chain evolution relation.

2.3 Continuous Parameter Markov Chain

In the previous section we concerned ourselves with stochastic processes the states of which assume discrete values. The order parameter (such as trial number) also takes discrete values. In the present section we shall be concerned with discrete state–continuous parameter stochastic processes that obey the Markovian assumption.

A simple example of discrete state–continuous time stochastic process is the Poisson process. To illustrate this simply, let $Y(t)$ denote the total number of events that occur between times 0 and t. Let us say that the first event occurs at $t = t_1$, second at $t = t_2$, and so on. At every instant the event occurs $Y(t)$ jumps abruptly from its previous value to the new value and stays constant until the next event occurs. The picture typically resembles a staircase function and clearly shows the discrete state–continuous time character of the process. The probability mass function for the process can be written as

$$\text{Prob}\{Y(t) = n\} = p_n(t) \tag{2.23}$$

and represents the probability that the variable $Y(t)$ assumes a value of n. Clearly, $p_n(t)$ is a function of time. Also since n can take only discrete values ($n = 0, 1, 2, \ldots$) we have

$$\sum_{n=0}^{\infty} p_n(t) = 1 \tag{2.24}$$

Let us now assume a discrete state–continuous parameter process and represent it as $X_N(t)$ where X assumes discrete values corresponding to $N = 0, 1, 2, \ldots$, etc. Note that while the time is continuous the changes in state occur only at discrete time points denoted by t_k where k takes the values $0, 1, 2, \ldots N$. Note also that the time ordering is essential and we have $t_0 < t_1 < \cdots t_N$. In describing such a system we would like to know the probability that the system moves to some arbitrary state j at the instant (t_n) conditioned on the previous values at t_{n-1}, t_{n-2}, etc. For systems subscribing to the Markovian assumption we need to know the details only at the previous time point. Thus we can write

$$\text{Prob}\{X(t_n) = j \mid X(t_{n-1}) = i_1, X(t_{n-2}) = i_2, \ldots X(t_0) = i_0\} = \text{Prob}\{X(t_n) = j \mid X(t_{n-1}) = i_1\} \quad (2.25)$$

The rhs of Equation (2.25) can be written more generally as

$$\text{Prob}\{X(t_2) = j \mid X(t_1) = i\} = p_{ij}(t_1, t_2) \qquad t_2 > t_1 \quad (2.26)$$

and is referred to as the transition probability. Clearly the transition probability does not depend on the values of $X(t)$ for times less than t_1. In the event $t_1 = 0$ and $t_2 = t$ the relation gets simplified to

$$\text{Prob}\{X(t) = j \mid X(0) = i\} = p_{ij}(t) \quad (2.27)$$

The function $p_{ij}(t)$ obviously satisfies the relation

$$0 \leqslant p_{ij} \leqslant 1 \qquad \text{for each} \quad i, j \text{ and } t$$

$$\sum_j p_{ij}(t) = 1 \quad (2.28)$$

Following the developments in the previous section, we shall refer to the Markov process as homogeneous (or stationary) if the transition probability $p_{ij}(t_1, t_2)$ depends only on the difference $(t_2 - t_1)$. We can thus write $p_{ij}(t_1, t_2) = p_{ij}(t_2 - t_1)$.

The joint probability mass function of a sequence of random variables $X(t_0), X(t_1), \ldots X(t_n)$ can be likewise expressed in terms of the transition probabilities. For this we define the joint probability mass function as

$$p(i_0, i_1, \ldots i_n) = \text{Prob}\{X(t_0) = i_0, X(t_1) = i_1, \ldots X(t_n) = i_n\} \quad (2.29)$$

where the rhs of Equation (2.29) can be simplified using the relationship between the joint, marginal and conditional distributions. The last could be identified with the appropriate transition probabilities to rewrite Equation (2.29) as

$$p(i_0, i_1, \ldots i_n) = p_{i_0 i_1(t_1 - t_0)}\, p_{i_1 i_2(t_2 - t_1)} \cdots p_{i_{n-1} i_n(t_n - t_{n-1})}\, p_{i_0} \quad (2.30)$$

where p_{i_0} refers to the probability mass function of $X(t_0)$.

It is noted from the general discussion above that after the state change has occurred the system remains in that state until some time when the next change occurs. Let us denote this *waiting time* for the state change to occur by t_w. Obviously our interest lies in estimating the waiting time t_w, given the state of the system. Let us suppose that the initial state (although not necessarily so) of the system at time t_0 is known to us as $\{X(t_0) = i_0\}$. To know the time taken for a change of state to occur (t_w), which is a random variable, let us arbitrarily divide the time scale into two parts $t_0 \rightarrow s$ and $s \rightarrow t$. We are now interested in finding the probability that the waiting time t_w exceeds a value of $(s + t)$ given the initial (or prior) state of the system. Mathematically we can write this relation as

$$\text{Prob}\{t_w > s + t \mid X(t_0) = i\} \tag{2.31}$$

To estimate this probability we should first know the probability that t_w exceeds s, given $X(t_0) = i$. If this probability is > 0, obviously no change has occurred until time s. We can then proceed further to calculate the probability that t_w now exceeds the sum $(s + t)$ given that the initial state is $X(t_0) = i$ and $t_w > s$. The implications of dividing the time scale into two arbitrary parts is now clear and Equation (2.31) can be written as

$$\begin{aligned}&\text{Prob}\{t_w > (s + t) \mid X(t_0) = i\}\\ &\qquad = \text{Prob}\{t_w > (s + t) \mid X(t_0) = i, t_w > s\}\ \text{Prob}\{t_w > s \mid X(t_0) = i\}\end{aligned} \tag{2.32}$$

In view of no change in state occurring at least until time s, we can write $X(t_0) = X(s) = i$. We then have

$$\begin{aligned}&\text{Prob}\{t_w > (s + t) \mid X(t_0) = i\}\\ &\qquad = \{\text{Prob}\ t_w > (s + t) \mid X(s) = i, t_w > s\}\ \text{Prob}\{t_w > s \mid X(t_0) = i\}\end{aligned} \tag{2.33}$$

or, in terms of functional relation,

$$g(s + t) = g(s)g(t) \tag{2.34}$$

where $g(t) = \text{Prob}\{t_w > t \mid X(t_0) = i\}$, $t > 0$. The above relation is satisfied if $g(t)$ has an exponentially decaying distribution. The waiting time for a state change to occur is therefore exponentially distributed and can be arbitrarily defined as $g(t) = \exp(-\lambda t)$ $(\lambda > 0, t > 0)$.

2.3.1 Chapman–Kolmogorov Equation

In the previous section we defined the transition probability $p_{ij}(t+\tau)$ which implies that given the initial state of the system as i, the state changes over to j at $(t+\tau)$. It is possible, however, that in moving from state i to state j, the system may exist in some arbitrary state k at time t. We then have

$$p_{ij}(t+\tau) = \sum_k \text{Prob}\{X(t+\tau) = j,\ X(t) = k \mid X(0) = i\} \qquad (2.35)$$

where the rhs can be written in terms of probability mass function giving

$$p_{ij}(t+\tau) = \sum_k \frac{\text{Prob}\{X(t+\tau) = j,\ X(t) = k,\ X(0) = i\}}{\text{Prob}\{X(0) = i\}}$$

Multiplying and dividing the rhs by $\text{Prob}\{X(0) = i, X(t) = k\}$ we have

$$p_{ij}(t+\tau) = \sum_k \frac{\text{Prob}\{X(0) = i,\ x(t) = k\}}{\text{Prob}\ X(0) = i} \times \frac{\text{Prob}\{X(t+\tau) = j,\ X(t) = k,\ X(0) = i\}}{\text{Prob}\{X(0) = i,\ X(t) = k\}} \qquad (2.36)$$

Employing now the definition of conditional probability, this relation simplifies to

$$p_{ij}(t+\tau) = \sum_k \text{Prob}\{X(t) = k \mid X(0) = i\} \times \text{Prob}\{X(t+\tau) = j \mid X(t) = k,\ X(0) = i\} \qquad (2.37)$$

Since for a Markov process knowledge of the last state alone is adequate, the second term in the above equation can be written as

$$p_{ij}(t+\tau) = \sum_k p_{ik}(t)\, p_{kj}(\tau) \qquad (2.38)$$

and holds for all i, j and $t \geqslant 0$, $\tau \leqslant 0$. The relation is known as the Chapman–Kolmogorov equation. The relation corresponds to Equation (2.17) for the discrete state–discrete time Markov process of the earlier section. It can also be written in terms of the transition probability matrices as

$$P(t+\tau) = P(t)P(\tau) \qquad (2.39)$$

which corresponds to Equation (2.20).

The differential forms of the Chapman–Kolmogorov equation are especially useful in treating the Markov chain. Equation (2.38) thus can be differentiated treating i and t as constants and j and τ as variables to obtain

$$\frac{dp_{ij}(t+\tau)}{d\tau} = \sum_k p_{ik}(t)\frac{d}{d\tau}p_{kj}(\tau) \tag{2.40}$$

where the partial derivative can be expanded to obtain

$$\begin{aligned}\left.\frac{d}{d\tau}p_{kj}(\tau)\right|_{\tau=0} &= \lim_{\Delta\tau\to 0}\frac{p_{kj}(\Delta\tau)-p_{kj}(0)}{\Delta\tau}\\ &= \lim_{\Delta\tau\to 0}\frac{p_{kj}(\Delta\tau)}{\Delta\tau} = \alpha_{kj}\end{aligned} \tag{2.41}$$

or

$$p_{kj}(\Delta\tau) = \alpha_{kj}(\Delta\tau) + O(\Delta\tau) \qquad k \neq j$$

Similarly for $k = j$ we get

$$\begin{aligned}\left.\frac{d}{d\tau}p_{kk}(\Delta\tau)\right|_{\tau=0} &= \lim_{\Delta\tau\to 0}\frac{p_{kk}(\Delta\tau)-p_{kk}(0)}{\Delta\tau}\\ &= \lim_{\Delta\tau\to 0}\frac{p_{kk}(\Delta\tau)-1}{\Delta\tau} = \alpha_{kk}\end{aligned} \tag{2.41}$$

or

$$p_{kk}(\Delta\tau) = 1 + \alpha_{kk}(\Delta\tau) + O(\Delta\tau)$$

where in Equations (2.40) and (2.41) we have used the initial conditions

$$\begin{aligned}p_{kj}(0) &= 0 \qquad k \neq j\\ p_{kk}(0) &= 1 \qquad k = k\end{aligned} \tag{2.42}$$

It is evident from Equations (2.40) and (2.41) that $\alpha_{kj} > 0$ while $\alpha_{kk} < 0$. Also using the relation $\sum_j p_{kj} = 1$ in Equation (2.40) it follows that $\sum_j \alpha_{kj} = -\alpha_{kk} = 0$. The α's are generally referred to as the transition densities and the matrix $p(\alpha)$ as the transition density matrix of the process.

Substituting Equation (2.41) in Equation (2.40) with τ set equal to zero, we get

$$\frac{\partial p_{ij}(t+\tau)}{\partial\tau} = \sum_k p_{ik}(t)\alpha_{kj} \tag{2.43}$$

or, in terms of transition probability matrix, as

$$\frac{\partial p(t)}{\partial \tau} = P(t)P(\alpha) \tag{2.44}$$

which admits solution in the form

$$P(t) = \exp[P(\alpha)t] \tag{2.45}$$

Similarly, treating i and t as variables and j and τ as constants we obtain the differential form

$$\frac{\partial P(t)}{\partial t} = \sum_k \alpha_{ik} p_{kj}(t) \tag{2.46}$$

or, in terms of matrix notation,

$$\frac{\partial p}{\partial t} = P(\alpha)P(t) \tag{2.47}$$

Equations (2.43) and (2.46) are known, respectively, as the forward and backward Kolmogorov equations.

2.3.2 Poisson, Birth, and Birth and Death Processes

In this section we shall be concerned with the general description of some of the processes that belong to the category of discrete state–continuous parameter Markov processes. Besides their theoretical significance, these processes are often invoked to explain a variety of physical phenomena encountered in practice. It is of interest therefore to know the general characteristics of these processes.

Poisson Process

In the example mentioned in the beginning of this chapter $Y(t)$ refers to the total number of events that occurred up to time t. Let us assume that the number of events between times t and $(t + s)$ are independent of the number of events that occurred up to time t. We shall also assume that the probability that a certain number of events occur depends only on the time interval (s) and is independent of the value of t. To simplify the situation, we shall make one more assumption concerning the number of events and specify that for an infinitesimally small interval of time the probability that only one event occurs is proportional to the interval of time, while that for more than one event is at least of the order of s [represented as $O(s)$]. Mathematically we can write this relation as

$$p_1(s) = \lambda s + O(s) \tag{2.48}$$

$$p_k(s) = O(s) \qquad k = 2, \ldots \tag{2.49}$$

Utilising this information in the general relation $\sum p_n(s) = 1$, $n = 0, 1, 2, \ldots$ we obtain

$$p_0(s) = 1 - \lambda s + O(s) \tag{2.50}$$

With these basic postulates let us now derive the differential equation describing the distribution of $p_n(s)$. For this purpose we start with $t = 0$ and on the time scale mark out the times t and $(t + s)$

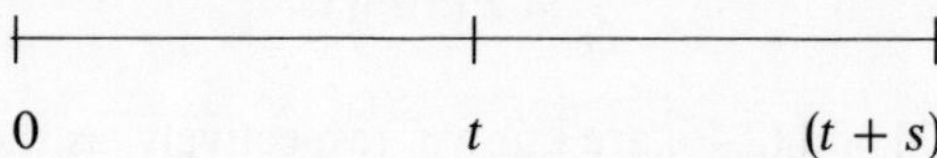

We denote $p_n(t + s)$ as the probability specifying that n events have occurred up to time $(t + s)$. A reflection on the problem indicates that these n events could have occurred in several ways, such as

(1) n events up to time t and no event between t and $t + s$,
(2) $(n - 1)$ events up to time t and one event between t and $t + s$,
(3) $(n - 2)$ events up to time t and two events between t and $t + s$, and so on.

If the n events would have occurred according to the above possibilities, then we can write the corresponding probabilities of steps (1)–(3) as

$$P(1) = \text{Prob}\{Y(t) = n\}\ \text{Prob}\{Y(s) = 0 \mid Y(t) = n\} \tag{2.51}$$

$$P(2) = \text{Prob}\{Y(t) = n - 1\}\ \text{Prob}\{Y(s) = 1 \mid Y(t) = n - 1\} \tag{2.52}$$

$$P(3) = \text{Prob}\{Y(t) = n - 2\}\ \text{Prob}\{Y(s) = 2 \mid Y(t) = n - 2\} \tag{2.53}$$

Equations (2.51)–(2.53) can be simplified as

$$p(1) = p_n(t)p_0(s) \quad = p_n(t)\{1 - \lambda s + O(s)\} \tag{2.54}$$

$$p(2) = p_{n-1}(t)p_1(s) = p_{n-1}(t)\{\lambda s + O(s)\} \tag{2.55}$$

$$p(3) = p_{n-2}(t)p_2(s) = p_{n-2}(t)\{O(s)\} \sim O(s) \tag{2.56}$$

In view of Equation (2.56) the probability of events $\leqslant (n-2)$ occurring up to time t is $O(s)$.

Since all these probabilities are mutually exclusive the total probability will be the sum of all these possibilities. We thus have

$$p_n(t+s) = p_n(t)\{1 - \lambda s\} + p_{n-1}(t)\{\lambda s\} + O(s) \qquad (2.57)$$

which can be rearranged as

$$\frac{p_n(t+s) - p_n(t)}{s} = -\lambda p_n(t) + \lambda p_{n-1}(t) + \frac{O(s)}{s} \qquad (2.58)$$

In the limit $s \to 0$ this relation simplifies to

$$\frac{dp_n(t)}{dt} = -\lambda[p_n(t) - p_{n-1}(t)] \qquad n \geqslant 1 \qquad (2.59)$$

For $n = 0$, from Equation (2.51) or (2.54) we have

$$p_n(t+s) = p_0(t)p_0(s) = p_0(t)[1 - \lambda s + O(s)] \qquad (2.60)$$

or

$$\frac{dp_0(t)}{dt} = -\lambda p_0(t) \qquad (2.61)$$

Equation (2.59) together with (2.61) constitutes the basic equation describing the Poisson process and can be solved using the methods of induction or characteristic function discussed in the earlier chapter. As the worked out example in Section 1.7 indicates, we can obtain

$$p_n(t) = \frac{\exp(-\lambda t)(\lambda t)^n}{n!}, \qquad n \geqslant 0 \qquad (2.62)$$

for the obvious initial conditions $p_n(0) = 0$, $p_0(0) = 1$. Also the noncentral moments of the distribution can be obtained using the generating function. It can be easily verified that the distribution has a mean and variance equal to λt. Thus $E[Y(t)] = \lambda t$ and the relation suggests that the events occur at a mean rate of λ per unit time.

The Poisson processes defined in Equation (2.62) possess some important properties which we would state without proof. The verification of these properties is a trivial task and would mostly require information presented in the earlier chapter. If $Y_1(t)$ and $Y_2(t)$ represent two independent Poisson processes with events occurring at mean rates of λ_1 and λ_2 per unit time, then the sum $Y(t) = Y_1(t) + Y_2(t)$ is also a

Poisson process with mean rate of $(\lambda_1 + \lambda_2)$. The subtraction $[Y_1(t) - Y_2(t)]$, however, does not form a Poisson process. The first two noncentral moments for the case can be easily obtained as

$$E[Y(t)] = (\lambda_1 - \lambda_2)t \qquad \text{and} \qquad E[Y^2(t)] = (\lambda_1 + \lambda_2)t + (\lambda_1 - \lambda_2)^2 t^2$$

A Poisson process can be decomposed into two or more processes each of which represents a Poisson process. To illustrate this let $Y(t)$ denote the total number of events that occurred at a mean rate per unit time of up to time t. If each event has a probability p of being monitored and if $Y_1(t)$ represents the actual number of events recorded, then $Y_1(t)$ forms a Poisson process with parameter λp. The remaining $Y_2(t)$ events also form another Poisson process with parameters $\lambda(1 - p)$. The situation could be generalised to a case where the Poisson process could be decomposed into n different Poisson processes with parameters λp_n, $n = 1, 2, \ldots$ and $\sum p_n = 1$.

If $Y(t)$ represents a Poisson process, then for some time $\tau < t$, the conditional probability that $Y(\tau) = m$ given that $Y(t) = n$ follows a binomial distribution law. Stated mathematically this implies the relation

$$\text{Prob}\{Y(\tau) = m \mid Y(t) = n\} = \binom{n}{m}(\tau/t)^m(1 - \tau/t)^{n-m} \tag{2.63}$$

Also, if $Y(t)$ and $Y(t + \tau)$ represent the total number of events that occurred up to t and $t + \tau$, the autocorrelation between them can be expressed as $(t/t + \tau)^{1/2}$.

Pure Birth Processes

The previous section was concerned with the general classification of the Poisson process where we note that events occur at a mean rate per unit time of λ. The parameter λ is constant and does not depend either on the number of events (n) that have occurred so far or the time t. In a general case it is possible to conceive of a situation where the parameter λ depends on the number of events that have occurred so far (n) and also on the time (t). Such processes are called as pure birth processes, and to reflect the dependence of λ on n we write this parameter as λ_n. Following the arguments for the case of the Poisson process, the probabilities of various events occurring can be written as

$$p_1(s) = \lambda_n s + O(s) \tag{2.71}$$

$$p_k(s) = O(s), \qquad k \geqslant 2 \tag{2.72}$$

$$p_0(s) = 1 - \lambda_n s + O(s) \tag{2.73}$$

It will be noticed that these equations for pure birth processes correspond to Equations (2.48)–(2.50) for the Poisson processes with $\lambda \rightarrow \lambda_n$. Proceeding analogously we can write the probability that n births have occurred up to time $(t + s)$ as

$$p_n(t + s) = p_n(t)\{1 - \lambda_n s\} + p_{n-1}\{\lambda_{n-1} s\} + O(s) \tag{2.74}$$

where the probabilities of all mutually exclusive ways of the birth of n species have been properly accounted for. Equation (2.74) can be rearranged to obtain the differential form

$$\frac{dp_n(t)}{dt} = -\lambda_n p_n(t) + \lambda_{n-1} p_{n-1}(t), \qquad n \geqslant 1 \tag{2.75}$$

and

$$\frac{dp_0(t)}{dt} = -\lambda_0 p_0(t), \qquad n = 0 \tag{2.76}$$

The set of differential-difference Equations (2.75)–(2.76) can be solved using the methods presented in Chapter 1, if the functional dependence of λ_n on n is known along with the initial conditions. A special case of interest arises when λ_n linearly varies with n. We can thus write $\lambda_n = n\lambda$. The special case is known as the Yule–Fury process. Substituting the functional dependence of λ_n on n in Equation (2.75) we can obtain the distribution $p_n(t)$ as

$$p_n(t) = e^{-\lambda t}(1 - e^{-\lambda t})^{n-1}, \qquad n \geqslant 1 \tag{2.77}$$

$$p_0(t) = 0, \qquad n = 0 \tag{2.78}$$

where we have used the initial conditions $p_1(0) = 1, p_n(0) = 0, n \neq 1$. The generating function corresponding to the distribution $p_n(t)$ can be written as

$$\psi(s, t) = \sum_{n=1}^{\infty} e^{-\lambda t}(1 - e^{-\lambda t})^{n-1} s^n \tag{2.79}$$

which, on readjustment and expansion in series, gives the necessary noncentral moments. The mean and variance of the process estimated this way can be written as

$$E\{Y(t)\} = e^{\lambda t}$$

and (2.80)

$$\mathrm{Var}\{N(t)\} = e^{\lambda t}\{e^{\lambda t} - 1\}$$

It is also possible to treat cases where λ is a function of time. In the governing equations for the Poisson process, we now have $\lambda(t)$ instead of λ alone. The governing equations [Equations (2.59) and (2.61)] can be solved for this case with the appropriate initial conditions to obtain the distribution of $p_n(t)$. The probability of n births in time t can be obtained as

$$p_n(t) = \frac{1}{n!}\left[\int_0^t \{\lambda(t)\,dt\}^n \exp\left\{-\int_0^t \lambda(t)\,dt\right\}\right], \qquad n \geqslant 1 \tag{2.81}$$

and

$$p_0(t) = \exp\left\{-\int_0^t \lambda(t)\,dt\right\} = \exp\{-m(t)\} \tag{2.82}$$

The generating function corresponding to Equation (2.80) can be written as

$$\psi(s, t) = \sum p_n(t)s^n = \exp[(mt)(s-1)] \tag{2.83}$$

which gives the mean as $E[Y(t)] = m(t)$.

It is possible, as in the Pólya process, to generalise the basic Poisson process by including the dependence of λ on both n and t. The governing equations now are similar to Equations (2.75) and (2.76) with λ_n and λ_{n-1} dependent also on t. If the functional dependence of λ on n and t is known, these equations can be solved for appropriate initial conditions to obtain the distribution $p_n(t)$ and moments of the process.

Birth and Death Process

In the previous section we were concerned with the pure birth process where the occurrence of an event leads to an increase in the net value of the system. It is possible, however, that the occurrence of an event would also lead to a decrease in the net value of the system. Thus in a system of n species the occurrence of an event such as reaction may lead to both an increase and a decrease in the total population n. We shall refer to these processes as birth and death processes and derive their basic governing equations in the present section.

The conditional probability that m births occur between times t and $(t + s)$ given that n births have occurred up to time t can be written for the pure birth process as $p(m, s \mid n, t)$ where for $m = 1$, 2 and 0, we have Equations (2.71)–(2.73).

Likewise, the conditional probability that m deaths occur during t to $(t + s)$ given that the system population is n at time t can be written as

$$q_1(t) = \mu_n s + O(s) \tag{2.84}$$

$$q_2(t) = O(s) \tag{2.85}$$

$$q_0(t) = 1 - \mu_n s + O(s), \qquad \mu_0 = 0 \tag{2.86}$$

In dealing with birth and death processes one needs to consider Equations (2.71)–(2.73) and (2.84)–(2.86) simultaneously. Let us stipulate that at any instant $(t + s)$ the total population is $\{Y(t + s) = n\}$. As before let us divide the total time scale into two parts $(0 - t)$ and $(t - t + s)$. The total number up to $(t + s)$ may have reached a value of n in several possible ways which are mutually exclusive. Hence we may have a total population of $(n - b + d)$ up to time t followed by b births and d deaths in the interval t to $t + s$. Only values of b and d corresponding to 0 and 1 are of interest since, according to Equations (2.74)–(2.73) and (2.84)–(2.86), the probability of b and d exceeding or equalling 2 is of the order of $O(s)$. We can thus have four possibilities corresponding to $(b, d) = (0, 0)$, $(1, 0)$, $(0, 1)$ or $(1, 1)$. The individual probabilities of these occurrences can be written as follows:

$$p_n(t)\{1 - \lambda_n s + O(s)\}\{1 - \mu_n s + O(s)\} \qquad \text{for} \quad (b, d) = (0, 0) \tag{2.87}$$

$$p_{n-1}(t)\{\lambda_{n-1} s + O(s)\}\{1 - \mu_{n-1} s + O(s)\} \qquad \text{for} \quad (b, d) = (1, 0) \tag{2.88}$$

$$p_{n+1}(t)\{1 - \lambda_{n+1} s + O(s)\}\{\mu_{n+1} s + O(s)\} \qquad \text{for} \quad (b, d) = (0, 1) \tag{2.89}$$

$$p_n(t)\{\lambda_n s + O(s)\}\{\mu_n s + O(s)\} \qquad \text{for} \quad (b, d) = (1, 1) \tag{2.90}$$

The probability that $Y(t) = n$ at $(t + s)$ can then be written as

$$\begin{aligned} p_n(t + s) = p_n(t)\{1 - (\lambda_n + \mu_n)s\} + p_{n-1}(t)\lambda_{n-1} s \\ + p_{n+1}(t)\mu_{n+1} s + O(s) \end{aligned} \tag{2.91}$$

The equation can be rearranged, and letting $s \to 0$ we get

$$\frac{dp_n(t)}{dt} = -(\lambda_n + \mu_n)p_n(t) + \lambda_{n-1}p_{n-1}(t) + \mu_{n+1}p_{n+1}(t), \qquad n \geqslant 1 \tag{2.92}$$

and

$$\frac{dp_0(t)}{dt} = -\lambda_0 p_0(t) + \mu_1 p_1(t) \tag{2.93}$$

Equations (2.92) and (2.93) constitute the basic relations for the birth and death processes. Note that, as in the pure birth process, the parameters λ_n and μ_n can depend on the population size at the instant as well as on the time. If the functional dependence of these parameters on n and t is known, then the equations can be solved subject to appropriate initial conditions. Thus, for the case of linear birth and death rates ($\lambda_n = n\lambda$ and $\mu_n = n\mu$), the equations can be solved to obtain the probability generating function

$$\psi(s,t) = \sum_{n=0}^{\infty} p_n(t)s^n = \left[\frac{\mu[1-\exp(\mu-\lambda)t] - [\mu - \lambda\exp(\mu-\lambda)t]s}{[\lambda - \mu\exp(\mu-\lambda)t] - \lambda[1-\exp(\mu-\lambda)t]s}\right]^{n_0} \tag{2.94}$$

where n_0 is the initial size of the system. Equation (2.94) can be expanded in series in s to obtain the distribution $p_n(t)$.

In the limiting case of $t \to \infty$ the time derivatives in Equations (2.92) and (2.93) can be equated to zero and the set of difference equations can be solved by the method of induction to obtain $p_n = \Gamma(n)/\Gamma j$, $n = 0, 1, 2, \ldots$, $j = 1, 2, \ldots n$, where

$$\Gamma(n) = \frac{\lambda_0 \cdot \lambda_1 \cdot \cdots \cdot \lambda_{n-1}}{\mu_1 \cdot \mu_2 \cdot \cdots \cdot \mu_n}$$

2.4 Continuous Parameter–Continuous State Processes

In the discussion so far, we considered Markov processes with discrete state space. $X(t)$ could therefore assume only integer values. It is possible, however, as in certain cases, that the random variables form a continuous state space. The present section considers such processes with continuous parameter. In view of the continuous state–space nature of such processes, we would have to use the probability density function or the distribution function as against the probability mass function used thus far. The continuous state–discrete time processes constitute a special case of these processes and will be briefly discussed in the next section.

The simple case where both the state and the order parameters are continuous is provided by the example of Brownian motion. In treating cases of this type we note one essential difference as against the cases considered so far. In the previous cases, we note that for a small change Δt in the order parameter, there is either no change in the state of the system or the state changes abruptly. In the present case, on the other hand, a small change Δt results in a corresponding small change Δx in the state of the system, with both x and t being continuous. In considering a stochastic process $X(t)$, we can visualise a continuous value depicting the variation of x with t. Obviously a complete description of such a process would require a knowledge of the probability density function $p(x_1, t_1, x_2, t_2, \ldots x_n, t_n)$, for which the conditional distribution can be written as

$$p(x_1, t_1; x_2, t_2; \ldots x_n, t_n)$$
$$= p(x_2, t_2; x_3, t_3; \ldots x_n, t_n \mid x_1, t_1) p(x_1, t_1) \quad (2.95)$$

or, on continuing, as

$$= p(x_3, t_3; x_4, t_4; \ldots x_n, t_n \mid x_1, t_1; x_2, t_2) p(x_1, t_1; x_2, t_2)$$

....

....

$$= p(x_n, t_n \mid x_{n-1}, t_{n-1}; x_{n-2}, t_{n-2}; \ldots x_1, t_1)$$
$$\times\, p(x_{n-1}, t_{n-1}; x_{n-2}, t_{n-2}; \ldots x_1, t_1) \quad (2.96)$$

If the process is Markovian in character then for successive times we have the conditional probability simplified thus:

$$p(x_n, t_n \mid x_{n-1}, r_{n-1}, \ldots x_1, t_1) = p(x_n, t_n \mid x_{n-1}, t_{n-1}) \quad (2.97)$$

Equation (2.96) implies that the conditional probability density at t_n can be uniquely determined if the corresponding density at t_{n-1} is given.

Knowledge of the densities at prior times is not necessary. The conditional probability $p(x_n, t_n \mid x_{n-1}, t_{n-1})$ is referred to as the transition probability. Equation (2.95) for the joint density can then be written in terms of transition probabilities as

$$p(x_1, t_1; x_2, t_2; \ldots x_n, t_n) = p(x_1, t_1)p(x_2, t_2 \mid x_1, t_1) \times p(x_3, t_3 \mid x_2, t_2) \ldots p(x_n, t_n \mid x_{n-1}, t_{n-1}) \tag{2.98}$$

The transition probability as defined above obeys the Chapman–Kolmogorov equation. Thus, while moving from state x_0 at t_0 to state x_1 at t_1, if the system passes through an intermediate state x at t, then we have

$$p(x_1, t_1 \mid x_0, t_0) = \int p(x_1, t_1 \mid x, t)p(x, t \mid n_0, t_0)\, dx \tag{2.99}$$

The summation over all possible states x as employed in the discrete space case now gets replaced by an integral for the continuous space.

To illustrate more clearly, let us take a specific example of a stochastic process $X(t)$ the probability density of which is given as $p(x_1, t_1, x_2, t_2, x_3, t_3)$. This probability density function can be written in terms of transition probabilities as

$$p(x_1, t_1; x_2, t_2; x_3, t_3) = p(x_1, t_1)p(x_2, t_2 \mid x_1, t_1)p(x_3, t_3 \mid x_2, t_2) \tag{2.100}$$

Let us now say that the system at x_1 at the instant t_1 moves over to state x_3 at time t_3. In doing so it moves through state x_2 at time t_2. The conditional probability $p(x_3, t_3 \mid x_1, t_1)$ can then be written as

$$p(x_3, t_3 \mid x_1, t_1) = \int p(x_2, t_2 \mid x_1, t_1)p(x_3, t_3 \mid x_2, t_2)\, dx_2 \tag{2.101}$$

Following the same line of argument, if the system at x_1 at instant t_1 moves to state x_2 at t_2, then the probability $p(x_2, t_2)$ can be written in terms of the transition probability as

$$p(x_2, t_2) = \int p(x_1, t_1)p(x_2, t_2 \mid x_1, t_1)\, dx_1 \tag{2.102}$$

The equation implies that, if the initial probability density $p(x_1, t_1)$ is known, the probability density at some time later could be related through the transition probability to the initial density. If this is so, then the central question concerns the estimation of the transition probability. We know that the transition probability follows the Chapman–Kolmogorov equation and it seems easier to employ the differential form of this equation. We shall therefore derive in the next section the differential form of the Chapman–Kolmogorov equation.

The transition probability as defined in Equation (2.97), or subsequently in Equations (2.99) and (2.101), depends on the time. For a stationary or time homogeneous Markov process, the transition probability depends only on the time difference. We can therefore write

$$p(x, t \mid x_0, t_0) = p_\tau(x|x_0) \qquad \text{where} \quad \tau = t - t_0 \tag{2.103}$$

The Chapman–Kolmogorov equation can be simply written for this case as

$$p_{\tau+\tau'} = \int p_{\tau'}(x_1|x)p_\tau(x|x_0)\,dx \tag{2.104}$$

where

$$\tau = t - t_0, \qquad \tau' = t_1 - t$$

In the limit $\tau \to 0$, Equation (2.103) can be reduced to

$$p_\tau(x|x_0) = p_0(x|x_0) = \delta(x - x_0) \tag{2.105}$$

2.4.1 Differential Forms of the Chapman–Kolmogorov Equation

In the present section we shall derive the differential forms of the Chapman–Kolmogorov equation. Given state x_0 at t_0, the transition probability that the state will be x at time $(t + \Delta t)$ can be written in terms of the Chapman–Kolmogorov equation as

$$p(x, t + \Delta t \mid x_0, t_0) = \int p(x, t + \Delta t \mid x - \Delta x, t)p(x - \Delta x, t \mid x_0, t_0)\,d(\Delta x) \tag{2.106}$$

where $(x - \Delta x)$ at time t could be considered as an intermediate state. Expanding the integral in Taylor's series and rearranging, we get

$$p(x, t + \Delta t \mid x_0, t_0) = \int p(Z)\, d(\Delta x) + \int \sum_{i=1}^{\infty} \frac{(-1)^i}{i} (\Delta x)^i \frac{\partial^i}{\partial x^i} p(Z)\, d(\Delta x) \qquad (2.107)$$

where

$$p(Z) = p(x + \Delta x, t + \Delta t \mid x, t) p(x, t \mid x_0, t_0) \qquad (2.108)$$

In order to evaluate the integrals in Equation (2.107) the common factor $p(x, t \mid x_0, t_0)$ from $p(Z)$ can be removed outside the integral sign to obtain

$$p(x, t + \Delta t \mid x_0, t_0) = p(x, t \mid x_0, t_0) \Bigg[\int p(x + \Delta x, t + \Delta t \mid x, t)\, d(\Delta x) + \int \sum_{i=0}^{\infty} \frac{(-1)^i}{i!} (\Delta x)^i \frac{\partial^i}{\partial x^i} [p(x + \Delta x, t + \Delta t \mid x, t)]\, d(\Delta x) \Bigg] \qquad (2.109)$$

Noting that the first integral in Equation (2.109) is unity, this equation can be rearranged to give

$$\operatorname*{Lim}_{\Delta t \to 0} \frac{p(x, t + \Delta t \mid x_0, t_0) - p(x, t \mid x_0, t_0)}{\Delta t} = \frac{\partial p(x, t \mid x_0, t_0)}{\partial t} = \sum \frac{(-1)^i}{i!} \frac{\partial^i}{\partial x^i} [a_i(x, t) p(x, t) \mid x_0, t_0)] \qquad (2.110)$$

$t > t_0; -\infty < x < \infty$ where

$$a_i = \lim_{\Delta t \to 0} \int (\Delta x)^i p(x + \Delta x, t + \Delta t \mid x, t)\, d(\Delta x)$$

Equation (2.110) is the differential form of the Chapman–Kolmogorov equation, also referred to as the forward Fokker–Planck equation or forward Kolmogorov equation. The equation, by virtue of Equation (2.105), has the obvious initial condition of

$$\operatorname*{Lim}_{t \to t_0} p(x, t \mid x_0, t_0) = p_0(x|x_0) = \delta(x - x_0) \qquad (2.111)$$

Also, as $x \to \infty$, the conditional distribution should tend to zero; we therefore have

$$\operatorname*{Lim}_{x \to \infty} p(x, t \mid x_0, t_0) = 0 \qquad (2.112)$$

Instead of using (x, t) as the variables and (x_0, t_0) as constants, if we were to begin with (x_0, t_0) as variables and (x, t) as constants, we would end up with an equation similar to Equation (2.110). The equation, now referred to as the backward Fokker–Planck or Kolmogorov equation, can be written as

$$\frac{\partial p(x, t \mid x_0, t_0)}{\partial t_0} = \sum_{i=1}^{\infty} \frac{1}{i!} \frac{\partial^i}{\partial x_0^i} [a_i(x_0, t_0) p(x, t \mid x_0, t_0)] \quad (2.113)$$

with initial condition given by Equation (2.111) and $\mathrm{Lim}_{x_0 \to \infty} p(x, t \mid x_0, t_0) = 0$. The two differential forms of the Chapman–Kolmogorov equation yield identical results for the transition probability and preference for one over the other is determined depending on the situation.

2.4.2 Master Equation

An especially useful differential version of the Chapman–Kolmogorov equation is obtained for a stationary Markov process, where it is known that the transition probability does not depend on the time but only on the time difference. We can then begin with the Chapman–Kolmogorov equation corresponding to the situation given by Equation (2.104). The transition probability can be rewritten for this case as

$$p(x_1, t + \tau' \mid x_0, t) = p_\tau(x_1 | x_0) \quad (2.114)$$

where if $\tau' \to 0$ we have $p_0(x_1|x_0) = \delta(x_1 - x_0)$ [Equation (2.105)]. If, however, τ tends to an infinitesimally small value other than zero, we can write $p(x_2, t + \tau \mid x_1, t)$ as in Equation (2.109) which can then be rearranged to give

$$p_{\tau'}(x_1|x_0) = (1 - \alpha_0 \tau')\, \delta(x_1 - x_0) + \tau' W(x_1|x_0) + O(\tau') \quad (2.115)$$

where we have made use of the relation $p_0(x_1|x_0) = \delta(x_1 - x_0)$, and $W(x_1|x_0)$ represents the transition probability per unit time to move from x_0 to x_1 and is clearly positive. The parameter α_0 is defined as

$$\alpha_0 = \int W(x_1|x_0)\, dx_1 \quad (2.116)$$

With this result let us now turn to the Chapman–Kolmogorov

Equation (2.104). Using Equation (2.115) we have

$$p_{\tau+\tau'}(x_1|x_0) = [1 - \alpha_0(x_1)\tau']\,p_\tau(x_1|x_0) + \tau' \int W(x_1|x)\,p_\tau(x|x_0)\,dx \tag{2.117}$$

Dividing by τ', and in the limit $\tau' \to 0$, we obtain

$$\frac{\partial p_\tau(x_1|x_0)}{\partial \tau} = \int [W(x_1|x)\,p_\tau(x,t) - W(x|x_1)\,p_\tau(x,t)]\,dx \tag{2.118}$$

Equation (2.118) represents the differential form of the Chapman–Kolmogorov Equation (2.104). As would be noticed it is valid for a stationary Markov process and is commonly referred to as the master equation.

To write the general form of Equation (2.118), let $X(t)$ denote a Markov process that is stationary. According to the discussion presented earlier we need to know only the first order probability density $p_1(x_1)$ and the transition probability $p_\tau(x_2|x_1)$. Let us now redefine a part of this process starting with some arbitrary time t_0 when the system is at a fixed value of, say, x_0. Alternatively, we may also define the initial probability density at t_0 by $p(x_0)$. Let this new process be referred to as $X'(t)$, $t > t_0$. The new process is clearly nonstationary; yet since it is a part of the original stationary process; the transition probability still depends only on the time difference. For clarity, these processes are referred to as homogeneous processes and indicate the nonstationary Markov processes with stationary transition probability. If $p_{t_1-t_0}(x_1|x_0)$ represents the transition probability to move from state x_0 at t_0 to state x_1 at t_1, we can write using Equation (2.102) the probability density of the next state as

$$p_1^*(x_1, t_1) = \int p(x_0)\,p_{t_1-t_0}(x_1|x_0)\,dx_0 \tag{2.119}$$

or, for a definite value of $X(t_0) = x_0$, as

$$p_1^*(x_1, t_1) = p_{t_1-t_0}(x_1|x_0) \tag{2.120}$$

Also, the transition probability from x_1 to x_2 can be defined as

$$p_1^*(x_2, t_2 \mid x_1, t_1) = p_{t_2-t_1}(x_2|x_1) \tag{2.121}$$

The asterisk in Equations (2.119) and (2.121) is used to distinguish them as referring to a nonstationary (homogeneous) process as against the normal notation used for a stationary Markov process.

An important point to note from this discussion is that by extracting a subensemble from the stationary Markov process, one realises a nonstationary process that obeys the relations presented above. The homogeneous process then usually occurs as a subensemble of some stationary Markov process (although the Weiner process is an exception). The nonstationary process in the limit $t \rightarrow \infty$, as one would expect, approaches the stationary process. In the limit $t \rightarrow \infty$, we would therefore expect

$$p_1^*(x_1, t_1) \longrightarrow p_1(x_1, 0) \tag{2.122}$$

or, by virtue of Equation (2.120),

$$p_1^*(x_1, t_1) \longrightarrow p_1(x_1) \longrightarrow p_{t_1 - t_0}(x_1|x_0) \tag{2.123}$$

The relation suggests that the transition probability in Equation (2.118) is identical to the distribution function $p_1(x)$ of the subensemble determined by the initial condition x_0. Equation (2.118) can then be written more generally and simply as

$$\frac{\partial p}{\partial t} = \int [W(x|x_1)p(x_1, t) - W(x_1|x)p(x, t)]\, dx_1 \tag{2.124}$$

If the random variable $X(t)$ possesses discrete states, then the equation can be written as

$$\frac{\partial p}{\partial t} = \sum_{n'} \{W_{nn'} p_{n'}(t) - W_{n'n} p_n(t)\}, \qquad W_{nn'} \geqslant 0, \quad n \neq n' \tag{2.125}$$

Equations (2.124) and (2.125) are the more common forms of the master equation frequently used in treating physical processes. The meaning is made especially clear in Equation (2.125) where we note the gain due to transitions from states n' (first term) and loss due to transitions to states n' (second term).

2.4.3 Alternative Forms of the Master Equation for Homogeneous Markov Processes: Continuous and Discrete State Space

In this section we present the alternative forms of the master equation for homogeneous Markov processes with either discrete or continuous state space. The only restriction imposed is that the state space be unbounded ($-\infty < x < \infty$). In view of the homogeneous nature of the Markov process we can talk in terms of time-dependent distribution function (or density function) as against the transition probability. Let $p^*(x, t \mid x_0)$ define the probability of finding the system in state x at time t given that it has been in state x_0 at zero time. Clearly, the initial condition is $p^*(x, 0 \mid x_0) = \delta(x - x_0)$. Further, we assume that the successive events are statistically independent and can be described by a time-dependent transition probability per unit time $W(x' + x)$. The parameter usually depends on some extensive property of the system such as its volume.

Let us now derive the equation for the evolution of $p^*(x, t \mid x_0)$. For this purpose the following changes may be considered: (1) No transition at all occurs, and (2) transition takes place. If no transition occurs during the time interval Δt, in view of the exponential distribution of the waiting time, we can write $\exp(-p(x_0)t)\,\delta(x - x_0)$ as its probability. In the event transition occurs, we have two possibilities:

(i) During the time $t = 0$ to $t = t_1$ the system moves over from state x_0 to x_1 according to the probability $p^*(x_1, t_1 \mid x_0)$. From $t = t_1$ to $t_1 + dt_1$ the system moves over from state x_1 to x and then remains at x for the rest of the time. Mathematically these steps could be represented as

$$\int dx_1 \int_{t_1=0}^{t_1=t} dt_1\, p^*(x_1, t_1 \mid x_0) W(x|x_1) \times \exp\{-p(x)(t - t_1)\} \quad (2.126)$$

(ii) It is possible that the system instead of following the course as in (i), stays in state x_0 up to the instant t_1. During the movement from t_1 to $t_1 + dt_1$, the system changes over from state x_0 to state x_1 according to the probability $W(x_1|x_0)$ and the subsequent evolution to state x from x_1 occurs according to $p^*(x, t - t_1 \mid x_1)$. The probability of this course of evolution can be written as

$$\int dx_1 \int_{t_1=0}^{t_1=t} dt_1 \exp[-p(x_0)t_1] W(x_1|x_0) \times p^*(x, t - t_1 \mid x_1) \quad (2.127)$$

Equations for the evolution of $p(x, t|x_0)$ are

$$p^*(x, t \mid x_0) = \exp[-p(x_0)t]\,\delta(x - x_0) + \int dx_1 \int_0^t dt_1\, p^*(x_1, t_1 \mid x_0) \times W(x|x_1) \times \exp[-p(x)(t - t_1)] \quad (2.128)$$

and

$$p^*(x, t \mid x_0) = \exp[-p(x_0)t]\, \delta(x - x_0)$$
$$+ \int dx_1 \int_0^{t_1} dt_1 \exp[-p(x_0)t_1] W(x_1|x_0)$$
$$\times\, h(x, t - t_1 \mid x_1) \quad (2.129)$$

Equations (2.128) and (2.129) can be reduced to the differential forms yielding

$$\frac{\partial p^*(x, t \mid x_0)}{\partial t} = -p(x)p^*(x, t \mid x_0) + \int W(x|x_1)\, p^*(x_1, t \mid x_0)\, dx_1 \quad (2.130)$$

and

$$\frac{\partial p^*(x, t \mid x_0)}{\partial t} = -p(x_0)p^*(x, t \mid x_0) + \int W(x_1|x_0)\, p^*(x, t \mid x_1)\, dx_1 \quad (2.131)$$

with the initial condition $p^*(x, t = 0 \mid x_0) = \delta(x - x_0)$.

Equations (2.130) and (2.131) are commonly referred to as the forward and backward forms of the master equation. Their solution yields the time-dependent probability distribution (density) function which can be subsequently used to obtain conditional averages according to the relation

$$\langle f(x)x_0 \rangle = \int f(x)p^*(x, t \mid x_0)\, dx = \psi(x_0, t) \quad (2.132)$$

The backward form of the master equation is especially suitable when our major interest lies in estimating some average quantities. The master equation can now be directly written in terms of the averaged quantity as

$$\frac{\partial \psi(x_0, t)}{\partial t} + p(x_0)\psi(x_0, t) = \int W(x_1|x_0)\psi(x_1, t)\, dx_1 \quad (2.133)$$

with the initial condition of $\psi(x_0, t = 0) = f(x_0)$. This integro-differential equation can now be solved directly to yield the desired results. In the next chapter we shall consider some applications involving this type of equation.

2.4.4 Applications of Master Equation

The conversion of the Chapman–Kolmogorov equation, that merely represents the Markovian character of the process, into a differential version, or more specifically into the master equation form, affords an easy mathematical formulation and interpretation. The master equation requires a knowledge of the quantities $[W(y|y')\Delta t]$ representing the probability of transition over a short period Δt. These are specific to the system and can be estimated independently. The Chapman–Kolmogorov equation, on the contrary, is a general symbolic relation valid for any Markov process and does not contain information about any specific system. This observation has an important implication: the master equation can be used to describe the long time behaviour and evolution of specific systems. It is this possibility that has made the master equation so useful. To illustrate this fact we shall consider a simple example the macroscopic equation of which can be described as

$$\frac{dn}{dt} = g(n) - r(n) \tag{2.134}$$

Specifying the functions $g(n)$ and $r(n)$ we know the transition probabilities which, when used in Equation (2.115), give the transition probability per unit time as

$$W(n|n') = g(n)\,\delta_{n,n'+1} + r(n)\,\delta_{n,n'-1} \tag{2.135}$$

Inserting Equation (2.135) in (2.125) we obtain the desired master equation. In general, solution of the master equation may present some problems, except for a handful of simple cases. The general methods of solution will be considered in the next chapter. However, to complete the discussion here we cite a simple example of a linear decay process where the transition probability per unit time can be obtained as

$$W_{nn'} = \gamma n'\,\delta_{n,n'-1} \qquad (n \neq n') \tag{2.136}$$

Substituting this in the master equation (2.125) we have

$$\frac{dp_n(t)}{dt} = \gamma(n+1)p_{n+1} - \gamma n p_n(t), \qquad p_n(0) = \delta_{n,n_0} \tag{2.137}$$

which possesses a simple solution for $p_n(t)$. Once $p_n(t)$ is known, all other quantities such as mean etc. can be calculated as per the relations presented in Chapter 1.

2.5 Continuous State–Discrete Parameter Markov Processes

It is now merely a matter of formality to discuss the present case to complete the discussion of Markov processes. In fact, some of the results for this case have already been presented in the previous section. In any case, the results in general can be obtained as a special case of the results presented earlier. We shall therefore only summarise the essential points.

For a Markov process with continuous state space $X_n(-\infty, \infty)$ for all possible n, we have the equation

$$\begin{aligned}\text{Prob}\{X_{n+1} = x \mid X_n = y, X_{n-1} = i, X_{n-2} = j, \ldots X_0 = k\} \\ = \text{Prob}\{X_{n+1} = x \mid X_n = y\} \quad (2.138)\end{aligned}$$

The conditional distribution gives the transition probability from state y to state x and the process is referred to as homogeneous if it is independent of n. For an m-step transition probability, we can write

$$\begin{aligned}\text{Prob}\{X_{n+m} \leqslant x \mid X_n = y, \ldots X_0 = k\} = \text{Prob}\{X_{n+m} \leqslant x \mid X_n = y\} \\ = p_m(y|x) \quad (2.139)\end{aligned}$$

For $m = 1$, this equation also defines the unit-step transition probability. As in the previous cases, the initial distribution $\text{Prob}\{X_0 \leqslant x\} = p_0(x)$ and the transition probability $p_m(y|x)$ uniquely determine the probability of the next state. The Chapman–Kolmogorov equation can also be analogously written as

$$p_{m+n}(y|x) = \int p_n(y|z)p_m(z|x)\,dz \quad (2.140)$$

where z refers to the intermediate state between x and y. The differential form of a similar equation has already been discussed in the earlier section.

REFERENCES

Bharucha-Reid, A. T., *Elements of the Theory of Markov Processes and Their Applications*, McGraw-Hill, New York, 1960.

Doob, J. L., *Stochastic Processes*, Wiley, New York, 1953.

Feller, W., *An Introduction to Probability Theory and Its Applications*, Vol. I, Wiley, New York, 1968.

Fisz, M., *Probability Theory and Mathematical Statistics*, Wiley, New York, 1963.

Gardiner, C. W., *Handbook of Stochastic Methods for Physics, Chemistry, and the Natural Sciences*, Springer-Verlag, Berlin-Heidelberg, 1983.

Howard, R. A., *Dynamic Probabilistic Systems, Vol. I, Markov Models*, Wiley, New York, 1971.

Karlin, S., *A First Course in Stochastic Processes*, Academic Press, New York, 1966.

King, R. P., *An Introduction to Stochastic Differential Equations—Abbreviated Lecture Notes*, Department of Chemical Engineering, University of Natal, Durban, South Africa, 1971.

Mortensen, R. E., *J. Stat. Phys.* **1-2**, 271 (1969).

Papoulis, A., *Probability, Random Variables and Stochastic Processes*, McGraw-Hill, New York, 1965.

Seinfeld, J. H. and Lapidus, L., *Mathematical Methods in Chemical Engineering*, Vol. III, Prentice-Hall, New Jersey, 1974.

Stratonovich, R. L., *SIAM J. Control* **4**, 362 (1966).

van Kampen, N. G., *Stochastic Processes in Physics and Chemistry*, North-Holland, Amsterdam, 1981.

PART II

Role of Internal Noise

3 Probabilistic Modelling of Simple Reaction Schemes

3.1 Introduction

THE REALISATION that, within the framework of stochastic processes, a phenomenon can be studied as a random process that can be changing both in time and space has enormously increased the scope of application of these processes. The present chapter is concerned with the construction of such a framework for chemically reacting systems. The basic mathematical techniques used for stochastic processes have already been summarised in Part I.

Before we begin on the formal description of these processes, let us consider the rationale of the formulation. Fundamental studies on chemical processes at the molecular level suggest that chemical reaction proceeds amongst discrete molecular species, and properties such as number of molecules and concentrations of species are not, in reality, continuous real valued functions of time. The commonly used deterministic (macroscopic) description of chemical processes in terms of differential equations, on the contrary, presupposes a continuous variation of these properties. The predictions from these macroscopic models are thus likely to be in error under certain conditions, necessitating a more realistic (stochastic) approach to such cases. This leads us logically to the question of identification of those situations for which macroscopic descriptions would be valid.

This can be done via the fundamental result derived from statistical mechanics (Landau and Lifshitz, 1958) which states that the macroscopic description is valid at thermodynamic equilibrium. It further postulates that, under this condition, if the average number of molecules is of the order of N, the fluctuations about this average will be of the order of $N^{1/2}$. It is important to note that this result holds for systems away from the points of macroscopic instability such as phase transitions or critical points. In the light of this fact one can think of two situations where the conventional macroscopic description would not be valid.

The first is clearly one where the number of molecules N, or the size of the system involved, is small. Here $N^{1/2}$ would not be negligible in relation to N, so that fluctuations would have to be considered. Most chemically reacting systems obviously do not fall in this category for N is typically in the range 10^{20}–10^{25} and the square root of the extent of fluctuations is therefore essentially inconsequential. There are situations, however, such as in diffusion controlled reacting systems, reactions in biological cells or diffusion–reaction in solid catalysts with pore dimensions of the size of molecules, where the macroscopic description would be inadequate. The second situation where conventional macroscopic description would be inadequate is near the points of instability. The square root law describing the fluctuations is no longer applicable, and as at equilibrium transitions the fluctuations tend to grow to produce observable effects. It is this type of situation that is more frequently encountered in chemically reacting systems and deserves consideration from the point of view of development of predictive theory.

The application of the mathematical tools developed for stochastic processes therefore seems justified for chemically reacting systems—at least for the situations elaborated above. The present chapter is concerned with the general theory for simple (elementary) reactions and the formulation of the basic framework for stochastic analysis. The development is restricted to isothermal homogeneous systems. More rigorous stochastic models (for specific reacting systems) will be considered in later chapters. The treatment presupposes acquaintance with the basic concepts of random processes and definitions of common terms such as mean, variance, distribution function, etc. These concepts, together with the mathematical background needed, have already been outlined in Chapters 1 and 2.

3.2 Stochastic Description of Chemical Reactions

3.2.1 The Birth and Death Description

Consider a molecule of a fluid subjected to a great number of random independent impulses owing to collisions with other molecules. This results in what is termed as Brownian motion. The idea of treating chemical reaction as a Brownian motion of particles seems to have been first exploited in the work of Kramers (1940) where an attempt was made to formulate macroscopic rate processes in terms of the molecular parameters. A theory of chemical fluctuations based on the notion of a "birth and death" process was also proposed (Delbruck, 1940) around the same time. This simple theory in essence keeps count of the total numbers of reactant and product molecules and uses the macroscopic rate law (of mass action) to model the transition rates between the possible states in molecule number space.

More quantitatively, let us consider a macroscopic system of volume V containing N species $(A_1, A_2, \ldots A_N)$ among which M chemical reactions are occurring. In a deterministic description of this system, we would average out the concentrations of the species in volume V at any time t, and neglecting the fluctuations about the average values would derive the deterministic equations giving their time evolution. It is also possible to take account of fluctuations in a simple way by assuming that the macroscopic mass balance equations as derived above define a Markov process in the number of molecular species space. If X_n denotes the number of molecules of species A_n, then the state of the system can be specified in terms of the probability $P(X_1, X_2, \ldots X_n; t)$ of having X_1 molecules of type A_1, X_2 molecules of type A_2, etc. in volume V at time t. The evolution of this probability, within the framework of the Markovian birth and death process, is governed by the stochastic master equation of the form

$$\begin{aligned}\frac{\partial P(X_1, X_2, \ldots X_N; t)}{\partial t} &= \frac{\partial P(\{X_N\}; t)}{\partial t} \\ &= \sum_{i=1}^{M} [W_i(\{X_n - v_{ni}\} \rightarrow \{X_n\}) . P(\{X_n - v_{ni}\}; t)] \\ &\quad - [W_i(\{X_n\} \rightarrow \{X_n + v_{ni}\}) . P(\{X_n\}, t)] \qquad (3.1)\end{aligned}$$

To explain the terms let us consider the death process. It represents the summation of the product of the transiiion probability per unit time $[W_i\{X_n\} \rightarrow \{X_n + v_{ni}\}]$ for the ith reaction $(i = 1, \ldots M)$ to occur given that the system is in state X_n and the probability $P(\{X_n\}; t)$ that the system is in state X_n at time t. The parameter v_{ni} represents the stoichiometric coefficient for species n in the ith reaction. It can be seen that for every reaction the birth and death contributions can be different, thus affecting the probability evolutions.

The transition probabilities W_i are defined in terms of the properties of the system and the molecules which take part in the ith reaction. For a homogeneous system this can simply be described as the probability that the ith reaction will occur in volume V within the next infinitesimal time $(t - t + \partial t)$ and can be written as

$$W_i(t)\,\partial t = k_i l_i(t)\,\partial t \tag{3.2}$$

where $l_i(t)$ represents a combinatorial factor expressing the number of ways in which the nth species necessary for the ith reaction can be selected from a population contained in volume V at time t. The parameter k_i represents the average probability that the particular combination of molecules would react accordingly in the next infinitesimal time ∂t, and is similar to the conventional (deterministic) rate constant used in the macroscopic model. It is necessary, however, to emphasise the distinction between the stochastic and deterministic rate constants; the two are strictly equivalent only for unimolecular reactions. A more elaborate discussion of the birth and death description of chemically reacting systems can be found elsewhere (McQuarrie, 1967; Oppenheim, Shuler and Weiss, 1977; van Kampen, 1975).

The application of the birth and death concept to chemical reactions essentially gives rise to a linear differential-difference equation with nonlinear coefficients. Some of the methods outlined in Chapter 1, such as the moment generating method, the method of characteristic equation, etc. can be applied to solve these equations. In general, only for a handful of cases is complete analytical solution of these equations possible. The exactly solvable cases have been summarised by McQuarrie (1967). For most other situations it is necessary to develop approximation schemes, and a detailed account of the principal methods used has been presented by Nicolis and Prigogine (1977).

3.2.2 Applications to Elementary Reactions

The present section will be devoted to the application of the birth and death concept to simple elementary reactions. The treatment will include unimolecular and bimolecular reactions and some simple extensions of these schemes.

Unimolecular Reactions

The simplest of the unimolecular reaction schemes is the process described by

$$A \longrightarrow B$$

This scheme has been analysed earlier by Bartholomay (1958, 1959). Let $X(t)$ represent the number of molecules of species A at any time t. The probability of transition from state (x) to $(x - 1)$ in the time interval $(t \rightarrow t + \Delta t)$ is given by $[kx\Delta t + O(\Delta t)]$ where k refers to the constant and $O(\Delta t)$ implies $[O(\Delta t)/\Delta t] \rightarrow 0$ as $t \rightarrow 0$. Also, it is postulated that the transition of state $x \rightarrow (x - j), j > 1$ in the interval Δt is at least of the order of $O(\Delta t)$, and that the reverse reaction occurs with zero probability. We can thus write a detailed balance of probabilities to obtain

$$P_x(t + \Delta t) - P_x(t) = k(x + 1)\Delta t P_{x+1}(t) - kx\Delta t P_x(t) + O(\Delta t) \tag{3.3}$$

which in the limit $\Delta t \rightarrow 0$ acquires the form

$$\frac{dP_x}{dt} = k(x + 1)P_{x+1} - kxP_x \tag{3.4}$$

The differential-difference Equation (3.4) can be converted into a linear partial differential equation using the generating function

$$G(s, t) = \sum_{x=0}^{\infty} P_x(t)s^x, \qquad |s| < 1 \tag{3.5}$$

to obtain

$$\frac{\partial G}{\partial t} = k(1 - s)\frac{\partial G}{\partial s} \tag{3.6}$$

The solution to Equation (3.6) subject to initial condition $G(s, 0) = s^{x_0}$ is

$$G(s, t) = [1 + (s - 1)\, e^{-kt}]^{x_0} \tag{3.7}$$

Once the generating function is obtained we can proceed to calculate the mean, variance, etc. by simply referring to their definitions (see Chapter 1):

$$E[X(t)] = (\partial G/\partial s)_{s=1} \tag{3.8}$$

$$\sigma^2[X(t)] = [(\partial^2 G/\partial s^2) + (\partial G/\partial s) - (\partial G/\partial s)^2]_{s=1} \tag{3.9}$$

The final results obtained are

$$E[X(t)] = x_0 \exp(-kt) \tag{3.10}$$

$$\sigma^2[X(t)] = x_0 e^{-kt}(1 - e^{-kt}) \tag{3.11}$$

It is observed that the expected value of $X(t)$ corresponds to the deterministic result (although this is only true strictly for unimolecular reactions). Additionally, higher moments such as σ^2 are obtained in such an analysis so that the effect of fluctuations on chemical kinetics can be included more confidently.

The case of unimolecular reactions where the initial condition to the system is not constant ($x_0 = x_0$ at $t = 0$) but is distributed has been solved by McQuarrie (1963) to show the effect of uncertainty in the initial condition. For large systems and assuming Poisson distribution he derived the equation to calculate the mean and variance. The equation for the mean is exactly identical to Equation (3.10), while that for variance contains a parameter p that characterises the distribution.

Among other unimolecular reactions analysed (McQuarrie, 1963) are the reversible and parallel reactions

$$A \rightleftharpoons B; \qquad \begin{array}{l} A \longrightarrow B \\ A \longrightarrow C \end{array}$$

Again, denoting the number of molecules of A at any time t by $X(t)$, the detailed balance equation for the reversible reaction can be written as

$$\frac{dP_x}{dt} = k_{-1}(x_0 - x + 1)P_{x-1} + k_1(x+1)P_{x+1} - [k_1 x + k_{-1}(x_0 - x)]P_x \tag{3.12}$$

where x_0 refers to the total number of molecules of species A and B. The case of parallel reaction requires consideration of two random variables $X(t)$ and $Y(t)$ representing, respectively, the number of molecules of species A and B. The governing master equation for this case can be written in a straightforward manner as

$$\frac{dP_{x,y}}{dt} = k_1(x+1)P_{x+1,y-1} + k_2(x+1)P_{x+1,y} - (k_1 x + k_2 y)P_{x,y} \tag{3.13}$$

Equations (3.12) and (3.13) can be solved in a manner analogous to that described for simple unimolecular reactions. The results obtained again indicate that the mean exactly coincides with the deterministic result, whereas the variance gives additional information regarding the effect of fluctuations.

The treatment of first-order reactions, essentially demonstrated for one-component systems, can be easily extended to multicomponent systems following first-order kinetics. Krieger and Gans (1960), Gans (1960) and Darvey and Staff (1966) have provided analyses for such multicomponent systems. The general result again brings out the equivalence between the calculated mean and the deterministic solution.

Bimolecular Reactions

The application of the general concept of stochastic description to bimolecular reactions of the type $2A \rightarrow B$ or $A + B \rightarrow C$ leads to second-order partial differential equations in the generating function, which in general are more difficult to solve than the ones encountered in unimolecular reactions. Renyi (1954) first considered the latter type of bimolecular reactions and provided solutions using Laplace transform techniques. McQuarrie, Jachimowski and Russell (1964) solved the equation using the generating function method, which appears to be practically more useful. For a typical bimolecular process of the type

$$2A \longrightarrow B$$

and with $X(t)$ again representing the random variable specifying the number of molecules of species A, the stochastic description can be completely defined by invoking the following considerations.

Since two molecules of A are involved in a reactive event, the probability of the transition $(x + 2) \rightarrow x$ in the interval $(t, t + \Delta t)$ is $\{1/2[k_1(x + 2)(x + 1)\Delta t] + O(\Delta t)\}$ where k_1 is a constant and $O(\Delta t)/\Delta t \rightarrow 0$ as $\Delta t \rightarrow 0$. The probability of any other transition $(x + j) \rightarrow x, j > 2$ in the time interval Δt is at least of the order of $O(\Delta t)$. In view of the irreversibility of the reactive process the probability of the transition $(x - j) \rightarrow x$, $j > 0$ is zero. Finally the probability of the transition $x \rightarrow x$ in the interval Δt is $\{(1 - 1/2kx[(x - 1)\Delta t] + O(\Delta t)\}$. With these assumptions we can write the following probability balance equation in the limit $\Delta t \rightarrow 0$:

$$\frac{dP_x}{dt} = \tfrac{1}{2}k_1(x + 2)(x + 1)P_{x+2} - \tfrac{1}{2}k_1 x(x - 1)P_x + O(\Delta t) \quad (3.14)$$

McQuarrie, Jachimowski and Russell (1964) solved this equation by using the separation of variations technique to obtain the mean:

$$\langle x \rangle = -\sum_{n=2}^{x_0} A_n T_n(t) \tag{3.15}$$

where x_0 is the initial number of A molecules and

$$A_n = \frac{1-2n}{2^n} \frac{\Gamma(x_0+1)\Gamma[(x_0-n+1)/2]}{\Gamma(x_0-n+1)\Gamma[(x_0+n+1)/2]}, \qquad n = 2, 4, \ldots x_0 \tag{3.16}$$

$$T_n = \exp\{-\tfrac{1}{2}k_1 n(n-1)t\} \tag{3.17}$$

Comparison of the stochastic mean with the deterministic result indicates that the two approach each other rapidly at increasingly higher values of x_0.

Bimolecular reactions of the type

$$A + B \longrightarrow C$$

have been analysed by McQuarrie, Jachimowski and Russell (1964). Noting that the condition $Y(t) - X(t) = Y_0 - X_0 = Z_0$ would always be satisfied, the following master equation can be written:

$$\frac{dP_x}{dt} = k_1(x+1)(Z_0+x+1)P_{x+1} - k_1 x(Z_0+x)P_x \tag{3.18}$$

Again the method of separation of variables can be used to give the first moment

$$\langle x \rangle = \sum_{n=1}^{x_0} \frac{(2n+Z_0)\Gamma(x_0+1)\Gamma(x_0+Z_0+1)}{\Gamma(x_0-n+1)\Gamma(x_0+Z_0+n+1)} T_n(t) \tag{3.19}$$

The second factorial moment is given by

$$\langle x(x-1) \rangle = \sum_{n=2}^{x_0} \frac{(n-1)(n+Z_0+1)(2n+Z_0)\Gamma(x_0+1)\Gamma(x_0+Z_0+1)}{\Gamma(x_0-n+1)\Gamma(x_0-Z_0+n+1)} T_n(t) \tag{3.20}$$

where

$$T_n(t) = \exp\{-n(n+Z_0)k_1 t\} \tag{3.21}$$

Equations (3.20) and (3.21) properly reduce to that for the pseudo first-order process when Z_0 tends to a large value.

The case of the more general type of bimolecular reaction

$$A + B \rightleftharpoons C + D$$

can likewise be solved using the method of separation of variables. The final results obtained are so complex as to be of little practical value. In the equilibrium limit, however, the reaction scheme yields the exact result (Dervey, Ninham and Staff, 1966). The probability balance is given by

$$\begin{aligned} P(a, b, c, d, t + \Delta t) = {} & k_1(a + 1)(b + 1)\Delta t \,.\, P(a + 1, b + 1, c - 1, d - 1; t) \\ & + k_2(c + 1)(d + 1)\Delta t \,.\, P(a - 1, b - 1, c + 1, d + 1; t) \\ & + [1 - (k_1 ab + k_2 cd)\Delta t] \,.\, P(a, b, c, d; t) + O(\Delta t) \end{aligned} \tag{3.22}$$

For a conservative system the four random variables (a, b, c, d) can be related through the initial conditions which are assumed to be given as $A(0) = \alpha$, $B(0) = \beta$, $C(0) = \gamma$ and $D(0) = \delta$. Clearly,

$$\alpha - a = \beta - b = c - \gamma = d - \delta \tag{3.23}$$

It is now possible to eliminate b, c, d using relation (3.23) to obtain the probability balance equation in terms of a single random variable a:

$$\begin{aligned} \frac{dP_x}{dt} = {} & k_1(a + 1)(\beta - \alpha + a + 1)P_{a+1} + k_2(\gamma + \alpha - a + 1)(\delta + \alpha - a + 1)P_{a-1} \\ & - \{k_1 a(\beta - \alpha + a) + k_2(\gamma + \alpha - a)(\delta + \alpha - a)\}P_a \end{aligned} \tag{3.24}$$

The master equations for several simplified cases such as

$$A + B \rightleftharpoons C$$

$$2A \rightleftharpoons C + D$$

$$2A \rightleftharpoons C$$

can be written analogously. These can then be transformed into differential-difference equations which can be solved in terms of hypergeometric series to obtain the mean and variance. McQuarrie

(1963) has summarised the equations for the stochastic mean, the deterministic result and the calculated variance for different reaction schemes. Simple calculations using these results indicate that when the variance approaches zero the stochastic and deterministic results also approach each other.

A fact apparent in the derivation of these equations is that not many bimolecular reactions can be solved exactly. Practical necessity therefore compels one to develop approximate procedures. A large number of approximation schemes are available in the literature, and we shall describe some of them in the next section. Of particular general importance is the approximation based on the Fokker–Planck description which enjoys great popularity, since explicit calculations with this description are often more convenient than with the rigorous master equation.

3.2.3 Approximations of the Master Equation

Four approximation schemes which can be of potential use in many nonlinear cases are outlined below. First, the Fokker–Planck approximation is described highlighting its limitations especially for nonlinear cases. This is followed by a general expansion method for the master equation. Alternative formulations such as Hamilton–Jacobi formulations are then discussed. Finally, a more recent method that solves the transport equations derived from the master equation is described.

Fokker–Planck Approximation

To derive the Fokker–Planck equation, the master Equation (3.1) can be written in a more convenient form as

$$\frac{\partial P(x,t)}{\partial t} = \int W(x \to x')P(x',t) - W(x' \to x)P(x,t)\,dx' \qquad (3.25)$$

where the transition probability $W(x \to x')$ can be expressed in terms of the difference $(x - x') = r$ as

$$W(x \to x') = W(x', r) \qquad (3.26)$$

The general master equation then becomes

$$\frac{\partial P(x,t)}{\partial t} = \int W(x-r,r)P(x-r,t)\,dr - P(x,t)\int W(x-r)\,dr \tag{3.27}$$

In view of the complexities involved in the solution of Equation (3.27), it can be simplified without sacrificing its main features. We make two approximations that seem rational. The first stems from the observation that the transition probabilities W vary sharply with the jump size $(x - x')$ and can be represented as a sharp peak. Second, the dependence of the transition probability on the starting value of x' itself is fairly weak, so that a relation of the type

$$W(x' + \Delta x, r) = W(x', r)$$

holds. Also, the probability $P(x, t)$ varies weakly with x, as in the case of the transition probability. In view of these assumptions, we can expand the first term in Equation (3.27) in Taylor's series to obtain

$$\frac{\partial P(x,t)}{\partial t} = \int W(x,r)P(x,t)\,dx - \int r\frac{\partial}{\partial x}[W(x,r)P(x,t)]\,dr$$
$$+ \frac{1}{2}\int r^2 \frac{\partial^2}{\partial x^2}[W(x,r)P(x,t)]\,dr - P(x,t)\int W(x,-r)\,dr \tag{3.28}$$

where only terms up to second order are retained. Further, defining

$$\alpha_i(x) = \int_{-\infty}^{\infty} r^i W(x,r)\,dr \tag{3.29}$$

Equation (3.28) becomes

$$\frac{\partial P(x,t)}{\partial t} = -\frac{\partial}{\partial x}[\alpha_1(x)P] + \frac{1}{2}\frac{\partial^2}{\partial x^2}[\alpha_2(x)P] \tag{3.30}$$

This is the familiar form of the Fokker–Planck equation originally derived by Planck. The derivation brings out clearly the relationship between the parameters α_1, α_2 and the transition probabilities $W(x, r)$. It is possible to include higher order terms in Taylor's series expansion and rewrite Equation (3.30) as

$$\frac{\partial P(x,t)}{\partial t} = \sum_{i=1}^{\infty} \frac{(-1)^i}{i!}\frac{\partial}{\partial x}\{\alpha_i(x)P\} \tag{3.31}$$

which is the well-known Kramers–Moyal expansion. If we multiply Equation (3.30) by x and sum we obtain

$$\frac{\partial\langle x\rangle}{\partial t} = \langle\alpha_1(x)\rangle \tag{3.32}$$

Also, if we neglect the fluctuations $\langle\alpha_1(x)\rangle$, the term α_1 can be related to the rate form in the macroscopic equation

$$\frac{dx}{dt} = Q(x) - R(x) \tag{3.33}$$

as

$$\alpha_i(x) = \frac{1}{i!}[Q(x) - (-1)^i R(x)] \tag{3.34}$$

Equation (3.31) arises as a result of the special case of the general property of Markov processes and always holds for such processes where $\alpha_i(x)$ is generally replaced by the ith moment of the transition probability

$$\alpha_i(x, t) = \frac{1}{i!}\lim_{\Delta t\to 0}\frac{1}{\Delta t}\int dx' P(x', t + \Delta t \mid x, t)(x' - x)^i \tag{3.35}$$

where $P(x', t + \Delta t \mid x, t)$ is the joint probability of finding the system in state x' at time $t + \Delta t$, given that it was in state x at time t.

The relative advantages of the Fokker–Planck equation over the corresponding master equation are:

(1) A differential equation is involved here as against the integro-differential master equation. The solution to the differential equations may also at times pose some problems, but in general they are always easier to handle.

(2) The Fokker–Planck equation can be constructed easily from sources other than the master equation. For instance, the coefficient α_1 is known if we have information regarding the macroscopic evolution [see Equation (3.34)]. Likewise the coefficient α_2 can be obtained from the equilibrium probability distribution which is the steady state solution of Equation (3.30):

$$P(x, \infty) = \frac{K}{\alpha_2(x)}\exp\left[2\int_0^2 \frac{\alpha_1(x)}{\alpha_2(x)}dx\right] \tag{3.36}$$

The equilibrium distribution for closed systems can be known from statistical mechanics, so that Equation (3.36) can be used to estimate α_2. Knowing α_1, α_2, one can construct a Fokker–Planck equation, thus avoiding complete knowledge of the transition probabilities. The chief advantage in such a formulation, however, is that, while it is valid for linear systems, in cases where the macroscopic rate laws are nonlinear the approach runs into difficulties. This arises primarily due to the neglect of fluctuations as was done in obtaining Equation (3.32). We shall return to this point later in Section 3.3.4, where a modified form of the Fokker–Planck equation that avoids this complication to some extent is derived.

Markov processes, the master equation of which obeys the Fokker–Planck equation (3.30), are also referred to as continuous processes. It would be appreciated that Equation (3.30) is always linear in the variable P, and its solution gives the familiar Gaussian distribution. The term "nonlinear Fokker–Planck equation" is used to denote the nonlinear dependence of the coefficient α_1 on x. The solution to these nonlinear equations may be non-Gaussian due to the dependence of the transport coefficients on the state of the system. Frequently, in the analysis of Fokker–Planck equations one is interested in quantities which are ensemble averaged properties such as the mean value of the macroscopic variables. While this can be easily found for linear equations, in nonlinear cases the mean values get coupled to the fluctuations of the macroscopic variables. This complicates the derivation of the transport equations describing the mean relaxation of the ensemble. In such instances, a need for fluctuation renormalisation of the transport equation arises, and over the years several renormalisation strategies have been proposed. While it is not out intention to go into the details of these techniques, a reference to some source material may be helpful: Zwanzig, Nordholm and Mitchell (1972), Mori and Fujisako (1972), Grabert and Wiedlinch (1980). As an alternative, one might use techniques such as eigenfunction expansion for the original master equation directly.

A simple method where the transition probabilities can be systematically expanded in terms of a suitably chosen parameter seems attractive. Such a method, due to van Kampen (1976, 1980), will be outlined in the next section.

Systematic Expansion of the Master Equation (*van Kampen, 1976, 1980*)

While the Fokker–Planck approximation of the master equation provides a convenient and useful method for linear processes, its use in nonlinear situations can sometimes lead to erroneous results. It is necessary therefore to look for alternative methods for the solution of the master equation. One such method is the systematic expansion of the master equation in some suitably chosen parameter. Obviously, it is necessary to select a parameter that appears in the master equation (i.e. in the transition probability), for larger values of which the fluctuations are small. In many instances, this parameter could simply be the size (Ω) of the system. The general expansion method consists of the following steps:

(1) Write the transition probability explicitly in terms of this parameter Ω.

(2) Anticipate the way in which the probability $P(x, t)$ will depend on this parameter.

(3) Employ (1) and (2) in the master equation and extract terms of higher order in Ω to obtain the macroscopic equations.

(4) The next order in Ω gives the linear Fokker–Planck equation for fluctuations.

(5) Collect results from (3) and (4) to obtain the final solution for $P(x, t)$ as defined in (2).

The method is discussed in greater detail by van Kampen (1976, 1980) who also applied it to a wide variety of situations. For the present purpose we shall exemplify it by considering the specific case of a chemical reaction

$$A \longrightarrow X, \qquad 2X \longrightarrow B$$

where, to begin with, it is assumed that the concentrations of species A and B are held constant. The macroscopic equation describing the change in the concentration of species X can be written as

$$\frac{dx}{dt} = k_1 x - k_2 x^2 \tag{3.37}$$

The master equation describing this situation can be formulated subject to the consideration detailed in Section 3.1.1. Further, using the notation

$$Ef(n) = f(n + 1), \qquad E^{-1}f(n) = f(n - 1) \tag{3.38}$$

the transformed master equation becomes

$$\frac{dP_x}{dt} = k_1\Omega(E^{-1} - 1)P + \frac{k_2}{\Omega}(E^2 - 1)x(x-1)P \tag{3.39}$$

where the number of molecules of A (assumed constant) is incorporated in the constant k_1, and the factor 1/2 arising from the consideration of the number of pairs $1/2x(x-1)$ is incorporated in the constant k_2. Note that the transition probabilities in Equation (3.39) have been expressed explicitly in terms of the parameter Ω representing the size of the system.

The next step calls for anticipation of the way in which the probability $P(x,t)$ itself would depend on the parameter Ω. It will be assumed that the probability $P(x,t)$ is a sharp peak located roughly at the values corresponding to the macroscopic part $\Omega\phi(t)$ with a width of the order of $\Omega^{1/2}$. Mathematically this can be represented by

$$P(x,t) = \Omega\phi(t) + \Omega^{1/2}y \tag{3.40}$$

The transformation (3.40) converts the variable x into the new variable y, and accordingly the probability distribution of x now becomes the probability distribution π of y:

$$P(x,t)\Delta x = \pi(y,t)\Delta y$$

or

$$\pi(y,t) = \Omega^{1/2}P(\Omega\phi(t) + \Omega^{1/2}y, t) \tag{3.41}$$

Equation (3.41), on appropriate differentiation, yields

$$\Omega^{1/2}\frac{\partial P}{\partial t} = \frac{\partial \pi}{\partial t} - \Omega^{1/2}\frac{d\phi}{dt}\frac{\partial \pi}{\partial y} \tag{3.42}$$

The operator E likewise can be restructured in terms of the new variable y:

$$E = \left(1 + \Omega^{-1/2}\frac{\partial}{\partial y} + \frac{\Omega^{-1}}{2}\frac{\partial^2}{\partial y^2} + \cdots\right) \tag{3.43}$$

Substitution of Equations (3.42) and (3.43) in (3.39) results in

$$\begin{aligned}\frac{\partial \pi}{\partial t} - \Omega^{1/2}\frac{d\phi}{dt}\frac{\partial \pi}{\partial y} &= k_1'\Omega\left(-\Omega^{-1/2}\frac{\partial}{\partial y} + \tfrac{1}{2}\Omega^{-1}\frac{\partial^2}{\partial y^2} - \cdots\right)\pi \\ &\quad + k_2\Omega\left(2\Omega^{-1/2}\frac{\partial}{\partial y} + 2\Omega^{-1}\frac{\partial^2}{\partial y^2} - \cdots\right) \\ &\quad \times (\phi + \Omega^{-1/2}y)(\phi + \Omega^{-1/2}y - \Omega^{-1})\pi \end{aligned} \tag{3.44}$$

where $k_1' = k_1/\Omega$. Frequently only terms up to second-order derivatives are retained and higher order terms are neglected.

Collecting terms of the order of $\Omega^{1/2}$ leads to the macroscopic equation

$$-\frac{d\phi}{dt} = -k_1' + 2k_2\phi^2 \tag{3.45}$$

while collection of terms of the order of Ω^0 leads to

$$\frac{\partial \pi}{\partial t} = 4k_2\phi\frac{\partial}{\partial y}y\pi + (k_1'/2 + 2k_2\phi^2)\frac{\partial^2 \pi}{\partial y^2} \tag{3.46}$$

Equation (3.46) represents the familiar Fokker–Planck equation which is linear in y. The solution to such an equation is known to be a Gaussian distribution. Calculation of the first and second moments therefore defines the fluctuations; this can be done by multiplying Equation (3.46) by y and y^2 to obtain (3.47) and (3.48), respectively.

$$\frac{d\langle y\rangle}{dt} = -4k_2\phi\langle y\rangle \tag{3.47}$$

$$\frac{d\langle y^2\rangle}{dt} = -8k_2\phi\langle y^2\rangle + k_1' + 4k_2\phi^2 \tag{3.48}$$

In both Equations (3.47) and (3.48), terms of the order of $\Omega^{1/2}$ have been neglected. Inclusion of a higher order term modifies (3.47) in that now an additional term containing σ^2 is involved. It would no more be possible to solve it explicitly and the mean of fluctuations cannot be obtained simply from its initial value but requires a knowledge of $\sigma^2(0)$ as well. Likewise it is possible to include higher order derivative terms in Equation (3.44). The equation, however, results in a nonlinear Fokker–Planck equation, and while one cannot expect to solve the complete equation it is possible to find the moments to any given order explicitly.

Hamilton–Jacobi Formulation

In another approach proposed by Kitahara (1976) the master Equation (3.25) is first converted into a Hamilton–Jacobi equation. For this purpose the general random variable X is scaled relative to the size of the system by defining $X = \Omega x$. The probability $P(X, t)$ is then expressed in terms of x as

$$P(X, t) = \psi(x, t) \tag{3.49}$$

Substituting this relation in the master Equation (3.25) leads to

$$\frac{1}{\Omega}\frac{\partial}{\partial t}\psi(x, t) = -H\left(x, \frac{1}{\Omega}\frac{\partial}{\partial x}\right)\psi(x, t) \tag{3.50}$$

where the Hamiltonian H is defined by

$$H(x, p) = \sum_r [1 - \exp(-rp)]w(x; r) \tag{3.51}$$

and

$$\Omega w(x; r) = W(X \to X + r) \tag{3.52}$$

Kubo, Matuso and Kitahara (1973) have assumed the exponential form for the time distribution function for Equation (3.49) in the limit $r \to \infty$. Expressing this as

$$\psi(x, t) \sim \exp[\Omega J(x, t)] \tag{3.53}$$

and substituting it in Equation (3.50) gives the familiar Hamilton–Jacobi equation for $J(x, t)$:

$$\frac{\partial J(x, t)}{\partial t} + H\left(x, \frac{\partial J(x, t)}{\partial x}\right) = 0 \qquad \text{where} \quad p = dJ/dx \tag{3.54}$$

The original master equation can thus be reduced to a new form which is much more convenient to handle.

The treatment for the one variable case can be easily generalised for the case of several random variables $(X_1 \ldots X_N)$ by rescaling them as $(X_i = x_i/\Omega, \ \ i = 1, \ldots N)$. The Hamiltonian H corresponding to Equation (3.51) for this general case can be written as

$$H(x_1, x_2, \ldots x_N, p_1, p_2, \ldots p_N) = \sum_{r_1, r_2, \ldots r_n} (1 - e^{-\sum_i r_i p_i}) \times w(x_1, x_2, \ldots x_N; r_2, r_2, \ldots r_N) \tag{3.55}$$

The corresponding multi-dimensional Hamiltonian–Jacobi equation would then be

$$\frac{\partial}{\partial t} J(\{x_i\}, t) + H\left(\{x_i\}, \left\{\frac{\partial J}{\partial x_i}\right\}\right) = 0 \tag{3.56}$$

The Hamiltonian–Jacobi equation can be solved subject to some initial condition

$$J(\{x_i\}, 0) = f(\{x_i\}) \tag{3.57}$$

The equation involves $x(t)$, $p(x, t)$ in an implicit form and it would be better to separate the dependencies. For this purpose we convert the problem into the $(x - p)$ domain related through t. This can be achieved if we start in the space $(x_1^0, x_2^0, \ldots x_N^0, p_1^0, p_2^0, \ldots p_N^0)$. From Equation (3.57)

$$\frac{\partial J}{\partial x_1}(\{x_i\}, 0) = \frac{\partial F}{\partial x_i}(\{x_i\}) = p_1^0 \tag{3.58}$$

where $i = 1, \ldots N$. The evolution of the $(x - p)$ domain is governed by the equations

$$\frac{\partial x_i(t)}{\partial t} = \frac{\partial H}{\partial p_i}(\{x_i\}, \{p_i\}) \tag{3.59}$$

and

$$\frac{\partial p_i(t)}{\partial t} = -\frac{\partial H}{\partial x_i}(\{x_i\}, \{p_i\}) \tag{3.60}$$

the solutions to which are given by

$$\{x_i\} = f_1(x_i^0; t) \tag{3.61}$$

$$\{p_i\} = f_2(x_i^0; t) \tag{3.62}$$

It is possible then to relate x_i and p_i using Equations (3.61) and (3.62):

$$p_i = r(x_i) = \frac{\partial J}{\partial x_i}, \qquad i = 1, \ldots N \tag{3.63}$$

Integration of this gives

$$J(x, t) = \int^x dx' r(x_i') \tag{3.64}$$

which is the desired solution. The general method is discussed in greater detail by Kitahara (1976) who has also applied it to situations where bistability or instability in the form of limit cycle prevails.

Systematic Expansion Method for Solving Transport Equations Derived from the Master Equation

A systematic expansion method for solving transport equations derived from the master equation has been recently proposed by Eder and Lackner (1983). The method, similar in its approach to the conceptually simpler van Kampen method, has several advantages, especially the accuracy of calculation of time dependent conditional average values of any general function $f(x)$ in the expansion parameter Ω. Conventionally, if $P(x, t \mid x_0)$ represents the probability of finding the system in state x at time t given that it was in state x_0 at $t = 0$, then all conditional averages can be calculated using the known relation

$$\langle f(x) \mid x_0 \rangle_t = \psi(x_0, t) = \int f(x) P(x, t \mid x_0)\, dx \tag{3.65}$$

with the initial condition

$$\psi(x_0, 0) = \int f(x)\, \delta(x - x_0)\, dx = f(x_0) \tag{3.66}$$

Substituting Equation (3.65) in the backward form of the master equation, the following integro-differential equation is obtained for the conditional average quantity $\psi(x_0, t)$, to be solved subject to initial condition (3.66):

$$\frac{\partial \psi(x_0, t)}{\partial t} + P_\Omega(x_0) \psi(x_0, t) = \int W_\Omega(x_0 \rightarrow x_1) \psi(x_1, t)\, dx_1 \tag{3.67}$$

Eder and Lackner (1983) preferred the backward form of the master equation since in general it is not possible to obtain conditional average quantities from the forward form unless the complete transition distribution function $P(x, t)$ is known [see Equation (3.66)].

The solution to Equation (3.67) is obtained using the expansion procedure. Following van Kampen the first step is to express the transition probability $W_\Omega(x_0 \rightarrow x')$ in a form

$$W_\Omega(x_0 \rightarrow x') = f(\Omega) \sum_{k=0}^{\infty} \frac{1}{\Omega^k} W_k[x_0, \Omega(x' - x_0)] \tag{3.68}$$

where $f(\Omega)$ is an arbitrary function and the transition probability is expressed in terms of the intensive variable x as against van Kampen's case where the extensive variable x/Ω is used. In the second step the desired function $\psi(x_1, t)$ is expanded in Taylor's series:

$$\psi(x_1, t) = \sum_{n=0}^{\infty} \frac{(x_1 - x_0)^n}{n!} \psi^n(x_0, t) \tag{3.69}$$

Using Equations (3.68), (3.69) and (3.67) we get

$$\frac{\partial \psi(x_0, t)}{\partial t} = \sum_{n=1}^{\infty} \frac{\beta_n(x_0)}{n!} \psi^n(x_0, t) \tag{3.70}$$

where

$$\beta_n(x_0) = \int W_\Omega(x_0 \to x_1)(x_1 - x_0)^n \, dx_1$$

or, using Equation (3.68),

$$\beta_n(x_0) = \frac{F(\Omega)}{\Omega^{n+1}} \sum_{k=0}^{\infty} \frac{\alpha_{n,k}(x_0)}{\Omega^k} \tag{3.71}$$

Redefining the variables in Equation (3.67) as

$$F(\Omega)t = \Omega^2 \tau, \qquad \psi(x, t) = \tilde{\psi}(x, \tau) \tag{3.72}$$

gives

$$\frac{\partial \psi(x, t)}{\partial \tau} = \sum_{n=1}^{\infty} \sum_{k=0}^{\infty} \frac{1}{\Omega^{n+k-1}} \frac{\alpha_{n,k}(x_0)}{n!} \psi^n(x_0, \tau) \tag{3.73}$$

where $\tilde{\psi}^n$ represents the nth derivative of $\tilde{\psi}$ with respect to x_0. Now we expand the parameter $\tilde{\psi}$ in powers of $1/\Omega$ yielding

$$\tilde{\psi}(x_0, \tau) = \sum_{l=0}^{\infty} \frac{1}{\Omega^l} \psi_l(x_0, \tau) \tag{3.74}$$

where ψ_l represents the Ω independent part of ψ.

Equation (3.74) can be written in a more convenient form as

$$\frac{\partial \psi_0}{\partial \tau} - \alpha_{1,0}(x_0) \frac{\partial \psi}{\partial x_0}(x_0. \tau) = 0 \tag{3.75}$$

$$\frac{\partial \psi_1}{\partial \tau} - \alpha_{1,0}(x_0) \frac{\partial \psi_1}{\partial x_0}(x_0, \tau) = \sum_{s=2}^{l+1} \sum_{k=1}^{s} \frac{\alpha_{k,s-k}(x_0)}{k!} \frac{\partial^k \psi_{l+1-s}}{\partial^k x_0}(x_0, \tau) \tag{3.76}$$

The solution to these equations can be obtained using the method of characteristics and the general solution subject to appropriate initial condition is

$$\langle f(x)\,|\,x_0\rangle_\tau = \psi(x_0, \tau)$$

$$= f(\bar{x}) + \frac{1}{\Omega}\left[f'(x)\phi_1(x_0, x) + f''(\bar{x})\frac{\sigma_{sc}^2(x_0, x)}{2}\right] + O\left[\frac{1}{\Omega^2}\right] \tag{3.77}$$

where the solution of Equation (3.75) gives

$$\psi(x_0, \tau) = F(\bar{x}(\tau)) \tag{3.78}$$

yielding

$$\bar{x}(\tau) = x_0(0) + G^{-1}[G(x_0) - \tau] \tag{3.79}$$

with $G^{-1}(x)$ representing the inverse of $G(x)$ defined as

$$G(x) = -\int \frac{dx_0}{\alpha_{1,0}(x_0)} \tag{3.80}$$

The parameters σ_{sc}^2 and ϕ_1 are defined in terms of $\bar{x}(\tau)$ as

$$\sigma_{sc}^2[x_0, \bar{x}] = \alpha_{1,0}^2(\bar{x}) \int_{x_0}^{\bar{x}} \frac{\alpha_{2,0}(z)}{\alpha_{1,0}^3(z)}\, dz \tag{3.81}$$

$$\phi_1[x_0, \bar{x}] = \alpha_{1,0}(x) \int_{x_0}^{\bar{x}} \left\{\frac{\alpha_{1,1}(z)}{\alpha_{1,0}^2(z)} + \frac{\alpha_{2,0}(z)}{2}\left[\frac{\alpha_{1,0}'(\bar{x}) - \alpha_{1,0}'(z)}{\alpha_{1,0}^3(z)}\right]\right\} dz \tag{3.82}$$

To illustrate the use of this method, let us consider the specific cases when $f(x) = x$ and x^2. According to Equation (3.77) the general solution is

$$f(x|x_0) = \langle x|x_0\rangle = \bar{x} + \frac{1}{\Omega}\phi_1(x_0, x) + O\left[\frac{1}{\Omega^2}\right] \tag{3.83}$$

$$f(x|x_0) = \langle x^2|x_0\rangle_\tau = \bar{x}^2 + \frac{1}{\Omega}\{\sigma_{sc}^2(x_0, \bar{x}) + 2x\phi_1(x_0, \bar{x})\}$$

$$+ O\left[\frac{1}{\Omega^2}\right] \tag{3.84}$$

Now defining the variance $\sigma^2 = \langle x^2\rangle - \langle x\rangle^2$ and using Equations (3.83) and (3.84) we obtain

$$\sigma^2[x_0, \tau] = \frac{1}{\Omega}\sigma_{sc}^2(x_0, \bar{x}) + O\left[\frac{1}{\Omega^2}\right] \tag{3.85}$$

making the meaning of the scaled variance quite clear. The use of the conventional van Kampen method would not give the additional function ϕ entering in Equations (3.83) and (3.84) unless higher order approximations are used. The present method gives the correct expansion of an average quantity without involving excessive computation and is therefore advantageous.

3.2.4 Modified Fokker–Planck Approximation

As described earlier, the conventional Fokker–Planck approximation for chemical reactions is obtained by truncating the Kramers–Moyal expansion of the master equation after the second term. In general, the resulting equation is much easier to handle than the original master equation and has gained considerable popularity. The approach is, however, known to have certain shortcomings which were generally stated in Section 3.3.3. In the present section we shall make these shortcomings explicitly clear by considering a specific example, and then suggest methods of overcoming them by appropriate modifications.

To begin with, we consider the case of a simple reversible reaction

$$2A \underset{k_{-1}}{\overset{k_1}{\rightleftharpoons}} B$$

The deterministic equation describing the evolution of species can be written as

$$\frac{dC_B}{dt} = -\frac{1}{2}\frac{dC_A}{dt} = k_1 C_A^2 - k_{-1} C_B \tag{3.86}$$

Also, if N_A and N_B represent, respectively, the number of particles of species A and B, then the general relation

$$N_A + 2N_B = N_A^0 + 2N_B^0 = N = V_C C \tag{3.87}$$

holds. Following the development in Section 3.2.1 the master equation for this situation can be written as

$$\frac{dP(N_B)}{dt} =$$

$$\frac{k_1}{V}(N - 2N_B + 2)(N - 2N_B + 1)P(N_B - 1) + k_{-1}(N_B + 1)P(N_B + 1)$$

$$-\frac{k_1}{V}(N - 2N_B)(N - 2N_B - 1)P(N_B) - k_{-1}N_B P(N_B) \tag{3.88}$$

The stationary solution to this equation can be easily found using the general methods described earlier:

$$P_{st}(N_B) = \frac{1}{Z}\frac{(N_{Be})^{N_B}(N - 2N_{Be})^{N-2N_B}}{N_B!\,(N - 2N_B)!} \tag{3.89}$$

where N_{Be} represents the equilibrium value of N_B and can be obtained from

$$\frac{N_{Be}}{(N - 2N_{Be})^2} = \frac{k_1}{k_{-1}V} = \frac{K}{V} \tag{3.90}$$

Let us now write the corresponding Fokker–Planck equation as given by Kramers–Moyal expansion:

$$\frac{\partial P(N_B)}{\partial t} = \sum_{n=1}^{\infty}\frac{(-1)^n}{n!}\frac{\partial^n}{\partial N_B^n}M_n(N_B)P(N_B) \tag{3.91}$$

The equation can be written more explicitly in the form

$$\frac{\partial \pi(C_B)}{\partial t} = \left\{-\frac{\partial}{\partial C_B}m_1(C_B) + \frac{1}{2V}\frac{\partial^2}{\partial C_B^2}m_2(C_B)\right\}\pi(C_B) \tag{3.92}$$

where the drift and diffusion terms are given by

$$m_1(C_B) = k_1(C - 2C_B)\left(C - 2C_B - \frac{1}{V}\right) - k_{-1}C_B \tag{3.93}$$

$$m_2(C_B) = k_1(C - 2C_B)\left(C - 2C_B - \frac{1}{V}\right) + k_{-1}C_B \tag{3.94}$$

and π represents the transformed probability related to P through volume V. The steady state solution to Equation (3.92) can be obtained using the relation (3.36). Clearly the solution to Equation (3.92) does not coincide with the continuum limit of the stationary probability of the master equation. Also, one observes from the master Equation (3.88) that, while it possesses the natural boundaries at $N_B = 0$ and $N_B = N/2$, the feature is not conserved in the Fokker–Planck approximation (3.92). The noise term therefore drives the system beyond the natural boundaries.

These shortcomings in the use of the Fokker–Planck equation are not important near the equilibrium state; however, the wings of the probability are considerably changed due to this approximation.

Calculations of parameters where the wings of the probability cannot be ignored (such as in the calculations of the mean passage time), cannot therefore be undertaken reliably using the Fokker–Planck approximation.

Grabert, Hanggi and Oppenheim (1983) took a more concrete view of this general situation and suggested a connection between the deterministic description in terms of the Onsager type of transport equations and the stochastic description in terms of the Fokker–Planck equation. Earlier, Grabert and Green (1979) and Grabert, Hanggi and Oppenheim (1980) had investigated such connection from a statistical mechanical viewpoint. The straightforward application of this theory to chemical reactions is somewhat incorrect due to the discrete state space character of chemical reactions. However, for large homogeneous systems, the discreteness should be of minor importance.

Essentially there are two steps in this scheme: first obtain the transport equations of the Onsager type, and second connect the transport laws to the Fokker–Planck equation. For a complete derivation of the transport equation, the general works of Grabert and Green (1979), Grabert, Graham and Green (1980), and Grabert, Hanggi and Oppenheim (1983) may be consulted, but for the sake of completeness we shall include the final form of the equation. Thus, we consider a homogeneous isothermal system occupying a volume V and containing an n-component ideal mixture of gases or liquids, among which the following reactions occur:

$$\sum_{i=1}^{n} \nu_i^{\alpha} X_i = 0 \qquad (\nu = 1, \ldots r) \tag{3.95}$$

The deterministic equation for this scheme would be

$$\frac{dC_i}{dt} = \sum_{\alpha=1}^{r} \nu_i^{\alpha}[k_{\alpha}\pi_-^{\alpha}(C_j) - k'_{\alpha}\pi_+^{\alpha}(C_j)] \tag{3.96}$$

where k_{α} and k'_{α} refer to volume independent forward and backward rate constants, ν_i is the stoichiometric coefficient of the αth reaction, and $\pi_-^{\alpha}, \pi_+^{\alpha}$ denote the products of the particle densities of reactants and products.

The deterministic law given by Equation (3.96) can be written in terms of the Onsager type of transport equation as

$$\frac{dC_i}{dt} = -\sum_{j=1}^{n} l^{ij} \frac{\partial g}{\partial C_j} \tag{3.97}$$

Here the parameters g and l^{ij} are functions of the intensive variables C_i and T only and are defined by

$$g(C_i, T) = G/V \tag{3.98}$$

$$l^i[\varepsilon_i, l^{ij}(\varepsilon_i, T)] = \frac{L^{ij}}{V} = \sum_{\alpha=1}^{r} v_i^\alpha v_j^\alpha \left(\frac{k'_\alpha}{k_B T}\right) \frac{\pi_+^\alpha(C_i) - \bar{K}_\alpha \pi_-^\alpha(C_i)}{\ln \pi_+^\alpha(C_i) - \ln \bar{K}_\alpha \pi_+^\alpha(C_i)} \tag{3.99}$$

where k_B is the Boltzmann constant and G the free energy of the system given by

$$\Delta G = G_0 + \sum_{i=1}^{n} N_i(\phi_i + k_B T \ln N_i) \tag{3.100}$$

Here G_0 refers to the free energy of the inert component such as solvent, and N_i is the number of particles of species i. The parameters ϕ and G_0 are temperature dependent. As a result of reaction the particle number of species i varies, and for the assumed stoichiometry can be written in terms of the extent of reaction:

$$\frac{dN_i}{dt} = \sum_{\alpha=1}^{r} v_i^\alpha \frac{d\lambda_\alpha}{dt} \tag{3.101}$$

where λ_α represents the progress variable for the αth reaction and can be simply obtained as

$$\frac{d\lambda_\alpha}{dt} = V(k_\alpha \pi_-^\alpha(C_i) - k' \pi^\alpha{}_+(C_i)) \tag{3.102}$$

It is therefore possible to know the variation in the particle number of species i and thus the variation of free energy G. The variation of free energy G could be expressed in an alternative form:

$$\frac{dG}{dt} = \frac{\partial G}{\partial N_i}\frac{\partial N_i}{\partial t} = (\mu_i)\frac{dN_i}{dt} = \frac{\partial G}{\partial \chi_i}\frac{\partial \lambda_i}{\partial t} = (\chi_i)\frac{d\lambda_i}{dt} \tag{3.103}$$

in terms of chemical potential or chemical affinities. Using these relations it is also possible to relate χ_i to μ_i, N_i, λ_i or vice versa.

It appears that the transport Equation (3.97) is a more formal and complicated way of writing the same deterministic equation. While this is indeed so, it exhibits a closer connection to quantities estimated during the stochastic description, making the relationship between the stochastic and deterministic formalisms even more explicit.

To move over to an equation of stochastic description using the deterministic transport law one could write the Fokker–Planck equation

$$\frac{\partial \pi(C_k)}{\partial t} = \sum_{ij} \frac{\partial}{\partial C_i} l^{ij}(C_k) \left[\frac{\partial g(C_k)}{\partial C_j} + \frac{k_B \tau}{V} \frac{\partial}{\partial C_j} \right] \pi(C_k) \tag{3.104}$$

For the bimolecular reaction considered earlier this equation can be written more explicitly as

$$\frac{\partial \pi(C_B)}{\partial t} = \left[-\frac{\partial}{\partial C_B} \alpha_1(C_B) + \frac{1}{2V} \frac{\partial^2}{\partial C_B^2} \alpha_2(C_B) \right] \pi(C_B) \tag{3.105}$$

where the drift and diffusion terms can be identified as

$$\alpha_1 = k_1(C - 2C_B)^2 - k_{-1} C_B + \frac{1}{2V} \frac{\partial m(C_B)}{\partial C_B} \tag{3.106}$$

$$\alpha_2 = 2k_B T l(C_B) = m(C_B) \tag{3.107}$$

and $m(C_B)$ is defined as

$$m(C_B) = 2k_{-1} \frac{C_B - K(C - 2C_B)^2}{\ln C_B - \ln K(C - 2C_B)^2} \tag{3.108}$$

The Fokker–Planck equation as obtained above preserves the natural boundaries $C_B = 0$, $C_B = C/2$ in consonance with the original master equation. Also, the diffusion term vanishes near the boundaries, clearly indicating that the noise characterised by nonlinear diffusion does not drive the system beyond the boundaries. The steady state probability distribution of the Fokker–Planck equation gives the same result as obtained from the master equation, viz.

$$\pi_{\text{st}} \propto \exp\left\{ -\frac{V}{k_B T} g(C_B) \right\} \tag{3.109}$$

The equation describes the free energy $g(C_B)$ as $-k_B T$ times the stationary probability distribution of the master equation or the associated Fokker–Planck equation. One can clearly see therefore the advantage in the use of the transport equation as against the deterministic equation, since the formal definitions of some of the parameters in the transport equations emerge more naturally from the stochastic description.

A comparison of the Fokker–Planck description constructed from the transport equations and by using the Kramers–Moyal expansion shows that the drift term in the former differs from that in the latter by a term $O(1/V)$. Thus

$$\alpha_1(C_B) = m_1(C_B) + O\left(\frac{1}{V}\right) \tag{3.110}$$

and the diffusion terms in the four approaches are obtained as

$$\alpha_2(C_B) = m_2(C_B) + \sum_{n=1}^{\infty} \frac{(-1)^n}{(n+1)!} m_{2+n}(C_B) \left[\frac{\chi(C_B)}{k_B T}\right]^n \tag{3.111}$$

The modified Fokker–Planck approximation has clear-cut advantages as shown above. Additionally, the approach is easily amenable to extensions to situations such as when nonideal mixtures are involved. The conventional birth–death type of master equation fails in such situations. Also, in systems where pressure and temperature fluctuations occur and where eventually conversion has to be taken into account, the approach holds promise.

3.2.5 Method of Continued Fraction

In this section we shall outline the method of continued fraction to obtain exact stationary solution and time-dependent solution to certain classes of stochastic processes described in terms of the master equation or the equivalent Fokker–Planck equation. It is known that the exact solution of the master equation is feasible only in a few cases, and even there the results of interest such as the spectral properties and correlation functions are difficult to obtain in explicit form using the exact solution. It is necessary therefore to devise schemes which would yield exact solutions that can be readily adapted for further use. The method of continued fraction is especially suitable in this regard (Hanggi and Talkner, 1978; Grossmann and Schranner, 1978; Hanggi, 1978). In the present section we shall describe this method by considering the specific example of a unit-step birth and death process, and subsequently extend it to cases where two or more step jumps occur.

The master equation governing the evolution of probability for a unit-step birth and death process derived in Chapter 2 can be written as

$$\frac{dp_0}{dt} = -\lambda_0 p_0(t) + \mu_1 p_1(t) \tag{3.112}$$

$$\frac{dp_i}{dt} = \lambda_{i-1} p_{i-1}(t) + \mu_{i+1} p_{i+1}(t) - (\lambda + \mu_i) p_i(t) \tag{3.113}$$

$i = 1, 2, \ldots$. We assume that the stationary solution to the physical system described as above is unique and positive. Even the absorbing states $\lambda_1 = \mu_i = 0$ are excluded from consideration. We define now the nearest neighbour transition function

$$\psi_i = \frac{p_i}{p_{i-1}}, \qquad i = 1, 2, \ldots \tag{3.114}$$

and using the definition in Equation (3.113) and rearranging it at steady state we get

$$\psi_{i,s} = \frac{\lambda_{i-1}}{(\lambda_i + \mu_i) - \mu_{i+1}\psi_{i+1,s}} \tag{3.115}$$

where the subscript s denotes the steady state condition. This equation can be written in the continued fraction form as with

$$a_{-1} = 1, \qquad a_0 = (\lambda_0 + \alpha_0)$$

$$a_1 = (\lambda_1 + \mu_1 + \alpha_1)(\lambda_0 + \alpha_0) - \mu_1 \lambda_0$$

$$b_1 = 0, \qquad b_2 = \beta_2 \lambda_0$$

and a_n and b_n follow the recursive relation

$$a_{n-1} = (n-1)a_{n-2} - \beta_{n-1}\alpha_{n-3}\frac{a_{n-2}a_{n-4}}{a_{n-3}} - (b_{n-1} + \mu_{n-1}a_{n-3})\left(\alpha_{n-3}\frac{b_{n-2} + \mu_{n-2}a_{n-4}}{a_{n-3}} + \lambda_{n-2}\right) \tag{3.135}$$

$$b_n = \beta_n[\alpha_{n-3}(b_{n-2} + \mu_{n-2}a_{n-4}) + \lambda_{n-2}a_{n-3}] \tag{3.136}$$

The stationary solution to the problem, as before, can be obtained as

$$p_{n,s} = p_{0,s}\frac{l_{n-1}}{m_n + m_{n-1}a_{n-2}\beta_{n+1}\psi_{n+1}} \tag{3.137}$$

$$\psi_{i,s} = \frac{\lambda_{i-1}}{(\lambda_i+\mu_i)-}\,\frac{\mu_{i+1}\lambda_i}{(\lambda_{i+1}+\mu_{i+1})-}\cdots$$

$$= \left(\frac{\lambda_{i-1}}{\mu_i}\right)\frac{1}{\left(1+\dfrac{\lambda_i}{\mu_i}\right)-}\,\frac{\lambda_i/\mu_i}{\left(-1+\dfrac{\lambda_{i+1}}{\mu_{i+1}}\right)-}\cdots = \left(\frac{\lambda_{i-1}}{\mu_i}\right)q_{i,s} \tag{3.116}$$

Similarly, using Equation (3.112),

$$\psi_i = \frac{\lambda_0}{\mu_i} \tag{3.117}$$

Equation (3.114) expresses the solution to the problem which can be extended up to the normalisation constant:

$$p_{n,s} = p_{0,s}^{(0)}\prod_{i=1}^{n}\psi_{i,s} \tag{3.118}$$

With the help of Equations (3.116) and (3.117), this can be written in a familiar form:

$$p_{n,s} = p_{0,s}\prod_{i=1}^{n}\frac{\lambda_{i-1}}{\mu_i} \tag{3.119}$$

Equation (3.119) represents the exact stationary solution to the problem.

In order to obtain the transient solution to the problem we define the Laplace transform

$$p_n(z) = \int_0^\infty e^{-zt}p_n(t)\,dt \tag{3.120}$$

which, when used in the governing equation, yields

$$\mu_1 p_1^{(n_0)}(z) = -\delta_{0,n_0} + (\lambda_0 + z)p_0^{(n_0)}(z) \tag{3.121}$$

$$\mu_{i+1}p_{i+1}^{(n_0)}(z) = -\lambda_{i-1}p_{i-1}^{(n_0)}(z) + (\lambda_i + \mu_i + z)p_i^{(n_0)}(z) - \delta_{i,n_0} \tag{3.122}$$

$i = 1, 2, \ldots$. $p_n^{(n_0)}(z)$ in these equations represents the Laplace transformation of the conditional probability $p(n, t \mid n_0, 0)$. Choosing the zero initial condition ($n_0 = 0$), we rewrite Equations (3.121) and (3.122) to yield the continued fraction as

$$p_0^{(0)}(z) = \cfrac{1}{\lambda_0 + z - \mu_1 \dfrac{p_1^{(0)}(z)}{p_0^{(0)}(z)}}$$

$$= \frac{1}{\lambda_0 + z -}\,\frac{\mu_1\lambda_0}{\lambda_1 + \mu_1 + z -}\,\frac{\mu_2\lambda_1}{\lambda_2 + \mu_2 + z -}\cdots$$

$$\times \cfrac{\mu_i\lambda_{i-1}}{\lambda_i + \mu_i + z - \mu_{i-1}\dfrac{p_{i+1}^{(0)}(z)}{p_i^{(0)}(z)}}\, p_0^{(0)}(z)$$

$$\times\, p_0^{(0)}(z) = \frac{1}{z+}\,\frac{\lambda_0}{1+}\,\frac{\mu_1}{z+}\,\frac{\lambda_1}{1+}\,\frac{\mu_2}{z+}\cdots \tag{3.123}$$

that converges uniformly.

Defining the transition function analogously to Equation (3.114) we have

$$\psi_n^{(0)}(z) = \frac{p_n^{(0)}(z)}{p_{n-1}^{(0)}(z)} \tag{3.124}$$

Using the definition in Equation (3.121) we obtain

$$\psi_{n+1}^{(0)}(z) = \frac{\lambda_n}{\lambda_{n+1} + \mu_{n+1} + z -}\,\frac{\mu_{n+2}\lambda_{n-1}}{\lambda_{n+2} + \mu_{n+2} + z -}\cdots \tag{3.125}$$

The final solution in the form of explicit continued fraction can be written following Equation (3.124) as

$$p_n^{(0)}(z) = p_0^{(0)}(z)\prod_{i=1}^{n}\psi_i^{(0)}(z)$$

$$= \frac{N_{n-1}/B_n(z)}{B_{n+1}(z)/B_n(z) -}\,\frac{\mu_{n+1}\lambda_n}{(\lambda_{n+1} + \mu_{n+1} + z) -}\cdots \tag{3.126}$$

where $B_n(z)$ satisfies the recursion relation

$$B_{n+1}(z) = (\lambda_1 + \mu_n + z)B_n(z) - \mu_n\lambda_{n-1}B_{n-1}(z) \tag{3.127}$$

If the initial condition is chosen such that n_0 is not equal to zero, the general solution reads:

$$p_i^{(n_0)}(z) = \frac{\mu_{n_0} B_i(z)}{B_{n_0}(z) m_1} \left(\frac{1}{B_{n_0+1}(z)/B_{n_0}(z)-} \; \frac{\mu_{n_0+1}(z)\lambda_{n_0}}{(\lambda_{n_0+1} + \mu_{n_0+1} + z)-} \cdots \right)$$

$$i = 0, 1, \ldots n_0$$

$$= \frac{N_{i-1} B_{n_0}(z)}{N_{n_0-1} B_i(z)} \left(\frac{1}{B_{i+1}(z)/B_i(z)-} \; \frac{\mu_{i+1}\lambda_i}{(\lambda_{i+1} + \mu_{i+1} + z)-} \cdots \right)$$

$$i = n_0, n_0 + 1, \ldots \quad (3.128)$$

The conditional probability $p(n, t \mid n_0)$ can be obtained by inverting these equations.

A straightforward extension of the method to more complex cases such as those involving two jumps is possible. Such cases are encountered in practice when one considers reaction schemes with bimolecular reaction steps. For these cases the stationary birth–death master equation can be formulated using a procedure analogous to that for Equation (3.112):

$$-(\lambda_0 + \alpha_0) p_{0,s} + \mu_1 p_{1,s} + \beta_2 p_{2,s} = 0 \quad (3.129)$$

$$\lambda_0 p_{0,s} - (\lambda_1 + \mu_1 + \alpha_1) p_{1,s} + \mu_2 p_{2,s} + \beta_3 p_{3,s} = 0 \quad (3.130)$$

$$\alpha_{i-2} p_{i-2,s} + \lambda_{i-1} p_{i-1,s} - (\lambda_i + \mu_i + \alpha_i + \beta_i) p_{i,s}$$

$$+ \mu_{i+1} p_{i+1,s} + \beta_{i+2,s} + p_{i+2,s} = 0, \qquad i = 2, 3, \ldots \quad (3.131)$$

Defining, as before, the nearest neighbour transition function $\psi_{i,s} = p_{i,s}/p_{i-1,s}$ and using in Equations (3.129) and (3.130) we get

$$\psi_{1,s} = (\lambda_0 + \alpha_0)/(\mu_1 + \beta_2 \psi_{2,s}) \quad (3.132)$$

$$\psi_{2,s} = \frac{(\lambda_0 + \alpha_0)(\lambda_1 + \mu_1 + \alpha_1) - \mu_1 \lambda_0}{\beta_2 \lambda_0 + (\lambda_0 + \alpha_0)(\mu_2 + \beta_3 \psi_{3,s})} \quad (3.133)$$

where $\psi_{3,s}$ can be obtained similarly, and the general relation for $\psi_{n,s}$ becomes

$$\psi_{n,s} = \frac{a_{n-1}}{b_n + a_{n-2}(\mu_n + \beta_{n-1} \psi_{n+1,s})} \quad (3.134)$$

where

$$l_n = a_0 a_1 a_2 \ldots a_n, \qquad n = 0, 1, 2, \ldots$$

$$m_n = (b_n + a_{n-2}\mu_n) m_{n-1} + \beta_n a_{n-1} a_{n-3} m_{n-2} \quad (3.138)$$

with $m_0 = 1$, $m_1 = \mu_1$. For the particular case when $\beta_i = 0$ for $i+ = 2, \ldots (N-1)$, we have the following well-known relation:

$$p_{n,s} = p_{0,s} \frac{a_{n-1}}{\mu_1 \mu_2 \ldots \mu_n}, \qquad n = 1, 2 \tag{3.139}$$

3.3 Limitations, Modifications and Alternatives to the Birth–Death Type of Formalism

The birth and death formalism, conventionally used to describe the stochastic formulation of chemical kinetics, is subject to a number of objections. The primary objections are: (1) the inadequacy of the approach to take account of spatial variations, and (2) the lack of rigorous microscopic justification of the Markovian master equation in the number of particle space (Oppenheim, Shuler and Weiss, 1969; Keiszer, 1972; Kurtz, 1972). The latter, while of considerable importance from a fundamental viewpoint, will not be considered here in any detail. The important conclusion may be stated: the conventionally used deterministic laws are a highly contracted form of the more rigorous Liouville equations where proper account is taken of the dynamics of interaction, including correlations, to obtain appropriate transport equations for the system. The conventional stochastic formulations also represent a highly contracted form of a more rigorous stochastic approach. Thus, for example, Keiszer (1972) demonstrated by considering a simple reversible isomerisation reaction scheme that the process could be classified into four different regimes on the basis of the relaxation times involved. In the first time region, which lasts up to the order of the mean free time, it is necessary to solve the full Liouville equation. In the subsequent regimes it is possible to contract the original rigorous equation, and it is only in the regime which occurs after the order of a microscopic chemical reaction time that the conventional stochastic theory becomes valid. The conventional stochastic theory is thus essentially coarse grained, and it is possible to involve some practical realities in the formulation provided the associated mathematical difficulties can be overcome.

One such, more rigorous, theory that takes account of the individual collisions between the molecules and their consequences has recently been proposed by Keizer and Conlan (1983). The theory clearly accounts for the fact that all collisions need not necessarily lead to reactive changes and in this sense is fine grained. Of course, to characterise collisions a proper reaction mechanism has to be taken into consideration and the general theory developed is applicable to unimolecular, bimolecular and mixed type of reactions. In the proper limits it is possible to obtain the conventional birth–death formulation and thus the general formulation represents a further step in the improvement of the stochastic description. Notwithstanding this and more such general formulations of stochastic processes for chemical reactions, the conventional stochastic theory, as used in this text, provides a first-order approximate description of chemical reactions, and is adequate for our purpose.

Using the conventional stochastic theory we shall consider in some detail the implications of the first objection, namely, the inadequacy of the birth–death formalism to account for spatial variation (such as due to diffusion). The homogeneous stochastic theory rests on the assumption that the processes which tend to distribute the particles uniformly (viz. diffusion) in a given volume of system are far too rapid in relation to the processes that alter the particle number (viz. chemical reactions). Clearly, this situation may not always hold true and spatial inhomogeneities can develop. Quantitative criteria measuring the relative importance of one over the other can be developed to ascertain the extent of uniformity. Thus, in a small volume δV of the system volume V, the probability that a molecule diffuses out of δV must exceed the probability that the molecule undergoes a reactive event within the volume δV, for the homogeneous model to apply. The probability of transport can be measured in terms of the extent of elastic collisions, and the reaction probability in terms of inelastic or reactive collisions. The quantification of this condition for typical chemical reaction schemes has been presented by Turner and Gillespie (1978) and discussed in terms of the phase–space formulation by Nicolis and Prigogine (1971).

The spatial inhomogeneities in a reacting mixture can be important and result under practical conditions of operation. While more elaborate theories that rigorously account for the basic physics involved can be formulated, the associated mathematical complexities soon render the use of these theories impractical. It is desirable therefore to provide a simple means of taking account of these inhomogeneities. Such a formalism can be easily developed if one discretises the space in terms of number of cells. Within each cell it is assumed that the reaction is homogeneous so that the usual concept of birth–death formalism holds. It is assumed that the cells are interconnected and the coupling of cells provides a means of accounting for the transport to and from the cell to the surroundings. Again this transport may be visualised in terms of a gain or loss to the cell. The two mechanisms of reaction and transport responsible for the generation and loss of species from a cell can thus be written in terms of the birth–death formalism.

To illustrate the point we shall consider a single stochastic variable X_k, where X represents, say, the composition in the kth cell. The state of the system, subdivided into k cells C_k, can then be described in terms of the evolution of the probability function $P(X_1, X_2, \ldots X_k, t)$. The multivariate master equation constructed will have the form

$$\frac{\partial P(\{X_k\}, t)}{\partial t} = \sum_{k=1}^{K} \{R_k p(\{X_k\}, t) + \tilde{D}[(X_k + 1)P(X_{k-1} - 1, X_k + 1) - X_k P(X_{k-1}, X_k)] + \tilde{D}[(X_k - 1)P(X_k + 1, X_{k+1} - 1) - X_k P(X_k, X_{k+1})]\} \quad (3.140)$$

where separate gain and loss contributions for the reaction and transport processes are explicitly written, and $\tilde{D}$ refers to the stochastic diffusivity that is related to the Fickian diffusivity by a relation obtained from kinetic theory arguments (Nicolis and Prigogine, 1971) for ideal mixtures:

$$\tilde{D} = D/\lambda l \quad (3.141)$$

where λ refers to the mean free path and l to the length of a mixing cell C_k. The choice of the volume of the mixing cell (λ^3) is arbitrary, but it is necessary to exercise discretion, especially since it is assumed that a homogeneous reaction that is described by the birth–death type of master equation occurs within it. It was made clear at the beginning of this section that such a formulation is only approximate since it ignores the molecular degrees of freedom such as velocity distribution, internal energy of states, etc. This imposes a lower limit on the size of the system for which the homogeneous stochastic description could be applied. As a reliable measure it is advisable to take l a few times the mean free path λ.

Equation (3.140) represents the master equation for the evolution of the probability and takes into account the gain and loss due to reaction and transport over all the cells C_k. The equation can be solved using the methods presented earlier. Alternatively, the stochastic simulation algorithm of Gillespie (1977) can be used. In general, when the number of cells considered is large, the solution to the equation becomes cumbersome.

It is possible to reduce the complexities still further by considering a two-cell system, one of volume ΔV and the other of the remaining volume $(V - \Delta V)$. It is assumed that within itself each cell is homogeneous and that the cells are connected to each other via a transport mechanism. This simple model is equivalent to the multi-cell model, where it is assumed that all cells except one are described in an averaged fashion. This kind of mean field approximation considerably reduces the attendant complexities in the multi-cell model, although now it seems applicable only to globally homogeneous situations. Such a two-cell system can be described in terms of the probability of finding X_1 molecules of type X in volume V_1 $(= V - \Delta V)$ and X molecules in volume ΔV at time t. The following final master equation can be written for situations involving diffusion only as the transport mechanism (Nicolis *et al.*, 1974, 1976; Malek-Mansour and Nicolis, 1975; Prigogine *et al.*, 1976):

$$\frac{\partial P_{\Delta v}(X,t)}{\partial t} = R_{\Delta v} P_{\Delta v}(X,t) + \hat{D}\langle X\rangle[P_{\Delta v}(X-1,t) - P_{\Delta v}(X,t)] + \hat{D}[(X+1)P_{\Delta v}(X+1,t) - XP_{\Delta v}(X,t)] \tag{3.142}$$

where the transport mechanism is written explicitly and $R_{\Delta v}$ represents the abbreviation for the gain and loss due to reaction in volume Δv. Equation (3.142) involves the mean of X which as usual is defined as $\sum XP_{\Delta v}(X,t)$.

The formulation as presented above reveals that the reduced stochastic description retains the nonlinearity; the essential features of chemical dynamics are therefore retained. Also, since attention in this theory is focussed on a small fluctuating volume Δv, the theory has a local character and in this sense provides a kind of stochastic theory for macroscopically homogeneous systems.

3.4 Langevin Equation

The previous sections were concerned with the description of stochastic processes in terms of the master equation and the methods that can be employed for its solution. The stochasticity in these systems arises essentially as a result of the discrete nature of the systems. The fluctuations in a chemical reaction arise because the process involves individual discrete reactive collisions. The fluctuations are therefore an intrinsic part of the system behaviour and generate what one may call an internal noise. The stochasticity as noted above should be distinguished from another form of stochasticity where the disturbance are external to the system. Thus, for instance, in the operation of a CSTR the input flow rate, heat transfer coefficient, etc. could vary about a certain mean value generating external disturbances on an otherwise deterministic system. A second example could be the occurrence of chemical reaction in the presence of diffusion and convection. When the flow field is turbulent the precise variation of the velocity components in the form of a functional relation is seldom known. However, the statistical properties of the flow can be known and incorporated in the phenomenological equation to take account of the stochasticity. The examples just presented, unlike those in the previous sections, involve an external source of disturbance imposed on an otherwise deterministic system.

The examples also indicate the form of the stochastic equation:

$$\frac{dx}{dt} = f(x) + \alpha L(t) \tag{3.143}$$

The disturbances are assumed to be rapidly varying, thus creating two time scales in Equation (3.143): One time scale represents the evolution of the macroscopic system, and the other the evolution of fluctuations in the macroscopic system. The dependent variables thus follow the conventional Newton's laws, whereas the stochastic evolution is determined by the fluctuations. Equation (3.143) as depicted here is called the Langevin equation after Langevin (1908) who first applied it to the case of Brownian motion. The equation is termed as linear if the dependence on the system variable x is linear and the noise term is additive. It is possible to conceive of a situation where the term $f(x)$ is not linear but the coefficient α of the fluctuating force is still constant.

Such an equation is generally referred to as quasilinear to express the constancy of the coefficient.

Equation (3.143) needs further qualification before it acquires any precise meaning. Since the fluctuating term $L(t)$ represents an average force, it is possible to write $\langle L(t)\rangle = 0$. Also, in view of the rapidity of its variation, it is possible to write $\langle L(t_1)L(t_2)\rangle = G\delta(t_1 - t_2)$. These two conditions define the first two moments. Further, it is assumed that all other higher cumulants vanish or that a relation of the type

$$\langle L(t_1)L(t_2)L(t_3)L(t_4)\rangle = \langle L(t_1)L(t_2)\rangle\langle L(t_3)L(t_4)\rangle + \cdots \quad (3.144)$$

is valid. This completely specifies the stochastic system and the fluctuating force defined in this way is called a white noise. The mathematical convenience of such a fluctuating force in a stochastic process is similar to that of a delta function in deterministic systems.

The presence of the fluctuating force in Equation (3.143) induces a spreading in the possible paths of evolution. The distribution of this defines a probability density, the evolution of which is governed by the Langevin equation. A relation between the linear Langevin equation and the equation of evolution of probability density was first obtained by Fokker (1914) and Planck (1917). This equation is usually referred to as the Fokker–Planck equation and we are already familiar with the general form of this equation. The question is: what relations exist between the coefficients of the Fokker–Planck equation and the corresponding Langevin equation? To deduce these we reexamine the definition of the coefficients α_1, α_2 of the Fokker–Planck equation

$$\frac{\partial P(x,t)}{\partial t} = \left[-\alpha_1(x)\frac{\partial}{\partial x} + \tfrac{1}{2}\alpha_2(x)\frac{\partial^2}{\partial x^2}\right]P \quad (3.145)$$

Let $P(x, t \mid x_0, t_0)$ represent the solution to Equation (3.145). Now take a small increment in t, $t = t_0 + \Delta t$, and compute for this small increment the moments of $(x - x_0) = \Delta x$. Taking the vanishing limit of Δt one can show that

$$\alpha_1(x_0) = \frac{\langle \Delta x\rangle}{\Delta t} \quad \text{and} \quad \alpha_2(x_0) = \frac{\langle(\Delta x)^2\rangle}{\Delta t} \quad (3.146)$$

It is possible to compute α_1 and α_2 from the original Langevin Equation (3.143) which can be written in the form

$$\Delta x = \int_t^{t+\Delta t} f[x(t')]\,dt' + \int_t^{t+\Delta t} L(t')\,dt' \quad (3.147)$$

or the average

$$\langle \Delta x \rangle = f[x(t)]\Delta t + O(\Delta t)^2 \tag{3.148}$$

which gives $\alpha_1(x)$ in the Fokker–Planck equation. Likewise one could write for the second coefficient

$$\begin{aligned}\langle (\Delta x)^2 \rangle = \Bigg\langle \Bigg[\int_t^{t+\Delta t} & f[x(t')^2]\, dt' \\ & + 2\int_t^{t+\Delta t} dt' \int_t^{t+\Delta t} dt_1 \langle f[x(t')]L(t_1)\rangle \\ & + \int_t^{t+\Delta t} dt' \int_t^{t+\Delta t} dt_1 \langle L(t')L(t_1)\rangle \Bigg\rangle\end{aligned} \tag{3.149}$$

The first term in this equation is of the order of $(\Delta t)^2$. In the second term, by expanding the function $f[x(t')]$ it can be verified that it does not contribute to α_2. The third term equals $\varepsilon\Delta t$. One can thus calculate α_2 as $\alpha_2 = \varepsilon$. Knowing both the coefficients it is now possible to write the Fokker–Planck equation corresponding to Equation (3.143) thus:

$$\frac{\partial P}{\partial t} = \left[-f(x)\frac{\partial}{\partial x} + \frac{\varepsilon}{2}\frac{\partial^2}{\partial x^2}\right] P \tag{3.150}$$

The method outlined above uses the ordinary rules of calculus and is unambiguous as long as the Langevin equation involved is linear or quasilinear. In the nonlinear Langevin case, however, there is some ambiguity. To illustrate this, let us consider a nonlinear Langevin equation

$$\frac{dx}{dt} = f(x) + \alpha(x)L(t) \tag{3.151}$$

where the function $f(x)$ is nonlinear and the coefficient of the fluctuating force depends on the system variable x. Remembering that $L(t)$ represents a random sequence of delta functions, every time it is reached the value of x is appropriately updated. It is this sudden jump in x that causes the difficulty, for it is not known what value of x is to be taken in calculating the coefficient $\alpha(x)$. If α is calculated on the basis of the value of x before the delta disturbance, then the corresponding Fokker–Planck equation obtained would be

$$\frac{\partial P(x,t)}{\partial t} = -\frac{\partial}{\partial x} f[x(t)]P + \frac{1}{2}\frac{\partial^2}{\partial x^2}\alpha(x)^2 P \tag{3.152}$$

However, if it is desired to calculate on the basis of the value of x that represents the mean of x before and after the delta disturbance, then the following Fokker–Planck equation is obtained:

$$\frac{\partial P(x,t)}{\partial t} = -\frac{\partial}{\partial x} f[x(t)] + \frac{1}{2}\frac{\partial}{\partial x}\alpha(x)\frac{\partial}{\partial x}\alpha(x)P \qquad (3.153)$$

The two approaches followed, respectively, by Ito and Stratonovich have given rise to the famous Ito–Stratonovich dilemma. Clearly, a distinction between the two exists only if α is dependent on x, i.e. for processes involving multiplicative noise. For linear or quasilinear Langevin equations with additive noise, the two approaches yield identical Fokker–Planck equations.

Considerable discussions on the Ito–Stratonovich formulations, the sources for the discrepancy and the correctness of the interpretations of Ito and Stratonovich exist in the literature (Mortensen, 1969; Gray and Caughey, 1965; Wong and Zakai, 1965; Ryter, 1978; West *et al.*, 1979; van Kampen, 1976, 1981). The major conclusion, and the one which is central to us here, is that the Stratonovich treatment of the stochastic differential equation should be followed in the analysis of dynamic systems. As further elaborated by West *et al.* (1979) this form leads to physically correct results and is consistent with the Fokker–Planck equation obtained from the master equation using Kramers–Moyal expansion (see Bedeaux, 1977; Hanggi and Talkner, 1978; Hanggi, 1978). The approach is also compatible with the path integral description of physical processes with fluctuations.

Appendix 3.A

Expansion of the Master Equation in Parameter Ω

In the present appendix, we outline the basic procedure for expanding the master equation in some suitably chosen parameter Ω as developed by van Kampen (1976). To illustrate, we begin with the master equation

$$\frac{\partial P}{\partial t} = \int [W(X|X')P(X',t) - W(X'|X)P(X,t)]\,dX' \qquad (1)$$

where $W(X|X')$ refers to the transition probability per unit time to move from state X' to X. The jump from X' to X is denoted as j $(= X - X')$. The transition probability $W(X|X')$ then can also be expressed in terms of the starting state X' and the unit jump j as $W(X|X') = W(X', j)$ where the latter can be written more explicitly as some function of the value of the state $(x' = X'/\Omega)$ and the unit jump as $f(x', j)$. Similar arguments suggest that $W(X'|X) = f(x, -j)$. The expression of transition probabilities written explicitly in terms of the functions $f(x', j)$ and $f(x, -j)$ is a necessity—for our next intention is to expand them in powers of Ω. This is therefore a crucial step for the expansion method to work. We now write the transition probability as

$$W(X|X') = f(\Omega)\{f_0(x'; j) + \Omega^{-1} f_1(x'; j) + \cdots\} \tag{2}$$

As the next step in the development of the expansion method we anticipate the dependence of the probability functions appearing in Equation (1) on the expansion parameter. The initial condition is $P(X, 0) = \delta(X - X_0)$. As the system evolves in time, for a large volume system, it is expected that the probability function still remains sharp with its width of the order of $\Omega^{1/2}$. Mathematically this can be expressed as

$$X(t) = \Omega\phi(t) + \Omega^{1/2} y \tag{3}$$

In view of this time-dependent transformation the original probability function changes over to a new function

$$P(X, t) = P(\Omega\phi(t) + \Omega^{1/2} y, t) = \pi(y, t) \tag{4}$$

which can be used to relate the differentials of P with respect to X and t to the differentials of y with respect to y and t.

The next step involves substituting Equations (2)–(4) in master Equation (1) to obtain the transformed equation in the probability function π as

$$\begin{aligned}\frac{\partial \pi(y, \tau)}{\partial \tau} - \Omega^{1/2} \frac{\partial \phi}{\partial \tau} \frac{\partial \pi}{\partial y} = {} & -\Omega^{1/2} \alpha_{1,0}(\phi) \frac{\partial \pi}{\partial y} - \alpha'_{1,0}(\phi) \frac{\partial}{\partial y} y\pi \\ & - \tfrac{1}{2}\Omega^{1/2} \alpha''_{1,0}(\phi) \frac{\partial}{\partial y} y^2 \pi + \tfrac{1}{2}\alpha_{2,0}(\phi) \frac{\partial^2 \pi}{\partial y^2} \\ & + \tfrac{1}{2}\Omega^{-1/2} \alpha'_{2,0}(\phi) \frac{\partial^2}{\partial y^2} y\pi - \frac{1}{3!}\alpha_{3,0}(\phi) \frac{\partial^3 \pi}{\partial y^3} \\ & - \Omega^{-1/2} \alpha_{1,1}(\phi) \frac{\partial \pi}{\partial y} + O(\Omega^{-1})\end{aligned} \tag{5}$$

where τ and the various α's are defined as

$$\tau = \Omega^{-1} f(\Omega) t \tag{6}$$

and

$$\alpha_{lk} = \int j^l f_k(x, j)\, dj \tag{7}$$

In view of the analogy of Equation (7) with the equation defining the moments the α's are also referred to as jump moments.

In Equation (5), collecting terms of the order of $\Omega^{1/2}$ leads to

$$\frac{d\phi}{d\tau} = \alpha_{1,0}(\phi) \tag{8}$$

which can be recognised as the macroscopic equation for the system. The initial condition for this equation can be obtained as $\phi(0) = X_0/\Omega = x_0$. Clearly the solution of the macroscopic equation is stable provided $\alpha'_{1,0} < 0$. It is also possible that the equation would possess more than one solution. An interesting case arises when $\alpha_{1,0} = 0$. Integration of Equation (8) then leads to $\phi(\tau) = \phi(0) =$ constant, implying that any small disturbance does not decay resulting in unstable situations. Equations possessing this property are referred to as diffusion equations. When $\alpha_{1,0} < 0$, the stability condition is violated even more strongly, leading to unstable situations. The expansion method in these types of situations becomes inapplicable.

Collecting terms of the order of $\Omega^{1/2}$ leads to the linear Fokker–Planck equation

$$\frac{\partial \pi}{\partial t} = -\alpha'_{1,0}(\phi) \frac{\partial}{\partial y} y\pi + \tfrac{1}{2}\alpha_{2,0}(\phi) \frac{\partial^2 \pi}{\partial y^2} \tag{9}$$

which possesses a solution that is Gaussian. It is sufficient therefore to know the first two moments to characterise the distribution and the equations for these can be easily obtained by multiplying Equation (9) by y (for first moment) and y^2 (for second moment). Averaging the solution of the resulting equations with appropriate initial conditions then gives the required moments:

$$\frac{\partial \langle y \rangle}{\partial t} = \alpha'_{1,0}(\phi)\langle y \rangle \tag{10}$$

$$\frac{\partial \langle y^2 \rangle}{\partial t} = 2\alpha'_{1,0}(\phi)\langle y^2 \rangle + \alpha_{2,0}(\phi) \tag{11}$$

REFERENCES

Barthalomay, A. F., *Bull. Math. Biophys.* **20**, 175 (1958).
Barthalomay, A. F., *Bull. Math. Biophys.* **21**, 363 (1959).
Bedeaux, D., *Phys. Lett.* **62A**, 10 (1977).
Darvey, I. G. and Staff, P. J., *J. Chem. Phys.* **44**, 990 (1966).
Darvey, I. G., Ninham, B. W. and Staff, P. J., *J. Chem. Phys.* **45**, 2145 (1966).
Delbruck, M., *J. Chem. Phys.* **8**, 120 (1940).
Edder, O. J. and Lackner, T., *Phys. Rev.* **28A**, 952 (1983).
Fokker, A. D., *Ann. Physik* **43**, 810 (1914).
Gans, P. J., *J. Chem. Phys.* **33**, 691 (1960).
Gillespie, D. T., *J. Stat. Phys.* **16**, 311 (1977).
Gillespie, D. T., *J. Phys. Chem.* **81**, 2340 (1977).
Grabert, H. and Green, M. S., *Phys. Rev.* **19A**, 1747 (1979).
Grabert, H., Graham, R. and Green, M. S., *Phys. Rev.* **21A**, 2136 (1980).
Grabert, H. and Weidlich, W., *Phys. Rev.* **21A**, 2147 (1980).
Grabert, H., Hanggi, P. and Oppenheim, I., *Physica* **117A**, 300 (1983).
Gray, A. M. Jr. and Caughley, T. K., *J. Math. and Phys.* **44**, 288 (1965).
Grossmann, S. and Schranner, R., *Z. Physik* **B30**, 325 (1978).
Hanggi, P., *Z. Naturforsch.* **33A**, 525 (1978).
Hanggi, P. and Talkner, P., *Phys. Lett.* **68A**, 9 (1978).
Keizer, J., *J. Chem. Phys.* **56**, 5775 (1972).
Keizer, J. and Conlan, F. J., *Physica* **117A**, 405 (1983).
Kitahara, K., *Adv. Chem. Phys.* **29**, 85 (1976).
Kramers, H. A., *Physica* **7A**, 284 (1940).
Krieger, J. M. and Gans, P. J., *J. Chem. Phys.* **32**, 247 (1960).
Kubo, R., Matuso, K. and Kitahara, K., *J. Stat. Phys.* **9**, 51 (1973).
Kurtz, T. G., *J. Chem. Phys.* **57**, 2976 (1972).
Landau, L. and Lifschitz, E. M., *Statistical Physics*, Pergamon, London, 1958.
Langevin, P., *Comptes Rendues* **146**, 530 (1908).
Malek-Mansour, M. and Nicolis, G., *J. Stat. Phys.* **13**, 197 (1975).
McQuarrie, D. A., *J. Chem. Phys.* **38**, 433 (1963).
McQuarrie, D. A., Jachimowski, C. J. and Russell, M. E., *J. Chem. Phys.* **40**, 2914 (1964).
McQuarrie, D. A., *Stochastic Approach to Chemical Kinetics*, Vol. 8 of Supplemental Review Series in Applied Probability, Methuen, London, 1967.
Mori, H. and Fujisaka, H., *Prog. Theor. Phys.* **49**, 764 (1972).
Mortensen, R. E., *J. Stat Phys.* **1**, 271 (1969).
Nicolis, G. and Prigogine, I., *Proc. Natl. Acad. Sci.* (*USA*) **68**, 2102 (1971).
Nicolis, G., Malek-Mansour, M., van Nypelseer, A. and Kitahara, K., *J. Stat. Phys.* **14**, 417 (1976).
Nicolis, G., Malek-Mansour, M., Kitahara, K. and van Nypelseer, A., *Phys. Lett.* **48A**, 217 (1974).
Nicolis, G. and Prigogine, I., *Self-Organisation in Non-Equilibrium Systems*, Wiley-Interscience, New York, 1977.
Oppenheim, I., Shuler, K. E. and Weiss, G., *J. Chem. Phys.* **50**, 460 (1969).
Oppenheim, I., Shuler, K. E. and Weiss, G., *Stochastic Processes in Chemical Physics: The Master Equation*, MIT Press, Cambridge, 1977.
Planck, M., *Sitzungsber. Preuss. Akad. Wiss. Phys. Math. Kl.* 325 (1917).
Prigogine, I., Nicolis, G., Herman, R. and Lam, T., *Collect. Phenom.* **2**, 103 (1976).
Renyi, A., *Magyar Ind. Akad. Alkalm. Mat. Int. Kozl.* **2**, 93 (1954).
Ryter, D., *Z. Physik* **B30**, 219 (1978).

Turner, J. S., *J. Phys. Chem.* **81**, 2379 (1977).
van Kampen, N. G., *Adv. Chem. Phys.* **34**, 145 (1976).
van Kampen, N. G., *Phys. Rep.* **24**, 171 (1976).
van Kampen, N. G., *Stochastic Processes in Physics and Chemistry*, North-Holland, Amsterdam, 1981.
West, B. J., Bulsara, A. R., Lindenberg, K., Seshadri, V. and Shuler, K. E., *Physica* **97A**, 211 (1979).
Wong, E. and Zakai, M., *Ann. Math. Stat.* **36**, 1560 (1965).
Zwanzig, R., Nordholm, K. S. J. and Mitchell, W. C., *Phys. Rev.* **5A**, 2650 (1972).

4 Probabilistic Modelling of Complex Reaction Schemes

4.1 Introduction

IN THE previous chapter we were concerned with the application of the general concepts of probability theory and Markov processes, developed in Chapters 1 and 2, to simple chemically reacting systems. In the present chapter we shall extend the application of these concepts to more complex situations. We note from the previous chapter that the master equation describing the evolution of a system can be solved exactly only for unit-step (birth and death type) linear processes, and introduction of even the slightest complexity renders the equation unsuitable for analytic treatment. It is necessary in such cases to resort to approximation schemes, and some of these have also been discussed in that chapter.

An approximation scheme, which is especially appealing and which logically attempts to solve the master eqûation, is the method of expansion of the master equation in some suitably chosen parameter of the system. This method has been discussed at some length in the previous chapter, but its application was necessarily restricted to simple unique steady state situations. In the present chapter we shall exemplify the use of this method in various other types of situations. We shall also be concerned with the restrictions in the use of this method to certain situations. More specifically the application of the method to the following situations will be considered: (1) multivariable cases, (2) diffusion processes, (3) systems involving bistability, and (4) the case of instability.

To recapitulate what we have so far established: The formulation of the master equation is based on the use of transition probabilities per unit time $W(X|X')$ which can be obtained for a specific system by alternative sources. Once $W(X|X')$ is known, the method envisages an explicit expression for it in terms of a suitably chosen parameter of the system Ω. This step is crucial for the expansion method to work. The next stage involves anticipation of the way $P(X, t)$ depends on the expansion parameter. One uses here the intuitive idea that, for a delta peak, to begin with $[P(X, 0) = \delta(X - X_0)]$, and at some time later the peak still remains sharp at some position that is $O(\Omega)$, while its width is of the order of $(\Omega^{1/2})$. Utilising these two steps in the original master equation and performing the necessary operations, one obtains a master equation in the transformed variable from which the microscopic law can be extracted as the lowest power in the expansion parameter Ω, while the next higher order gives the linear noise approximation as a Fokker–Planck equation describing the evolution of fluctuations. The Fokker–Planck equation is easily used to compute the moments of the fluctuating component, and the equation being linear avoids the Stratonovich–Ito conflict. The method is therefore neat, and one is tempted to generalise this single variable case to multivariable cases. Indeed the method also works in such situations, and from a multivariable master equation, employing a similar procedure, one can obtain the corresponding macroscopic equations as well as the multivariable Fokker–Planck equation describing the fluctuating components of the variables. The method is best illustrated by considering an example of interactions on a catalyst surface (Tambe, Kulkarni and Doraiswamy, 1985a).

4.2 The Case of the Catalyst Surface

Let us consider the case of a catalyst surface where the following sequence of steps occurs:

$$A_p + p[Z] \rightleftharpoons p[AZ]$$

$$B_q + q[Z] \rightleftharpoons q[BZ]$$

$$m[AZ] + n[BZ] \longrightarrow [A_m B_n] + (m + n)Z$$

Z in this sequence refers to vacant sites and p, q, m and n are arbitrary coefficients. In one case it is assumed that the sequence follows a conservation law for the total number of sites on the surface:

$$L = (Z + AZ + BZ) \tag{4.1}$$

For constant numbers of species A and B this suggests AZ and BZ as the two variables for the system. Further, assuming that $p = q = m = n = 1$, the macroscopic equations determining the evolution of the system can be written in suitable form as (Tambe, Kulkarni and Doraiswamy, 1985a)

$$\frac{d[AZ]}{dt} = k_1 p_A(L - AZ - BZ) - k_{-1}[AZ] - k_3[AZ][BZ] \tag{4.2}$$

$$\frac{d[BZ]}{dt} = k_2 p_B(L - AZ - BZ) - k_{-2}[BZ] - k_3[AZ][BZ] \tag{4.3}$$

Subject to proper choice of the volume parameter and scaling relation for t, these equations become

$$\frac{d[AZ]}{dt} = [a_{11}L - a_{12}(AZ + BZ)] - a_2[AZ] - \frac{a_3}{L}[AZ][BZ] \tag{4.4}$$

$$\frac{d[BZ]}{dt} = [b_{11}L - b_{12}(AZ + BZ)] - b_2[BZ] - \frac{b_3}{L}[AZ][BZ] \tag{4.5}$$

where the various coefficients are defined as follows:

$$\begin{gathered} a_{11} = k_1 p_A, \qquad a_{12} = pa_{11}, \qquad b_{11} = k_2 p_B, \qquad b_{12} = qb_{12} \\ a_2 = k_{-1}, \qquad b_2 = k_{-2} \qquad \text{and} \qquad a_3 = b_3 = k_3 L \end{gathered} \tag{4.6}$$

The joint probability density function $P(AZ, BZ, t)$ satisfies the master equation

$$\begin{aligned} \frac{dP(AZ, BZ, t)}{dt} = {} & a_{11}L(E_{AZ}^{-1} - 1)P + a_{12}(E_{AZ} - 1)(AZ + BZ)P \\ & + a_2(E_{AZ} - 1)(AZ)P + \frac{a_3}{L}(E_{AZ} - 1)[(AZ)(BZ)]P \\ & + b_{11}L(E_{BZ}^{-1} - 1)P + b_{12}(E_{BZ} - 1)(AZ + BZ)P \\ & + b_2(E_{BZ} - 1)(BZ)P + \frac{b_3}{L}(E_{BZ} - 1)[(AZ)(BZ)]P \end{aligned} \tag{4.7}$$

Defining new transformed variables

$$[AZ] = L\phi(t) + L^{1/2}\tilde{x}$$
$$[BZ] = L\psi(t) + L^{1/2}y \tag{4.8}$$

the changed probability function $\pi(\tilde{x}, y, t)$ can be obtained as

$$\frac{\partial \pi}{\partial t} - L^{1/2}\frac{\partial \phi}{\partial t}\frac{\partial \pi}{\partial \tilde{x}} - L^{1/2}\frac{\partial \psi}{\partial t}\frac{\partial \pi}{\partial y}$$
$$= a_{11}L\left(-L^{-1/2}\frac{\partial}{\partial \tilde{x}} + \frac{L^{-1}}{2}\frac{\partial^2}{\partial \tilde{x}^2}\right)\pi$$
$$+ a_{12}L\left(L^{-1/2}\frac{\partial}{\partial \tilde{x}} + \frac{L^{-1}}{2}\frac{\partial^2}{\partial \tilde{x}^2}\right)(\phi + L^{-1/2}\tilde{x} + \psi + L^{-1/2}y)\pi$$
$$+ a_2L\left(L^{-1/2}\frac{\partial}{\partial \tilde{x}} + \frac{L^{-1}}{2}\frac{\partial^2}{\partial \tilde{x}^2}\right)(\phi + L^{-1/2}\tilde{x})\pi$$
$$+ a_3L\left(L^{-1/2}\frac{\partial}{\partial \tilde{x}} + \frac{L^{-1}}{2}\frac{\partial^2}{\partial \tilde{x}^2}\right)[(\phi + L^{-1/2}\tilde{x})(\psi + L^{-1/2}y)]\pi$$
$$+ b_{11}L\left(-L^{-1/2}\frac{\partial}{\partial y} + \frac{L^{-1}}{2}\frac{\partial^2}{\partial y^2}\right)\pi$$
$$+ b_{12}L\left(L^{-1/2}\frac{\partial}{\partial y} + \frac{L^{-1}}{2}\frac{\partial^2}{\partial y^2}\right)[\phi + \psi + L^{-1/2}(\tilde{x} + y)]\pi$$
$$+ b_2L\left(L^{-1/2}\frac{\partial}{\partial y} + \frac{L^{-1}}{2}\frac{\partial^2}{\partial y^2}\right)(\psi + L^{-1/2}y)\pi$$
$$+ b_3L\left(L^{-1/2}\frac{\partial}{\partial y} + \frac{L^{-1}}{2}\frac{\partial^2}{\partial y^2}\right)[(\phi + L^{-1/2}\tilde{x})(\psi + L^{-1/2}y)]\pi \tag{4.9}$$

Equating terms of the orders $L^{1/2}$ and L^0 in Equation (4.9) gives the following relations for the evolution of the macroscopic and fluctuating components of the original variables AZ and BZ:

$$\frac{d\phi}{dt} = a_{11}(1 - \phi - \psi) - a_2\phi - a_3\phi\psi \tag{4.10}$$

$$\frac{d\psi}{dt} = b_{11}(1 - \phi - \psi) - b_2\psi - a_3\phi\psi \tag{4.11}$$

$$\begin{aligned}\frac{\partial \pi}{\partial t} &= [a_{12}(\tilde{x} + y) + a_2\tilde{x} + a_3(\phi y + \psi\tilde{x})]\frac{\partial \pi}{\partial \tilde{x}} \\ &+ [b_{12}(\tilde{x} + y) + b_2 y + b_3(\phi y + \psi\tilde{x})]\frac{\partial \pi}{\partial y} \\ &+ \tfrac{1}{2}[a_{11}(1 + \phi + \psi) + a_2\phi + a_3\phi\psi]\frac{\partial^2 \pi}{\partial \tilde{x}^2} \\ &+ \tfrac{1}{2}[b_{11}(1 + \phi + \psi) + b_2\psi + b_3\phi\psi]\frac{\partial^2 \pi}{\partial y^2}\end{aligned} \tag{4.12}$$

Equations (4.10) and (4.11) representing the macroscopic part of the system exactly correspond to the original deterministic Equations (4.2) and (4.3), while the multivariate Fokker–Planck Equation (4.12) governs the evolution of fluctuations, viz.

$$\frac{d\langle\tilde{x}\rangle}{dt} = -[a_{12}\langle\tilde{x} + y\rangle + a_2\langle\tilde{x}\rangle] + a_3[\phi\langle y\rangle + \psi\langle\tilde{x}\rangle] \tag{4.13}$$

$$\begin{aligned}\frac{d\langle\tilde{x}^2\rangle}{dt} &= -2[a_{12}\langle\tilde{x}^2 + y^2\rangle] + a_2\langle\tilde{x}^2\rangle \\ &+ a_3[\phi\langle y^2\rangle + \psi\langle\tilde{x}^2\rangle] \\ &+ a_{11}(1 + \phi + \psi) + a_2\phi + a_3\phi\psi\end{aligned} \tag{4.14}$$

Similar equations can be written for the species y. Equations (4.10) and (4.11) can be solved subject to appropriate initial conditions. These equations being nonlinear are difficult to solve analytically; however, at steady state it is generally possible to obtain solutions for ϕ and ψ. Let (ϕ_{st}, ψ_{st}) be one such solution for arbitrary values of p, q, m and n. We can substitute these values of ϕ and ψ in Equations (4.12) and (4.13) to calculate the variations of fluctuations at steady state. Equation (4.12), however, involves $\langle y\rangle$ which can be eliminated from an equation for $\langle y\rangle$ similar to Equation (4.13):

$$\langle y\rangle = \frac{[a_{12} - b_{12}]\langle\tilde{x}\rangle}{(a_{12} - b_{12} + a_2 - b_2)} \tag{4.15}$$

Substitution of $\langle y\rangle$ in Equation (4.12) in terms of $\langle\tilde{x}\rangle$ from Equation (4.15) gives an exponential decay of $\langle\tilde{x}\rangle$. The results thus indicate that around the steady state (ϕ_{st}, ψ_{st}) the fluctuations would die out exponentially. This implies that the average values of AZ and BZ also represent their macroscopic part. This result was obtained earlier by Grechanikov and Yablonskii (1982) using the Monte Carlo simulation.

The foregoing analysis for the specific case of $p = q = m = n = 1$ can be generalised for any value of the stoichiometric coefficient (Tambe, Kulkarni and Doraiswamy, 1985a). Also, it is possible to relax the restriction that the total number of sites L on the surface is constant. The removal of this restriction would imply the breaking down of the relation given by Equation (4.1), which would now be valid only at the local time t:

$$L(t) = Z + (AZ) + (BZ) \tag{4.16}$$

This necessitates consideration of Z also as a variable, and the evolution of probability distribution, subject to the reaction sequence considered, would now take the form

$$\begin{aligned}\frac{\partial P}{\partial t}(Z, AZ, BZ, t) = {} & k_1 A_p (E_Z^p E_{AZ}^{-p} - 1)Z^p - k_{-1}(E_Z^{-p} E_{AZ}^p - 1)[AZ]^p \\ & + k_2 B_q (E_Z^q E_{BZ}^{-q} - 1)Z^q \\ & + k_{-2}(E_Z^{-q} E_{BZ}^q - 1)[BZ]^q \\ & + k_3 (E_{AZ}^m E_{BZ}^n E_Z^{-(m+n)} - 1)[AZ]^m [BZ]^n \end{aligned} \tag{4.17}$$

Following the general methodology and defining

$$\begin{aligned} Z &= L_m \phi_1 + L_m^{1/2} x_1 \\ AZ &= L_m \phi_2 + L_m^{1/2} x_2 \\ BZ &= L_m \phi_3 + L_m^{1/2} x_3 \end{aligned} \tag{4.18}$$

where L_m refers to the maximum number of sites on the surface, Equation (4.17) can be rewritten as follows:

$$\begin{aligned} & \frac{\partial \pi}{\partial t} - L_m^{1/2} \frac{\partial \phi_1}{\partial t} \frac{\partial \pi}{\partial x_1} - L_m^{1/2} \frac{\partial \phi_2}{\partial t} \frac{\partial \pi}{\partial x_2} - L_m^{1/2} \frac{\partial \phi_3}{\partial t} \frac{\partial \pi}{\partial x_3} \\ & = a_{11} \left[p L_m^{-1/2} \frac{\partial}{\partial x_1} + \frac{p^2}{2!} L_m^{-1} \frac{\partial^2}{\partial x_1^2} - p L_m^{-1/2} \frac{\partial}{\partial x_2} + \frac{p^2 L_m^{-1}}{2!} \frac{\partial^2}{\partial x_2^2} \right] L_m \\ & \quad \times [\phi_1^p + p\phi_1^{p-1} L_m^{-1/2} x_1 + [p(p-1)/2!]\phi_1^{p-2} L_m^{-1} x_1^2 + \cdots + L_m^{-p/2} x_1^p] \end{aligned}$$

$$+ a_{12}\left[-pL_m^{-1/2}\frac{\partial}{\partial x_1} + \frac{p^2}{2!}L_m^{-1}\frac{\partial^2}{\partial x_1^2} + pL_m^{-1/2}\frac{\partial}{\partial x_2} + \frac{p^2L_m^{-1}}{2!}\frac{\partial^2}{\partial x_2^2}\right]L_m$$

$$\times [\phi_2^p + p\phi_2^{p-1}L_m^{1/2}x_2 + [p(p-1)/2!]\phi_2^{p-2}L_m^{-1}x_2^2 + \cdots + L_m^{-p/2}x_2^p]$$

$$+ b_{11}\left[qL_m^{-1/2}\frac{\partial}{\partial x_1} + \frac{q^2}{2!}L_m^{-1}\frac{\partial^2}{\partial x_1^2} - qL_m^{-1/2}\frac{\partial}{\partial x_3} + \frac{q^2}{2!}L_m^{-1}\frac{\partial^2}{\partial x_3^2}\right]L_m$$

$$\times [\phi_1^2 + q\phi_1^{q-1}L_m^{-1/2}x_1 + [q(q-1)/2!]\phi_1^{q-2}L_m^{-1}x_1^2 + \cdots + L_m^{-q/2}x_1^q]$$

$$+ b_{12}\left[-qL_m^{-1/2}\frac{\partial}{\partial x_1} + \frac{q^2}{2!}L_m^{-1}\frac{\partial^2}{\partial x_1^2} - qL_m^{-1/2}\frac{\partial}{\partial x_3} + \frac{q^2}{2!}L_m^{-1}\frac{\partial^2}{\partial x_3^2}\right]L_m$$

$$\times [\phi_3^q + q\phi_3^{q-1}L_m^{-1/2}x_3 + [q(q-1)/2!]\phi_3^{q-2}L_m^{-1}x_3^2 + \cdots + L_m^{-q/2}x_3^q]$$

$$+ f_1\left[mL_m^{-1/2}\frac{\partial}{\partial x_2} + \frac{m^2}{2!}L_m^{-1}\frac{\partial^2}{\partial x_2^2} + nL_m^{-1/2}\frac{\partial}{\partial x_3} + \frac{n^2}{2!}L_m^{-1}\frac{\partial^2}{\partial x_3^2}\right.$$

$$\left. -(m+n)L_m^{-1/2}\frac{\partial}{\partial x_1} + \frac{(m+n)^2}{2!}L_m^{-1}\frac{\partial^2}{\partial x_1^2}\right]L_m$$

$$\times [\phi_2^m\phi_3^n + L_m^{-1/2}(n\phi_2^m\phi_3^{n-1}x_3 + m\phi_3^n\phi_2^{m-1}x_2)]$$

$$+ L_m\left[\frac{m(m-1)}{2!}\phi_2^{m-2}\phi_3^n x_2^2 + \frac{n(n-1)}{2!}\phi_3^{n-2}\phi_2^m x_3^2 + mn\phi_2^{m-1}\phi_3^{n-1}x_2x_3\right]$$

$$+ L_m^{-1.5} \quad \text{terms} \tag{4.19}$$

In this equation $\pi(x_1, x_2, x_3, t)$ is the transformed probability distribution corresponding to $P(Z, AZ, BZ, t)$, the rate terms have been expanded in the binomial series and the various groups are defined as

$$a_{11} = \frac{k_1 p_A}{L_m^{p-1}}, \qquad a_{12} = \frac{k_{-1}}{L_m^{p-1}}, \qquad b_{12} = \frac{k_{-2}}{L_m^{q-1}}$$
$$f_1 = \frac{k_3}{L_m^{m+n-1}}, \qquad b_{11} = \frac{k_2 p_B}{L_m^{q-1}} \tag{4.20}$$

Collecting all terms of the order $L^{1/2}$ together, we get

$$-\frac{d\phi_1}{dt} = a_{11}p\phi_1^p - a_{12}p\phi_2^p + b_{11}q\phi_1^q - b_{12}q\phi_3^q - f_1 m(m+n)\phi_2^m\phi_3^m \tag{4.21}$$

$$-\frac{d\phi_2}{dt} = -a_{11}p\phi_1^p + a_{12}p\phi_2^p + f_1 m\phi_2^m\phi_3^n \tag{4.22}$$

$$-\frac{d\phi_3}{dt} = f_1 n\phi_2^m\phi_3^n - b_{11}q\phi_1^q + b_{12}q\phi_3^q \tag{4.23}$$

Likewise, collecting all terms of the order L^0 together, we obtain the Fokker–Planck equation

$$\begin{aligned}
\frac{\partial\pi}{\partial t} = \frac{\partial}{\partial x_1}\{&a_{11}p^2\phi_1^{p-1}x_1 - a_{12}p^2\phi_2^{p-1}x_2 + b_{11}q^2\phi_1^{q-1}x_1 \\
&- b_{12}q^2\phi_3^{q-1}x_3 - f_1(m+n)[n\phi_2^m\phi_3^{n-1}x_3 + m\phi_3^n\phi_2^{m-1}x_2]\} \\
&+ \frac{\partial}{\partial x_2}\{-a_{11}p^2\phi_1^{p-1}x_1 + a_{12}p^2\phi_2^{p-1}x_2 \\
&\qquad + f_1 m[n\phi_2^m\phi_3^{n-1}x_3 + m\phi_3^n\phi_2^{m-1}x_2]\} \\
&+ \frac{\partial}{\partial x_3}\{f_1 n[n\phi_2^m\phi_3^{n-1}x_3 + m\phi_3^n\phi_2^{m-2}x_2] - b_{11}q^2\phi_1^{q-1}x_1 \\
&\qquad + b_{12}q^2\phi_3^{q-1}x_3\} \\
&+ \frac{1}{2}\frac{\partial^2}{\partial x_1^2}\{a_{11}p^2\phi_1^p + b_{11}q^2\phi_1^q + b_{12}q^2\phi_3^q + a_{12}p^2\phi_2^p + f_1(m+n)^2\phi_2^m\phi_3^n\} \\
&+ \frac{1}{2}\frac{\partial^2}{\partial x_2^2}\{a_{11}p^2\phi_1^p + a_{12}p^2\phi_2^p + f_1 m^2\phi_2^m\phi_3^n\} \\
&+ \frac{1}{2}\frac{\partial^2}{\partial x_3^2}\{f_1 n^2\phi_2^m\phi_3^n + b_{11}\phi^q q^2 + b_{12}q^2\phi_3^q\}
\end{aligned} \tag{4.24}$$

The mean of fluctuation is given by

$$\begin{aligned}
\frac{\partial\langle x_1\rangle}{\partial t} = -\{&a_{11}p^2\phi_1^{p-1}\langle x_1\rangle - a_{12}p^2\phi_2^{p-1}\langle x_2\rangle \\
&+ b_{11}q^2\phi_1^{q-1}\langle x_1\rangle - b_{12}q^2\phi_3^{q-1}\langle x_3\rangle \\
&- f_1(m+n)[n\phi_2^m\phi_3^{n-1}\langle x_3\rangle + m\phi_3^n\phi_2^{m-1}\langle x_2\rangle]\}
\end{aligned} \tag{4.25}$$

$$\begin{aligned}
\frac{\partial\langle x_2\rangle}{\partial t} = \{&a_{11}p^2\phi_1^{p-1}\langle x_1\rangle - a_{12}p^2\phi_2^{p-1}\langle x_2\rangle \\
&- f_1 m[n\phi_2^m\phi_3^{n-1}\langle x_3\rangle + m\phi_3^n\phi_2^{m-1}\langle x_2\rangle]\}
\end{aligned} \tag{4.26}$$

$$\frac{\partial\langle x_3\rangle}{\partial t} = \{b_{11}q^2\phi_1^{q-1}\langle x_1\rangle - b_{12}q^2\phi_3^{q-1}\langle x_3\rangle$$

$$- f_1 n[n\phi_2^m\phi_3^{n-1}\langle x_3\rangle + m\phi_3^n\phi_2^{m-1}\langle x_2\rangle]\} \tag{4.27}$$

The equations describing the mean of the fluctuations in the variables are all linear in the fluctuation terms, and can be easily solved around the macroscopic steady states. As evident from the structure of these equations, the mean of the fluctuations represents an exponential decaying function with time. At steady state one could therefore expect the fluctuations to die out completely. The macroscopic part of the system thus also represents the average value of the variable. The result, already known for the special case of $p = q = m = n = 1$, has been generally shown here to be valid for any arbitrary set of values of the stoichiometric coefficients, both when the total number of sites on the surfaces is constant or otherwise.

Example 4.1 *Effect of fluctuations on critical slowing down in autocatalytic reactions*

Tambe, Kulkarni and Doraiswamy (1985c) have analysed the case of autocatalytic reaction ($X_1 + X_2 \rightleftarrows 2X_1$) in a CSTR that is modelled by the macroscopic rate equations

$$\frac{dx_1}{dt} = \frac{F}{V}(x_{10} - x_1) + k_1x_1x_2 - k_2x_1^2 \tag{1}$$

$$\frac{dx_2}{dt} = \frac{F}{V}(x_{20} - x_2) - k_1x_1x_2 + k_2x_1^2 \tag{2}$$

where x_{10} and x_{20} refer to the input concentrations of species x_1 and x_2, respectively. To obtain the relaxation time, the Jacobian corresponding to Equations (1) and (2) can be written to obtain a characteristic equation for the process and its eigenvalues as

$$\lambda_1 = F/V = k_f, \qquad \lambda_2 = [(k_1 + 2k_2)x_{1s} + k_f - k_1x_{2s}] \tag{3}$$

Note that λ_2 can approach zero for a certain set of parameter values, implying that the systems will relax infinitely slowly to reach the stationary state. The typical case of critical slowing down is indicated in Figure 4.1.

The master equation corresponding to this situation can be written

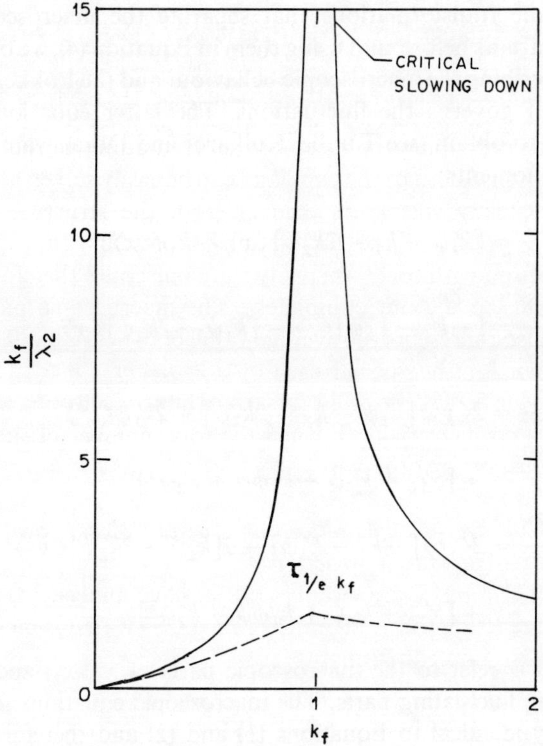

FIGURE 4.1 Critical slowing down in deterministic and stochastic models.

following Tambe, Kulkarni and Doraiswamy (1985c) as

$$\frac{dP}{dt} = \Bigg\{ V(E_{x_1}^{-1} - 1)\left[x'_{10} + \frac{k'_1}{V^2} x_1 x_2 \right] + (E_{x_1} - 1)\left[k_f x_1 + \frac{k'_2}{V} x_1(1 - x_1) \right]$$

$$+ V(E_{x_2}^{-1} - 1)\left[x'_{20} + \frac{k'_2}{V^2} x_1(1 - x_1) \right]$$

$$+ (E_{x_2} - 1)\left[k_f x_2 + \frac{k'_1}{V} x_1 x_2 \right] \Bigg\} P(x_1, x_2, t) \tag{4}$$

where

$$x'_{10} = \frac{k_f x_{10}}{V}, \qquad x'_{20} = \frac{k_f x_{20}}{V}, \qquad k'_1 = k_1 V, \qquad k'_2 = k_2 V \tag{5}$$

Defining the transformations that separate the macroscopic and fluctuating parts as before, and using them in Equation (4), we obtain the equation describing the macroscopic behaviour and the Fokker–Planck equation that governs the fluctuations. The latter equation can be decomposed to obtain (see Tambe, Kulkarni and Doraiswamy, 1985c) the various moments:

$$\frac{d\langle y\rangle}{dt} = [k_1'\psi - k_f - 2k_2'\phi]\langle y\rangle + k_1'\phi\langle z\rangle \tag{6}$$

$$\frac{d\langle z\rangle}{dt} = [-k_f - k_1'\phi]\langle z\rangle - [k_1'\psi - 2k_2'\phi]\langle y\rangle \tag{7}$$

$$\frac{d\langle y^2\rangle}{dt} = 2\langle y^2\rangle[k_1'\psi - k_f - 2k_2'\phi] + 2k_1'\phi\langle z^2\rangle + [x_{10}' + k_f\phi + k_1'\phi\psi + k_2'\phi^2] \tag{8}$$

$$\frac{d\langle z^2\rangle}{dt} = 2\langle z^2\rangle[-k_f - k_1'\phi] - 2[k_1'\psi - 2k_2'\phi]\langle y^2\rangle + [x_{20}' + k_f\psi + k_1'\phi\psi - k_2'\phi^2] \tag{9}$$

where ϕ and ψ refer to the macroscopic parts of x_1, x_2, and y, z the corresponding fluctuating parts. The macroscopic equations for ϕ and ψ are exactly identical to Equations (1) and (2) and therefore display similar feature of critical slowing down. The eigenvalues are similar to Equation (3) with x_1 and x_2 replaced by ϕ and ψ.

The fluctuation Equations (6)–(9) can be easily solved around the steady states (ϕ_s, ψ_s) to obtain the time variation of fluctuations. The calculation of the eigenvalues for these equations gives

$$\lambda_1 = k_f, \qquad \lambda_2 = k_f - k_1'\psi_{st} + (k_1' + 2k_2')\phi_{st} \tag{10}$$

Comparison of Equations (3) and (10) reveals that for a critical value of $k_f = k_{fc}$ these equations identically show the phenomenon of critical slowing down. The stochastic analysis thus additionally gives information regarding the slowing down of fluctuations.

The complete transient solution obtained by numerical methods for a set of parameter values is sketched in Figure 4.2. As evident from the figure, both the deterministic and stochastic methods yield identical results for this case.

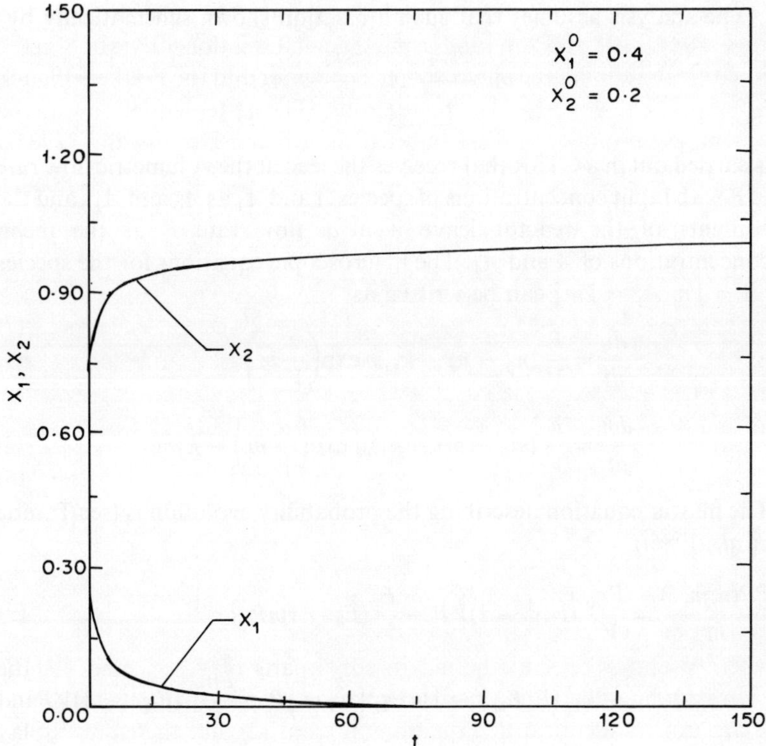

FIGURE 4.2 Plot of x_1 and x_2 profiles for the deterministic and stochastic models.

Example 4.2 *Fluctuations in reactions with nonsystemic type of autocatalytic feedback* (*Tambe* et al., *1985f*)

As a second example we examine the simple case of an autocatalytic reaction scheme analysed by Ravikumar, Kulkarni and Doraiswamy (1984). The simple model analysed considered a nonsystemic type of autocatalytic feedback and differs from the conventional systemic feedback in that, while in the latter the rate of reaction is directly affected due to change in concentration of the product species, in the former the product affects the rate through the rate constant of the reaction. For more details concerning the differences between the systemic and nonsystemic types of feedback, reference may be made to the review by Franck (1978).

The analysis assumes that such a reaction shown schematically by

$$A \xrightarrow{k_1} A_1 \xrightarrow{k_2} A_2$$

is carried out in a CTSR that receives the feed at the volumetric flow rate of F with input concentrations of species A and A_1 as A_0 and A_{10} and the contents of the reactor leave it at a flow rate F at the mean concentrations of A and A_1. The macroscopic equations for the species $[A = Vn, A_1 = Vm]$ can be written as

$$\frac{dn}{dt} = \frac{F}{V}(n_0 - n) - k_{10} n \exp\left(\frac{\alpha}{V} m\right) \tag{1}$$

$$\frac{dm}{dt} = \frac{F}{V}(m_0 - m) + k_{10} n \exp\left(\frac{\alpha}{V} m\right) - k_2 m \tag{2}$$

The master equation describing the probability evolution is (see Tambe *et al.*, 1985f).

$$\begin{aligned}\frac{\partial P(n, m, t)}{\partial t} &= \frac{Fn_0}{V^2}(E_n^{-1} - 1)VP + \frac{F}{V}(E_n - 1)nP \\ &\quad + k_{10}(E_n E_m^{-1} - 1)n \exp\left(\frac{\alpha}{V} m\right)P + \frac{Fm_0}{V^2}(E_m^{-1} - 1)VP \\ &\quad + \frac{F}{V}(E_m - 1)mP + k_2(E_m - 1)mP \end{aligned} \tag{3}$$

To solve Equation (3) using the systematic expansion procedure, we redefine the system variables thus:

$$n = V(\phi + V^{-1/2}x) \tag{4}$$

$$m = V(\psi + V^{-1/2}y) \tag{5}$$

where ϕ and ψ correspond to the macroscopic part of the variables n and m, and x and y measure the fluctuating contributions. In view of the large volume system envisaged, the fluctuating contributions are taken as proportional to the square root of the size of the system.

Employing the transformations (4) and (5) in Equation (3) we obtain the following equations describing the evolution of the probability. In view of the changed variables the probability $P(n, m, t)$ changes over to the new probability $\pi(x, y, t)$ and the following master equation results:

$$\frac{\partial\pi(x, y, t)}{\partial t} - V^{1/2}\frac{\partial\pi}{\partial x}\frac{\partial\phi}{\partial t} - V^{1/2}\frac{\partial\pi}{\partial y}\frac{\partial\psi}{\partial t}$$

$$= \frac{Fn_0}{V^2}\left(-V^{-1/2}\frac{\partial}{\partial x} + \frac{V^{-1}}{2}\frac{\partial^2}{\partial x^2}\right)V\pi$$

$$+ \frac{F}{V}\left(V^{-1/2}\frac{\partial}{\partial x} + \frac{V^{-1}}{2}\frac{\partial^2}{\partial x^2}\right)V(\phi + V^{-1/2}x)\pi$$

$$+ k_{10}\left(V^{-1/2}\frac{\partial}{\partial x} + \frac{V^{-1}}{2}\frac{\partial^2}{\partial x^2} - V^{-1/2}\frac{\partial}{\partial y} + \frac{V^{-1}}{2}\frac{\partial^2}{\partial y^2}\right)V$$

$$\times \{[\phi + V^{-1/2}x]\exp[\alpha(\psi + V^{-1/2}y)]\}\pi +$$

$$\frac{Fm_0}{V^2}\left(-V^{-1/2}\frac{\partial}{\partial y} + \frac{V^{-1}}{2}\frac{\partial^2}{\partial y^2}\right)V\pi$$

$$+ k_2\left(V^{-1/2}\frac{\partial}{\partial y} + \frac{V^{-1}}{2}\frac{\partial^2}{\partial y^2}\right)V(\psi + V^{-1/2}y)\pi \quad (6)$$

Collection of terms of order $V^{1/2}$ leads to the following equations describing the macroscopic variables:

$$\frac{-d\phi}{dt} = -\frac{Fn_0}{V^2} + \frac{F}{V}\phi + k_{10}\phi\exp\left(\frac{\alpha}{V}\psi\right) \quad (7)$$

$$\frac{-d\psi}{dt} = -\frac{Fm_0}{V^2} + \frac{F}{V}\psi - k_{10}\phi\exp\left(\frac{\alpha}{V}\psi\right) + k_2\psi \quad (8)$$

It is apparent that Equations (7) and (8) are identical to the deterministic set of equations which have been solved over a parametric space by Ravikumar, Kulkarni and Doraiswamy (1984). Similarly collection of terms of the order of V^0 leads to the following Fokker–Planck equation describing the fluctuations:

$$\frac{\partial\pi}{\partial t} = \frac{\partial}{\partial x}\left\{\frac{F}{V}x + k_{10}x\exp\left(\frac{\alpha}{V}\psi\right) + k_{10}\phi\frac{\alpha}{V}y\exp\left(\frac{\alpha}{V}\psi\right)\right\}\pi$$

$$+ \frac{1}{2}\frac{\partial^2}{\partial x^2}\left\{\frac{Fn_0}{V^2} + \frac{F}{V}\phi + k_{10}\phi\exp\left(\frac{\alpha}{V}\psi\right)\right\}\pi$$

$$+\frac{\partial}{\partial y}\left\{\frac{F}{V}y - k_{10}x\exp\left(\frac{\alpha}{V}\psi\right) - k_{10}\phi\frac{\alpha}{V}y\exp\left(\alpha\frac{\psi}{V}\right) + k_2 y\right\}\pi$$

$$+\frac{1}{2}\frac{\partial^2}{\partial y^2}\left\{\frac{Fm_0}{V} + \frac{F}{V}\psi + k_{10}\phi\exp\left(\frac{\alpha}{V}\psi\right) + k_2\psi\right\}\pi \tag{9}$$

The mean of fluctuations $\langle x\rangle$ and $\langle y\rangle$ can be obtained from this equation by calculating the first moment as

$$\frac{d\langle x\rangle}{dt} = -\left[\frac{F}{V} + k_{10}\exp\left(\frac{\alpha}{V}\psi\right)\right]\langle x\rangle - \left[k_{10}\phi\frac{\alpha}{V}\exp\left(\frac{\alpha}{V}\psi\right)\right]\langle y\rangle \tag{10}$$

$$\frac{d\langle y\rangle}{dt} = \left[k_{10}\exp\left(\frac{\alpha}{V}\psi\right)\right]\langle x\rangle - \left[\frac{F}{V} - k_{10}\phi\frac{\alpha}{V}\exp\left(\frac{\alpha}{V}\psi\right) + k_2\right]\langle y\rangle \tag{11}$$

The macroscopic Equations (7) and (8) have been solved earlier by Ravikumar, Kulkarni and Doraiswamy (1984). For a set of parameter values ($Fn_0/V^2 = 1$, $Fm_0/V^2 = 0$, $F/V = 1$, $\alpha/V = 6$, $k_2 = 8.0$, $k_{10} = 0.25$) it was shown that increasing the value of α has an influence on the trajectory reaching the final state and a critical value of α ($= 23.83$) exists for which long transients develop. This feature of the system is preserved even in the stochastic formulation as would be evident from the calculation of eigenvalues of the coefficients matrix. These can be readily obtained as

$$\beta_{1,2} = \frac{-(a_1 + a_3)}{2} \pm \tfrac{1}{2}[(a_1 - a_3)^2 + 4a_2a_4]^{1/2} \tag{12}$$

where a_1 and a_4 are the coefficients of the term $\langle x\rangle$ and a_2 and a_3 are the coefficients of the term $\langle y\rangle$ in Equation (10) and (11), respectively. Clearly, a_1, a_2 and a_4 in these equations are nonnegative while a_3 alone can take positive or negative values depending on the value of α for other fixed values of parameters. Also Equations (10) and (11) guarantee that the fluctuations $\langle x\rangle$ and $\langle y\rangle$ will dampen out as long as $a_3 > 0$. The system is therefore stable if $a_3 > 0$. For $a_3 < 0$ and $|a_3| > |a_1|$ the trace changes its sign indicating a change in the stability of the system. The critical value of α noted in the deterministic analysis can therefore be calculated *a priori* if other parameters of the system are known.

Thus, as long as α is less than the critical value, the fluctuations in $\langle x\rangle$ and $\langle y\rangle$ will dampen out and the predictions from the deterministic and stochastic formulations would lead to identical results. One could similarly calculate the second moments as

$$\frac{d\langle x^2\rangle}{dt} = -2\left[\frac{F}{V} + k_{10}\exp\left(\frac{\alpha}{V}\psi\right)\right]\langle x^2\rangle - 2k_{10}\phi\frac{\alpha}{V}\exp\left(\frac{\alpha}{V}\psi\right)\langle xy\rangle$$
$$+\left[\frac{Fn_0}{V^2} + \frac{F}{V}\phi + k_{10}\phi\exp\left(\frac{\alpha}{V}\psi\right)\right] \tag{13}$$

$$\frac{d\langle y^2\rangle}{dt} = -2\left[\frac{F}{V} - k_{10}\phi\frac{\alpha}{V}\exp\left(\frac{\alpha}{V}\psi\right) + k_2\right]\langle y^2\rangle$$
$$+ 2k_{10}\exp\left(\frac{\alpha}{V}\psi\right)\langle xy\rangle + \left[\frac{Fm_0}{V^2} + \frac{F}{V}\psi + k_{10}\phi\exp\left(\frac{\alpha}{V}\psi\right) + k_2\psi\right] \tag{14}$$

and

$$\frac{d\langle xy\rangle}{dt} = \left[k_{10}\exp\left(\frac{\alpha}{V}\psi\right)\right]\langle x^2\rangle - \left[k_{10}\phi\frac{\alpha}{V}\exp\left(\frac{\alpha}{V}\psi\right)\right]\langle y^2\rangle$$
$$-\left[\frac{2F}{V} + k_{10}\exp\left(\frac{\alpha}{V}\psi\right) - k_{10}\phi\frac{\alpha}{V}\exp\left(\frac{\alpha}{V}\psi\right) + k_2\right]\langle xy\rangle \tag{15}$$

These relations can be used, for instance, to calculate the equilibrium fluctuations as

$$\langle x^2\rangle_e = \left[\frac{l_8 l_3}{l_2} + \frac{l_9 l_5 l_3}{l_2 l_4} + \frac{l_9 l_6}{l_4}\right] \Big/ \left[l_7 + \frac{l_1 l_8}{l_2} + \frac{l_5 l_9 l_1}{l_4 l_2}\right] \tag{16}$$

$$\langle xy\rangle = \frac{l_3}{l_2} - \frac{l_1}{l_2}\langle x^2\rangle_e \tag{17}$$

$$\langle y^2\rangle_e = \frac{l_5 l_3}{l_2 l_4} + \frac{l_6}{l_4} - \frac{l_5 l_1}{l_2 l_4}\langle x^2\rangle_e \tag{18}$$

which, together with solutions to Equations (10) and (11), give

$$\langle x\rangle = \frac{\langle x_0\rangle[\beta_2 + a_1] + a_2\langle y_0\rangle}{\beta_2 - \beta_1}\exp(\beta_1 t)$$
$$+ \frac{\langle x_0\rangle(a_1 + \beta_1) + a_2\langle y_0\rangle}{\beta_1 - \beta_2}\exp(\beta_2 t) \tag{19}$$

Here $\langle x_0 \rangle$ and $\langle y_0 \rangle$ are the initial values of the fluctuations; and β_1, β_2, the two eigenvalues determined by Equations (12), can be used to construct the autocorrelation function as

$$\langle x(0)x(t) \rangle = \frac{1}{\beta_2 - \beta_1} \{[(a_1 + \beta_2) e^{\beta_1 t} - (a_1 + \beta_1) e^{\beta_2 t}] \langle x^2 \rangle_e + [e^{\beta_1 t} - e^{\beta_2 t}] \langle xy \rangle_e\} \tag{20}$$

where the factors $\langle x^2 \rangle_e$ and $\langle xy \rangle_e$ are evaluated at the steady state. (ϕ_{st}, ψ_{st}) can be obtained by solving Equations (7) and (8). Clearly, Equation (20) involves both the exponential time factors. A similar relation can be easily derived for the variable y. The typical variation of autocorrelation function for three different values of α (6, 12 and 18) is shown in Figure 4.3(a,b,c).

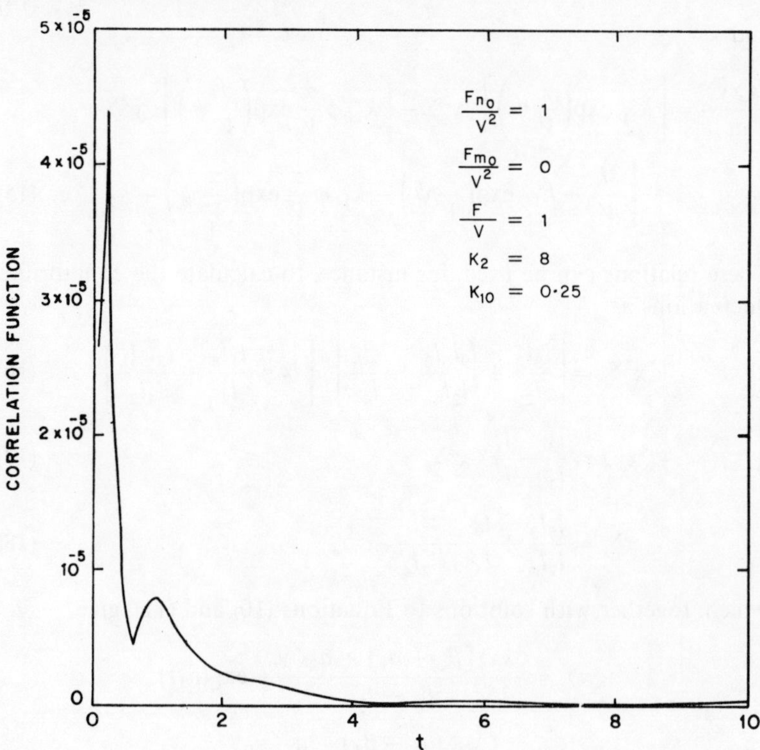

FIGURE 4.3a Plot of correlation function versus t ($\alpha = 6$).

As the value of α approaches the critical value ($\alpha = 23.83$) it is seen that the correlation function approaches zero faster, indicating rapidly varying processes for these higher values of α.

Knowledge of the autocorrelation function helps us to calculate the fluctuation spectrum (s) and transient diffusion rates (D) thus:

$$s(w) = \frac{1}{2\pi} \int_0^\infty \langle x(0)x(t')\rangle \cos wt' \, dt' \tag{21}$$

$$D = \int dt' \langle x(0)x(t)\rangle \tag{22}$$

The deterministic analysis of the autocatalytic reacting system indicates a unique stable solution for a set of parameter values

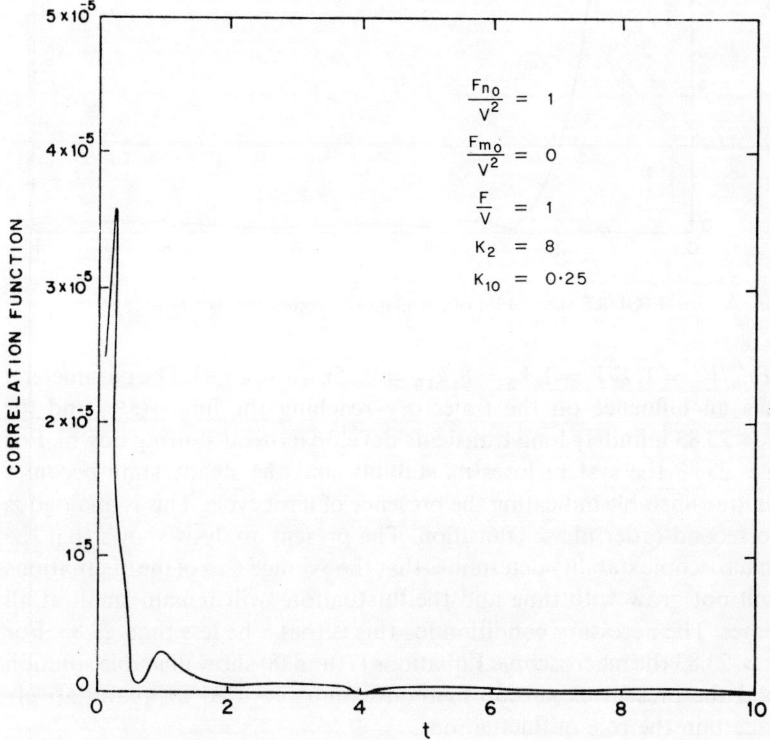

FIGURE 4.3b Plot of correlation function versus t ($\alpha = 12$).

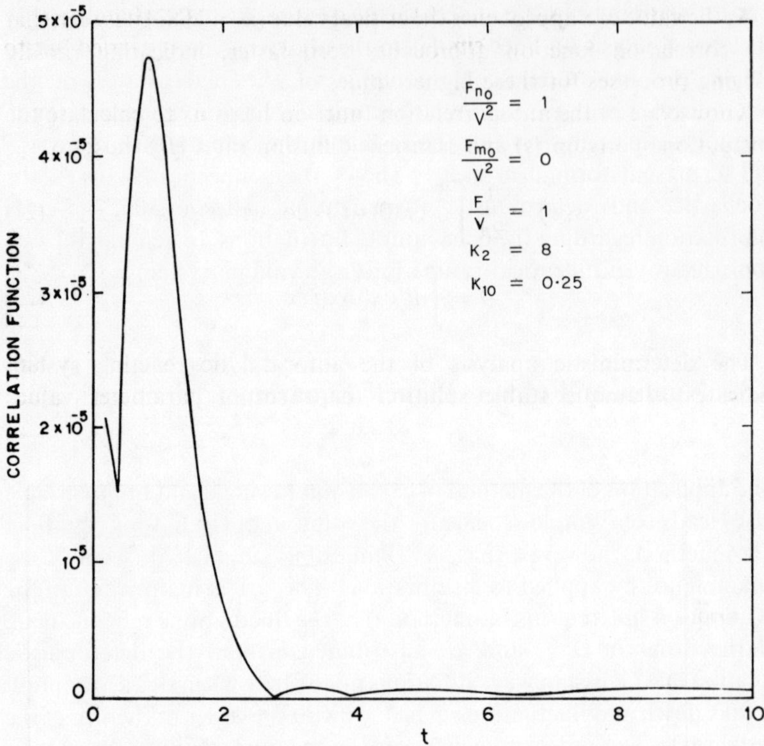

FIGURE 4.3c Plot of correlation function versus t ($\alpha = 18$).

($Fn_0/V^2 = 1, F/V = 1, k_2 = 8, k_{10} = 0.25$), if $\alpha < 23.83$. The parameter α has an influence on the trajectory reaching the final state, and for $\alpha = 23.83$ infinitely long transients develop (critical slowing down). For $\alpha > 23.83$ the system loses its stability and the steady state becomes unique unstable indicating the presence of limit cycle. This is analogous to second-order phase transition. The present analysis shows that the macroscopic stability determines that the average $\langle x \rangle$ of the fluctuations will not grow with time and the fluctuations will remain small at all times. The necessary condition for this is that α be less than 23.83. For $\alpha > 23.83$ the macroscopic Equations (7) and (8) show unstable solution and the present expansion method cannot be used to quantitatively ascertain the role of fluctuations.

For a different set of parameter values ($k_1 V/F = 0.05$, $k_2 V/F = 1$,

$\alpha = 15$), the deterministic analysis indicates the existence of multiplicity of states. Analysis on lines similar to the above has been carried out to obtain the mesostates corresponding to the macrostates of the deterministic analysis. As long as the initial fluctuations remain small, the stochastic analysis yields the results of the deterministic analysis.

The present formalism clearly shows the connection between the stochastic and deterministic formulations. Additionally, it gives information regarding the behaviour of fluctuations. In general, the two approaches yield identical results for large volume systems.

4.3 Nonlinear Fokker–Planck Equation

The application of the method of expansion to single and multivariable cases leads to a simple scheme for the solution of the master equation. The method, however, has its limitations in that it cannot be unambiguously applied to all situations. This arises mainly due to the assumption inherent in this method that the fluctuations remain small, of the order of $\Omega^{1/2}$, and are distributed around the deterministic solution in a Gaussian way. Situations might arise when this assumption breaks down and fluctuations in fact grow to the order of the size of the system (Ω). The separation between the macroscopic and fluctuating parts as envisaged in the expansion method is then not applicable and recourse to other methods becomes necessary. Typical of such situations is a system possessing absorbing boundaries or one near phase transition and critical points. In cases of this type, the asymptotic expansion of the master equation, as will be shown subsequently, leads to a nonlinear Fokker–Planck equation as its most dominant term and should be used as an approximation to the master equation. The subject is dealt with in Section 4.3.1.

Section 4.3.2 deals with the Fokker–Planck equation derived from the Langevin equation. We might recall here from the discussion in Chapter 2 that these stochastic equations are more convenient for treating external fluctuations, the statistical characteristics of which can be identified. While the Langevin type of equations have been used for this purpose, the ease of their formulation has often tempted us to use them for treating internal fluctuations as well. This has often caused considerable confusion, for the internal noise cannot be defined *a priori* and in fact information regarding it must emerge naturally from the system. In other words, it would appear that formulations such as the master equation should alone be used when problems with internal noise are to be analysed. Some reflection, however, suggests a simple way out. The idea is to derive the Fokker–Planck equations both from the master equation and the generalised Langevin equation and make them equivalent. The Fokker–Planck equation as derived from the master equation would contain the necessary information regarding the internal fluctuations which should be interpreted as a fluctuating function in the Langevin equation. In other words, through the intermediacy of the Fokker–Planck equation, we can make the Langevin equation consistent with the master equation. As would be seen in Section 4.3.2, the fluctuating function to be used in the Langevin equation so as to make it consistent with the master equation is a simple variant of the reaction functions in the macroscopic description.

This approach should have great practical appeal in engineering sciences where we often start with a macroscopic description of the system. This equation can be suitably appended with the fluctuating function that is determined by the methodology mentioned above, to yield a stochastic equation that is consistent with the master equation. Note that the only information required is the macroscopic description. The resulting Langevin equation can now be used to account for the effect of fluctuations. The equation is much more convenient and is more easily formulated. We can thus avoid knowledge of transition probability functions required in the master equation approach.

Having ascertained the importance of the nonlinear Fokker–Planck equation in the previous sections, in Section 4.3.4 we resort to some of the simple methods in use for obtaining solutions to the equations.

4.3.1 Reduction of Master Equation to Fokker–Planck Equation

The application of the method of expansion to the master equation leads to a situation where we recover the macroscopic law in the lowest order in the expansion parameter and a linear Fokker–Planck equation describing the fluctuations results in the next higher order approximation. Referring to Appendix 3.A in Chapter 3, the general form of this macroscopic equation is given by

$$\frac{\partial \phi}{\partial \tau} = \alpha_{1,0}(\phi) \tag{4.28}$$

which possesses an asymptotically stable solution only if the condition $\alpha'_{1,0} < 0$ is satisfied. In the examples considered thus far, $\alpha_{1,0}(\phi)$ has always been greater than zero and the condition $\alpha'_{1,0} < 0$ has always been met. An interesting variation occurs when $\alpha_{1,0}(\phi) = 0$. Integration of the macroscopic equation (4.28) now suggests that $\phi(\tau) =$ constant $= \phi(0)$. Any slight variation in the initial value therefore does not decay and the solution is rendered unstable. In fact, the variance of fluctuation as indicated by the linear Fokker–Planck equation takes the form

$$\frac{d\langle y^2\rangle}{d\tau} = \alpha_{2,0}(\phi) \tag{4.29}$$

indicating that the variance grows with time. The fluctuations would therefore attain the order of the macroscopic part (Ω) in time

$$\tau = \Omega/\alpha_{2,0} \tag{4.30}$$

and the expansion method that assumes the fluctuations to be small fails after this time. In other words, the anticipation that P is a smoothly varying function of X with a width of the order of $\Omega^{-1/2}$ is not justified. However, we could modify this by assuming a width of the order Ω, the variable X being thus scaled as $x = X/\Omega$. Using this definition in the master equation and going through the further steps as indicated in Appendix 3.A, we finally obtain the following transformed master equation:

$$\begin{aligned}\frac{\partial P(X,t)}{\partial \tau} &= \left\{-\frac{\partial}{\partial x}\alpha_{1,1}(x)P + \frac{1}{2}\frac{\partial^2}{\partial x^2}\alpha_{2,0}(x)P\right\}\\ &\quad + \Omega^{-1}\left\{\frac{1}{2}\frac{\partial^2}{\partial x^2}\alpha_{2,1}(x)P - \frac{1}{3!}\frac{\partial^3}{\partial x^3}\alpha_{3,0}(x)P - \frac{\partial}{\partial x}\alpha_{2,1}(x)P\right\}\\ &\quad + O(\Omega^{-2})\end{aligned} \tag{4.31}$$

The first bracket represents the major contribution yielding the approximation

$$\frac{\partial P(x,t)}{\partial \tau} = -\frac{\partial}{\partial x}\alpha_{1,1}(x)P + \frac{1}{2}\frac{\partial^2}{\partial x^2}\alpha_{2,0}(x)P \tag{4.32}$$

where τ is now defined as $\tau = f(\Omega)\Omega^{-2}t$.

In contrast to the previous situation where the dominant term leads to a macroscopic law, the present scheme leads to a nonlinear Fokker–Planck equation. We might recall that a nonlinear Fokker–Planck equation was derived from the master equation on the assumption of small jumps and slow variations of $W(x,t)$ and $P(x,t)$ with respect to x in the previous chapter. The equation rederived here fixes it as the first term in the systematic expansion of the master equation in powers of Ω^{-1}, that has the important property of having $\alpha_{1,0} = 0$. The special class of master equation that is characterised by this condition is referred to as the diffusion equation.

The question of deriving a Fokker–Planck equation from the associated master equation is a little more involved than what can be appreciated from the above discussion. A Markovian process with a discrete stochastic variable is described by a master equation (see Chapter 2), while for a continuous stochastic variable the Fokker–Planck formulation is employed (see Chapter 3). An equivalent description of the Fokker–Planck formulation can be given in terms of the Langevin equation or the mathematically more rigorous stochastic differential equation. The master equation provides a natural way to account for the internal fluctuations arising as a result of the interactions amongst the discrete constituents of the system. The Fokker–Planck or the equivalent stochastic differential equation, on the other hand, provides a natural way to extend the phenomenological equation by adding the external stochastic parameter. The difficulty arises when the Fokker–Planck formulation is used to treat the internal noise in the system. While such a formulation has been extensively used for this purpose with good success, the lack of clarity regarding its relation to the master equation has been a major source of confusion. This brings to mind the first question concerning the relation between the master equation and the Fokker–Planck equation. Is it possible at all to express the discrete state space master equation as an equivalent continuous state space Fokker–Planck equation? Also, if this is possible, then what is the structure and range of validity of the Fokker–Planck equation?

As seen in the previous chapter the master equation can be transformed into a Fokker–Planck equation using the Kramers–Moyal expansion. The discrete variables in the master equation are scaled using a large characteristic parameter, such as its volume. The discrete number X in the original formulation appears as a quasi-continuous quantity ($x = X/V$) in the transformed equation for large volume. The large characteristic parameter also justifies neglect of higher order terms which when truncated after the second term yields the nonlinear Fokker–Planck equation. Doubts regarding the systematic character of truncating the Kramers–Moyal expansion were raised by van Kampen (1976) who, as we already know, postulated a certain dependence of P on Ω. The assumption implied is that P is Gaussian in the limit $\Omega \to \infty$ and that the fluctuations are distributed around the deterministic solution with a width of Ω.

Clearly such a postulation is valid only for $\Omega \to \infty$. For large but finite Ω, it is known and proved rigorously (Kurtz, 1971) that the fluctuations of the Markov process do not necessarily coincide with the fluctuations around the deterministic solution calculated using $\Omega \to \infty$, as in van Kampen's method. This implies that the essential characteristics of the master equation may be lost for some systems when the expansion method is employed. The contention of van Kampen arrived at on the basis of his postulation regarding the dependence of P on Ω, that the nonlinear Fokker–Planck equation is inconsistent, is therefore suspect and the question has been reexamined by Horsthemke and Brenig (1977) who have rigorously derived a nonlinear Fokker–Planck equation as an asymptotic approximation of the master equation in the limit of large but finite Ω.

The method, based on the theory of continuous Markov processes and employing the Ito form of the stochastic differential equation, defines an intensive variable $x = X/\Omega$, with the transition probability per unit time appropriately scaled as $w = W/\Omega$. This implies a corresponding reduction in the jump of the quantity x and the new probability function p is related to P as $p(x, t) = P(X, t)$. For large Ω the quantity x can be considered as quasi-continuous which in the thermodynamic limit ($X \to \infty$, $\Omega \to \infty$, $X/\Omega =$ finite) becomes a continuous quantity. It is shown mathematically (Arnold, 1973; Gichman and Skorochod, 1971) that the Markov process so defined would possess continuous state space character provided the condition

$$\lim_{t\to s}\frac{1}{t-s}\int p(s, x_1 \mid t, x_2)\, dx_2 = 0 \qquad (x_2 - x_1) > 0 \tag{4.32}$$

is satisfied. Additionally, if the system possesses the first two truncated differential moments, then the Markov process is referred to as a diffusion process and takes the following form of the nonlinear Fokker–Planck equation:

$$\frac{\partial p}{\partial t} = -\frac{\partial}{\partial x}\alpha_1 p + \frac{1}{2}\frac{\partial^2}{\partial x^2}\alpha_2 p \tag{4.33}$$

where p refers to the conditional probability density. Alternatively, the nonlinear Fokker–Planck equation such as Equation (4.33) necessarily represents an asymptotic form of a quasi-continuous Markov process which in the limit of large Ω can be represented as a continuous process.

The differential moments α_1, α_2 can be obtained from the general definition as

$$\alpha_n = \sum_j \left(\frac{j}{\Omega}\right)^n \Omega w(j, x, t) \tag{4.34}$$

w is usually available in a canonical form

$$w = w^0 + \frac{1}{\Omega} w^1 + O(\Omega^{-2}) \tag{4.35}$$

We therefore have α_n as

$$\alpha_n = \alpha_n^0 + \frac{1}{\Omega}\alpha_n^1 + \cdots \tag{4.36}$$

The coefficients α_1 and α_2 can be immediately recognised as equal to $\alpha_{1,1}$ and $\alpha_{2,0}$ defined previously for the expansion method. Equation (4.33) is therefore identical to Equation (4.31) derived previously by modifying the expansion method.

It thus seems that the nonlinear Fokker–Planck equation (or the equivalent Langevin or stochastic differential equation) is the most general asymptotic representation of the master equation for large Ω. The error involved in approximating the master equation by the Fokker–Planck equation is of the order of Ω^{-1} since the neglected higher order moment terms are of this order.

In Table 4.1 we present representative forms of the basic equations that approximate the original master equation as obtained using different methods. The deterministic equation is also included for the sake of comparison. To indicate the essential differences between the predictions of the different methods, Table 4.2 lists the basic reaction schemes analysed. The second column of the table gives the basic deterministic equation along with the associated master equation for the system. The third column of the table defines the various coefficients which can be used to construct the nonlinear Fokker–Planck equation and the linear noise approximation.

The stationary solution of the master equation for the first case can be readily obtained as $P_{st}(X, \infty) = \delta_{x,0}$, in view of zero being an absorbing boundary of the master equation. The corresponding solution for the nonlinear Fokker–Planck equation can be obtained subject to Equation (4.32) as

$$p_{st}(x) = \frac{(k_1 a + k_3 b + k_2 x)^{2V(2k_1 a + k_2/V)/k_2}}{[(k_1 a + k_3 b)x + k_2 x^2]\exp(2Vx)} \,(constant) \tag{4.37}$$

Clearly $p_{st}(x)$ diverges as $x \to 0$ and is therefore nonnormalisable. A careful study reveals that $x = 0$ being an absorbing boundary will be reached in finite time. We note that both the master equation and the Fokker–Planck equation predict large fluctuations that take the system away from the deterministic solution. Under these conditions one would expect that the expansion method of van Kampen would not work. Indeed, its application would lead to the second nontrivial state of the deterministic equation with the fluctuations around it distributed in a Gaussian way. The simple example of an absorbing boundary therefore reveals that the application of the expansion method to such cases may have nontrivial consequences.

In the second case in Table 4.2 we encounter a situation where both the master equation as well as the nonlinear Fokker–Planck equation predict no stationary solution, while the expansion method yields a stationary solution with fluctuations distributed around it in a Gaussian way.

The third case considers another nonlinear model, the macroscopic equation of which indicates that the detailed balance property is not obeyed. The stationary solution of the master equation for this case can be readily obtained and used to calculate the first two noncentral moments as per the prescription in Chapter 1. The moments obtained

TABLE 4.1
Fokker–Planck equations from the master equation

	Basic equation	Remarks
Master equation	$\partial p/\partial t = \{W(X\|X')p(X',t) - W(X'\|X)p(X,t)\}$	Rigorous equation taking account of internal fluctuations
Deterministic equation	$dX/dt = Q(X) - R(X)$	Fluctuations completely ignored
Expansion method	(i) Macroscopic eq.: $d\phi/dt = \alpha_{1,0}(\phi)$ $\alpha_{1,0} = Q(\phi) - R(\phi)$ (ii) Linear noise approximation: $\partial\pi/\partial t = -\frac{\partial}{\partial y}(\alpha'_{1,0}, p) + \frac{1}{2}\frac{\partial^2}{\partial y^2}(\alpha_{2,0}, p)$ where $p(x,t) = \pi(y,t)$, $X = \Omega\phi + \Omega^{\frac{1}{2}}y$, $\alpha_{2,0} = Q(\phi) + R(\phi)$	p is assumed to be Gaussian with the fluctuations distributed around the deterministic solution with a width of $\Omega^{\frac{1}{2}}$. Strictly valid as $\Omega \to \infty$. For large but finite Ω, it may not provide accurate representation for certain situations such as systems with absorbing boundary, diffusion processes, phase transition and near critical point. Suitable modification required to take care of these special situations.
Fokker–Planck approximation (Kramers–Moyal expansion)	$\frac{\partial p}{\partial t} = -\frac{\partial}{\partial x}\alpha_{1,0}p + \frac{1}{2}\frac{\partial^2}{\partial X^2}\alpha_{2,0}P$ $x = X/\Omega$	As already elaborated in Chapter 3
Fokker–Planck approximation (asymptotic expansion)	$\frac{\partial p}{\partial t} = -\frac{\partial}{\partial x}\alpha_{1,1}p + \frac{1}{2}\frac{\partial^2}{\partial x^2}\alpha_{2,0}p$	Valid for any situation including phase transition and critical points

TABLE 4.2
Reaction schemes with their deterministic, master and Fokker–Planck equations

Reaction scheme	Deterministic and master equations	Coefficient of Fokker–Planck equation
1. $A + X \rightleftharpoons 2X$	$\frac{dx}{dt} = (k_1a - k_3b)x - k_2x^2$	$\alpha_{1,1} = (k_1a - k_3b)x - k_2x^2 + \frac{k_2x}{V}$
$B + X \longrightarrow C$	$\frac{dp}{dt} = \{k_1a(E-1)x + k_3b(E^{-1}-1)x\}p + k_2(E-1)x(x-1)p$	$\alpha_{2,0} = (k_1a + k_3b)x + k_2x^2$
		$\alpha'_{1,0} = k_1a - k_3b - 2k_2x$
2. $A + 2X \longrightarrow 3X$	$\frac{dx}{dt} = k_1ax^2 - k_2bx + k_3d^2$	$\alpha_{1,1} = k_1ax^2 - k_2bx + k_3d^2 - k_1ax/V$
$X + B \longrightarrow AC$	$\frac{dp}{dt} = k_1a(E-1)x(x-1)p + k_2b(E^{-1}-1)xp + k_3d^2(E-1)p$	$\alpha_{2,0} = k_1ax^2 + k_2bx + k_3d^2$
$2D \longrightarrow X + D$		$\alpha'_{1,0} = 2k_1ax - k_2b$
3. $A \longrightarrow X$	$\frac{dx}{dt} = k_1a - k_2x^2$	$\alpha_{1,1} = -k_1a - 2k_2x/V + 2k_2x^2$
$2X \longrightarrow B$	$\frac{dp}{dt} = k_1a(E-1)p + k_2(E^{-1}-1)^2x(x-1)p$	$\alpha_{2,0} = k_1a + 4k_2x^2$
		$\alpha'_{1,0} = -2k_2x^2$

from the master equation completely coincide with those from the nonlinear Fokker–Planck equation and are given as

$$\langle x \rangle = x_s + (8V)^{-1} + O(V^{-2}) \tag{4.38}$$

$$\langle \delta x\, \delta x \rangle = \tfrac{3}{4} x_s V^{-1} + O(V^{-2}) \tag{4.39}$$

The result clearly indicates that the stationary distribution is not centered around the deterministic solution, as in the expansion method.

The application of the nonlinear Fokker–Planck equation to different situations such as systems with absorbing boundaries, systems with phase transitions and critical points, as indicated by the examples considered above, suggests that the method yields results that are identical to the results of the original master equation. The van Kampen expansion method cannot be used for these types of situations and modification of the original method is necessary.

The method of obtaining the correct form of the nonlinear Fokker–Planck equation as demonstrated for the single variable case can be generalised to obtain a multivariable Fokker–Planck equation from the multivariable master equation. The typical form of the equation is

$$\frac{\partial P(x,t)}{\partial t} = -\frac{\partial}{\partial x_i}\boldsymbol{\alpha}_{1,p} + \frac{1}{2}\frac{\partial^2}{\partial x_i\, \partial x_j}\boldsymbol{\alpha}_2 p \tag{4.40}$$

where $\boldsymbol{\alpha}_1$ and $\boldsymbol{\alpha}_2$ represent matrices involving the arguments x and summation over all the variables is implied. The general procedure is similar to that illustrated in Table 4.2 (see also Appendix 5.A).

4.3.2 Fokker–Planck Equation from Langevin Equation

It was mentioned in the introduction of Chapter 3 that conventionally two different approaches, the master equation approach and stochastic differential equation approach, have been used to treat the fluctuations in physical and chemical systems. It has been made clear here that the former approach naturally accounts for internal fluctuations while the latter suits the need when external disturbances are to be accounted. Despite this fact, the stochastic differential equations have often been used to treat systems with internal noise. This arises naturally, for in most engineering sciences we often start with the macroscopic description of the system which when suitably appended with a noise term, such as white noise, leads to the stochastic differential equation. In generalised form this equation becomes

$$\frac{dx}{dt} = G[x(t)] + \varepsilon F[x(t), f(t)] \tag{4.41}$$

where $f(t)$ is a fluctuating function, G and F are arbitrary functions of x, and t and ε represent the strength of the fluctuations of (dx/dt). In many instances the fluctuating function F is merely dependent on time as, for example, when

$$F[x(t), f(t)] = f(t) \tag{4.42}$$

We then refer to Equation (4.41) as representing the stochastic differential equation with additive noise. In certain other instances the fluctuating function may involve dependence on both x and t and can be decomposed as

$$F[x(t), f(t)] = g[x(t)]f(t) \tag{4.43}$$

The stochastic differential Equation (4.41) with this kind of fluctuating function is referred to as an equation with multiplicative noise. In view of the separable dependence of F on x and t, mathematically this form is equivalent to the equation with additive noise, for a simple transformation such as

$$\frac{dv}{dt} = \frac{dx/dt}{g[x(t)]} \tag{4.44}$$

can transform a multiplicative noise to the additive noise. The most complex form of the function F would involve its dependence on x and t which cannot be put in a separable form as in Equation (4.43). The stochastic differential Equation (4.41) is then said to have a nonadditive noise term.

To see how one proceeds to obtain a Fokker–Planck equation from the various Langevin equations, let us begin with the simple example of a case with additive noise. The equation in this case would typically have the form

$$\frac{dx}{dt} + Ax(t) = f(t) \tag{4.45}$$

where A represents some phenomenological constant. The fluctuating force $f(t)$ is defined in terms of its statistical properties and it is generally assumed to be a delta correlated Gaussian random noise with zero mean. The properties of $f(t)$ are defined as

$$\langle f(t) \rangle = 0 \tag{4.46}$$

$$\langle f(t) f(t') \rangle = 2D\, \delta(t - t') \tag{4.47}$$

where higher moments of $f(t)$ vanish and $2D$ represents the variance of the fluctuation. With these properties of $f(t)$, the solution to Equation (4.45) in the limit of large time shows the following asymptotic properties:

$$\operatorname*{Lim}_{t \to \infty} \langle x(t) \rangle = 0, \qquad \lim_{t \to \infty} \langle x^2(t) \rangle = \frac{D}{A} \tag{4.48}$$

To construct an equivalent Fokker–Planck equation, we now need to construct an equation that describes the evolution of the conditional probability density $p(x, t \mid x_0)$ that postulates that the system would lie in the range of x and $x + dx$ at time t given that it has been at x_0 in the previous stage. This equation can be obtained from a knowledge of the first two moments of the distribution $[\Delta x(t) = x(t + \Delta t) - x(t)]$:

$$\begin{aligned} \lim_{\Delta t \to 0} \frac{[\Delta x(t)]}{\Delta t} &= \alpha_1(x) \\ \lim_{\Delta t \to 0} \frac{[\Delta x(t)]^2}{\Delta t} &= \alpha_2(x) \end{aligned} \tag{4.49}$$

as

$$\frac{\partial p}{\partial t} = -\frac{\partial}{\partial x} \alpha_1(x) p + \frac{1}{2} \frac{\partial^2}{\partial x^2} \alpha_2(x) p \tag{4.50}$$

For the particular example in Equation (4.45) we have $\alpha_1 = -Ax$ and $\alpha_2 = 2D$, yielding

$$\frac{\partial p}{\partial t} = A \frac{\partial}{\partial x}(x, p) + D \frac{\partial^2 p}{\partial x^2} \tag{4.51}$$

with the initial condition $p(x, t \mid x_0) = \delta(x - x_0)$. The equation with multiplicative term can be written as

$$\frac{dx}{dt} = Ax + B(x) f(t) \tag{4.52}$$

Making use of the transformation

$$y = \int^{x} \frac{dx}{B(x)} \tag{4.53}$$

we can convert Equation (4.52) into a simple form as

$$\frac{dy}{dx} = \frac{Ax}{B(x)} + f(t) \tag{4.54}$$

for which, from the previous case, the Fokker–Planck equation can be written as

$$\frac{\partial p(y,t)}{\partial t} = -\frac{\partial}{\partial y}\frac{Ax}{B(x)}p(y,t) + D\frac{\partial^2 p(y,t)}{\partial y^2} \tag{4.55}$$

Noting now the relations $p(x,t) = p(y,t)$ and $dy/dx = p(y,t)/B(x)$, Equation (4.55) can be written as

$$\frac{\partial p(x,t)}{\partial t} = -\frac{\partial y}{\partial x}\frac{\partial}{\partial y}Ax(y)p(x,y,t) + D\frac{\partial y}{\partial x}\frac{\partial x}{\partial y}B(x)p(x,t) \tag{4.56}$$

or

$$\frac{\partial p(x,t)}{\partial t} = -\frac{\partial}{\partial x}[Ax + DB(x)B'(x)]p(x,t) + D\frac{\partial^2}{\partial x^2}B^2(x)p(x,t) \tag{4.57}$$

yielding the desired form of the Fokker–Planck equation.

In the examples treated thus far, the fluctuating function has been dependent on time t alone. In a general case where functional dependence on both x and t is involved, the procedure for obtaining the Fokker–Planck equation is no longer unambiguous.

To illustrate the difficulty let us write Equation (4.41) as

$$dx = G[x(t),t]\,dt + F[x(t),t]\,dw(t) \tag{4.58}$$

where $dw(t)$ represents the derivative of a Wiener process that implies white noise. On integration of Equation (4.58), we have

$$x(t) = x(0) + \int_0^t Gx(t),t\,dt + \int_0^t F[x(t),t]\,dw(t) \tag{4.59}$$

The last integral in Equation (4.59) is with respect to a Wiener process and presents no difficulty if F is a function of t alone. However, when the integral F also depends on the state x that is a random process, then a difficulty in integration arises. The rules of calculus formulated in this regard by Stratonovich and Ito differ and lead to different results. This implies that two different Fokker–Planck equations associated with the same Langevin Equation (4.58) can be formulated. The rules proposed by Stratonovich are the ordinary rules of calculus with which we are already familiar, while Ito prescribes mathematically more rigorous rules. The Fokker–Planck equation corresponding to the stochastic differential equation can be generally written as

$$\frac{\partial p}{\partial t} = -\frac{\partial}{\partial x}\alpha_1(x)p + \frac{1}{2}\frac{\partial^2}{\partial x^2}\alpha_2(x)p \tag{4.60}$$

where now the coefficients α_1 and α_2 differ depending on whether one prefers to use the Stratonovich or the Ito prescription. Omitting the detailed mathematics involved, we can identify these coefficients for the two prescriptions:

$$\alpha_1 = G[x(t), t], \qquad \alpha_2 = F^2[x(t), t] \qquad \text{(Ito)} \tag{4.61}$$

$$\alpha_1 = G[x(t), t] + \tfrac{1}{2}F\frac{dF}{dx}, \qquad \alpha_2 = F^2[x(t), t] \qquad \text{(Stratonovich)} \tag{4.62}$$

The coefficients in Equations (4.61) and (4.62) can also be used to define the backward form of Fokker–Planck equation which can be generally written as

$$-\frac{\partial p}{\partial t} = \alpha_1(x)\frac{\partial p}{\partial x} + \tfrac{1}{2}\alpha_2(x)\frac{\partial^2 p}{\partial x^2} \tag{4.63}$$

The realisation that more than one form of Fokker–Planck equation exists for the same stochastic differential equation raises the important question of which one to choose for practical application. The question has been examined in detail by West *et al.* (1979) who applied the treatments of Ito and Stratonovich to stochastic equations involving nonadditive noise (as would be noticed from the definitions of α_1 and α_2, if the function $F[x(t), t]$ depends only on t, then the two approaches lead to identical results). The major conclusion has been that the Ito treatment of such systems always results in solutions that are quantitatively incorrect or at times physically unrealisable. This suggests that, despite the mathematical rigour of the Ito treatment, for most practical physical processes one should follow the Stratonovich prescription.

Based on the discussion presented above it is clear that given a Langevin equation we can obtain the corresponding Fokker–Planck equation which would rightly describe the physical situation. One should bear in mind, however, that an explicit form of the fluctuations function $F[x(t), t]$ can be known only if the disturbance is external in nature so that we can identify its statistical characteristics. Our main intention here, however, is to see if we could use it to describe the internal fluctuations. From a fundamental viewpoint such a treatment is questionable and the only way to account for them would be to resort to the master equation.

From a practical viewpoint, however, it is convenient to construct an equivalent form of the master equation in terms of a Langevin equation. This can be easily achieved since Fokker–Planck equations from both the master equation and the Langevin equation can be constructed and made equivalent. The procedure also allows identification of the fluctuating function. To illustrate the relation between the master equation and the Langevin equation, we begin with the master equation for the probability density, viz.

$$\frac{\partial p(x,t)}{\partial t} = \int dx'\{W(x|x')p(x',t) - W(x'|x)p(x,t)\} \tag{4.64}$$

The equation can be reduced to the Fokker–Planck equation as discussed in the previous section (Section 4.3.1) using the Kramers–Moyal expression:

$$\frac{\partial p(x,t)}{\partial t} = \sum_{n=1}^{\infty} \frac{(-1)^n}{n!} \frac{\partial}{\partial x} \alpha_n(x) p(x,t) \tag{4.65}$$

where the various moments (α's) are defined as

$$\alpha_n(x) = \int dx'(x' - x)^n W(x'|x) \tag{4.66}$$

We are interested in associating a Langevin equation that is consistent with this description of the master equation. Towards this end Bedeaux (1977) has constructed an equation of the form

$$\frac{dx}{dt} = G[x(t)] + \sum_{m=2}^{\infty} F_m[x(t)] f_m(t) \tag{4.67}$$

where F_m is an arbitrary function of x and f_m the fluctuating function.

Note that the stochastic Equation (4.67) has a multiplicative noise. Based on the development presented earlier in this section, we can associate the following Fokker–Planck equation with Equation (4.67) where the Stratonovich prescriptions are used:

$$\frac{\partial p(x,t)}{\partial t} = -\frac{\partial}{\partial x} G[x(t)] + \sum_{n,m=2}^{\infty} \frac{(-1)^m}{m!} \left\{\frac{\partial}{\partial x} F_m(x)\right\}^m p(x,t) \quad (4.68)$$

Equations (4.65) and (4.68) are now made equivalent by identifying the α's in (4.65) with the corresponding terms in (4.68). As is often assumed, if the fluctuating function is a zero centred Gaussian noise, then its higher order moments vanish. We then need to consider only the case of $m = 2$ and Equation (4.68) simplifies to

$$\frac{\partial p(x,t)}{\partial t} = -\frac{\partial}{\partial x}\left[G[x(t)] + \frac{1}{4}\left\{\frac{\partial}{\partial x} F_2^2(x)\right\}\right] p(x,t) + \frac{1}{2}\left\{\frac{\partial^2}{\partial x^2}[F_2^2(x)p(x,t)]\right\} \quad (4.69)$$

To identify $F_2[x(t)]$ let us take a specific example

$$\frac{dx}{dt} = Q(x) - R(x) = G(x) \quad (4.70)$$

which possesses the following master equation:

$$\frac{\partial p(x,t)}{\partial t} = Q(x-1)p(x-1,t) + R(x+1)p(x+1,t) - [Q(x) + R(x)]p(x,t) \quad (4.71)$$

The associated Fokker–Planck equation obtained through Kramers–Moyal expansion, Equation (4.65), possesses the differential moments α given by Equation (4.66) which can be simplified as

$$\alpha_n(x) = [Q(x) + (-1)^n R(x)] \quad (4.72)$$

We then obtain

$$\alpha_1(x) = Q(x) - R(x)$$

$$\alpha_2(x) = Q(x) + R(x)$$

Comparing this result with Equation (4.69) we can immediately relate the fluctuating function as

$$F_2(x) = Q(x) + R(x) \tag{4.73}$$

The following Langevin equation can then be written:

$$\frac{dx}{dt} = Q(x) - R(x) + \{Q(x) + R(x)\} f(t) \tag{4.74}$$

which would have the Fokker–Planck equation given by Equation (4.69) that is consistent with the master equation. The concept used here is best illustrated by considering an example.

Example 4.3

For a chemical reaction $X + Y \rightleftarrows 2X$ with unit rate constants and no external disturbances, find the stationary probability density using the Fokker–Planck descriptions.

We are already aware that for a situation of the type considered in this example, we should formulate a master equation which would adequately represent the fluctuations in the number densities n_x and n_y of the molecules of species X and Y. Before we formulate the master equation we note from the reaction stoichiometry that the number $n_x + n_y = \text{constant} = \text{n}$. This is therefore a single variable problem, and the evolution of the probability density $p(x, t)$ can be written as

$$\frac{\partial p(x,t)}{\partial t} = (E_x - 1)nn_x p + 2(E_x^{-1} - 1)n_x(n_x - 1)p \tag{1}$$

where E_x, E_x^{-1} refer to the difference operator. The master equation can be reduced to the Fokker–Planck equation as indicated in Section 4.3.1 as

$$\frac{\partial P(x,t)}{\partial t} = -\frac{\partial}{\partial x}\alpha_{11}(x)p + \frac{1}{2}\frac{\partial^2}{\partial x^2}\alpha_{20}(x)p \tag{2}$$

where

$$\alpha_{11} = x(n - 2x) + \frac{n}{4\Omega} \qquad \text{and} \qquad \alpha_{20} = \text{n}x/\Omega \tag{3}$$

For stationary solution, equating the rhs of Equation (2) to zero and integrating twice, we obtain

$$p(x, \infty) = \frac{K_1}{\sqrt{x}} \left\{ \exp\left[-\frac{2\Omega}{n}\left(x - \frac{n}{2}\right)^2 \right] \right\}$$
$$\times \left[1 + K_2 \left(\frac{n}{2\Omega}\right)^{1/2} \int^{(2\Omega/n)^{1/2}[(n/2)-x]} \left[\frac{n}{2} + \left(\frac{n}{2\Omega}\right)^{1/2} x' \right]^{1/2} e^{x'^2} \right] dx' \tag{4}$$

where K_1 and K_2 are constants of integration, which can be evaluated by the boundary and normalisation conditions. The closed nature of the system stipulates that the probability current across the boundaries $x = 0$ and $x = n$ should vanish at all times. Thus, for $0 < x < n$, if we define $p(x, t)\, dx = Q(x, t)$, then we should have $dQ(x, t)/dt = 0$ for $x = n$ and 0. Clearly this condition is satisfied by Equation (4) only if $K_2 = 0$. We thus have

$$p(x, \infty) = \frac{K_1}{\sqrt{x}} \exp\left[-\frac{2\Omega}{n}\left(x - \frac{n}{2}\right)^2 \right] \tag{5}$$

where the constant K_1 can be found from the normalisation condition

$$\int_0^x p(x, t)\, dx = 1 \tag{6}$$

Let us now formulate the stochastic differential equation corresponding to the reaction under study, the macroscopic equation of which can be written as

$$\frac{dx}{dt} = x(n - x) - x^2 = Q(x) - R(x) \tag{7}$$

We note that the fluctuating function $F[x(t)]$ in Equation (4.73) can be identified as

$$\alpha_2 = F^2[x(t)] = nx \tag{8}$$

The stochastic equation associated with the deterministic equation is then

$$dx = [x(n - x) - x^2]\, dt + \frac{nx}{\Omega} dw \tag{9}$$

The two Fokker–Planck equations associated with the stochastic equation are

$$\frac{\partial p(x,t)}{\partial t} = \left\{ -\frac{\partial}{\partial x}[x(n-x) - x^2] + \frac{1}{2}\frac{\partial^2}{\partial x^2}\frac{nx}{\Omega} \right\} p(x,t) \quad \text{(Ito)} \quad (10)$$

and

$$\frac{\partial p(x,t)}{\partial t} = \left\{ -\frac{\partial}{\partial x}\left[x(n-x) - x^2 + \frac{n}{4\Omega}\right] + \frac{1}{2}\frac{\partial^2}{\partial x^2}\frac{nx}{\Omega} \right\} p(x,t)$$

$$\text{(Stratonovich)} \quad (11)$$

We can immediately recognise that Equation (11) is identical to Equation (2). The Stratonovich prescription therefore yields the result already obtained. Let us now investigate Equation (10) obtained using the Ito prescription.

At stationary state integration of Equation (10) leads to

$$p(x,\infty) = \frac{K_1}{x}\left\{ \exp\left[-\frac{2\Omega}{n}\left(x - \frac{n}{2}\right)^2 \right] \right\}$$

$$\times \left[1 + B\left(\frac{n}{2\Omega}\right)^{1/2} \int^{(2\Omega/n)^{1/2}[(n/2)-x]} \exp(x'^2)\,dx' \right] \quad (12)$$

The chief problem in this equation arises when we try to estimate the constants of integration. A quick trial will reveal that it is impossible to satisfy the boundary and normalisation conditions simultaneously. If we choose the constant K_2 by satisfying boundary condition $dQ(x,t)/dt = 0$ ($x = n$ or 0), then K_1 cannot be obtained from the normalisation condition and vice versa. The discrepancy arises due to the application of the Ito prescription which leads to physically unrealistic results.

4.3.3 Methods of Solution

Sections 4.3.1 and 4.3.2 were concerned with the derivation of the Fokker–Planck equation that is equivalent to a given master equation or the stochastic Langevin equation. We also note from the discussion in these sections that a properly formulated Fokker–Planck equation retains all the essential features of the original equation and has the advantage of ease of handling for obtaining the solution. The discussion also makes clear the relationship between the Langevin, Fokker–Planck and the master equations; starting with any one, we can derive the other equations that are consistent with the original one. In the present section we shall consider some of the methods more commonly employed in analysing these equations. The authoritative treatise of Risken (1984) detailing the methods of solution and applications of the Fokker–Planck equation should be consulted for further details. Other source material devoted to this aspect include the reviews of van Kampen (1976) and Hanggi and Thomas (1982), the handbook of stochastic methods by Gardiner (1983) and the book by van Kampen (1981).

It is clear that the macroscopic behaviour of a large class of nonequilibrium systems can be adequately described either in terms of the Fokker–Planck, Langevin or master equation. Often these equations are difficult to solve and one of the central problems has been to devise new mathematical methods for obtaining easier solutions to these equations. A large body of literature attempting to provide such alternatives exists. These include the cumulant expansion technique (Kubo, 1963; van Kampen, 1974), the continued fraction method (Hanggi, Rossel and Trautmann, 1978; Haag and Hanggi, 1979; Grossmann and Schranner, 1978), the methods based on the projection operator techniques (Haken, 1978; Mori, Morita and Mashiyama, 1980; Gardiner, 1983; Chaturvedi and Shibata, 1979; Grossmann and Schranner, 1978; Chaturvedi, 1983), and the information-theoretic approach (Czajkowski, 1981; Kociszewski, 1981; Owedyk, 1983, 1984). Other references include van Kampen (1976); Kubo, Matsuo and Kitahara (1973); Janssen (1974); Matheson, Walls and Gardiner (1975); Gortz (1976); Matsuo (1977); van Kampen (1977); Oppenheim, Shuler and Weiss (1977); Gardiner and Chaturvedi (1977); Matsuo, Lindenberg and Shuler (1978); Caroli and Roulet (1979); Weidlich and Hagg (1980); Lemarchand (1980); Gortz and Walls (1976); Haag, Weidlich and Aber (1977); and Haag (1978).

The list is by no means complete, but is adequate to give sufficient details of the procedures and to exemplify them by considering specific cases. For the present purpose we shall only broadly outline some of these methods. Other methods such as the expansion technique, the continued fraction method, etc. have already been discussed in Chapter 3.

Path Integral Method

In obtaining a solution to the Fokker–Planck equation we are interested in obtaining the transition probability of the underlying continuous Markov process. The idea of representing this transition probability by means of a weighted sum over all possible alternative trajectories of the system is appealing and the discrete path sum representation closely follows the original concept proposed by Wiener (1930). Onsager and Machlup (1953) also initiated an alternative version of the path integral, where the probability density of the stochastic process was expressed in terms of a dissipation function called the Onsager–Machlup function. Path integral methods have since been successfully used and a good exposition of these can be found in the books by Bharoocha-Reid (1972), Muhlschlegel (1978), Feynman and Hibbs (1965) and a review by Dashen (1979).

To discuss the method we begin with a simple stochastic equation with additive noise; the Fokker–Planck equation corresponding to this can be immediately identified as per the prescription in the previous section and the Fokker–Planck operator can be written as

$$\mathscr{L}p = \left[-\frac{\partial}{\partial x}\alpha_1(x) + \varepsilon\frac{\partial^2}{\partial x^2} \right] p \tag{4.75}$$

This operator in general is not Hermitian and it is convenient to transform it into a Hermitian form as

$$\bar{\mathscr{L}} = p_{\text{eq}}^{-1/2}\mathscr{L}p_{\text{eq}}^{1/2} = \varepsilon\frac{\partial^2}{\partial x^2} - \frac{\alpha_1^2(x)}{4\varepsilon} - \tfrac{1}{2}\alpha_1^1(x) \tag{4.76}$$

where p_{eq} denotes the equilibrium probability distribution function defined as

$$p_{\text{eq}} = \exp\frac{1}{\varepsilon}\int^x \alpha_1(x)\,dx \tag{4.77}$$

In view of the transformation the changed probability distribution can be written as

$$p(x,t) = \exp\left(\frac{1}{2\varepsilon}\int^x \alpha_1(x)\,dx\right)G(x,t) \tag{4.78}$$

where $G(x, t)$ satisfies the equation

$$\frac{\partial}{\partial t} G(x,t) = \bar{\mathscr{L}} G(x,t) \tag{4.79}$$

The operator is now Hermitian and the formal solution to Equation (4.79) subject to initial condition $G(x, 0) = \delta(x)$ can be obtained using the Wiener–Feynman–Kac formula as

$$G(x,t) = \int_{x=0}^{x} ux(t) E\left\{ \exp - \int_0^t \left[\frac{\alpha_1^2(x)}{4\varepsilon} + \tfrac{1}{2}\alpha_1'(x) \right] dx \right\} \tag{4.80}$$

where $E\{\,.\,\}$ denotes the expectation value and u the integration over a path. The transition probability can then be obtained referring to Equation (4.78):

$$p(x,t) = \int_0^{\infty} ux(t) \exp\left(- \int_0^t \left\{ \frac{1}{4\varepsilon} \left[\frac{dx}{dt} - \alpha_1(x) \right]^2 + \tfrac{1}{2}\alpha_1^1(x(t)) \right\} dt \right) \tag{4.81}$$

A formula similar to the above can be obtained for a system described by a stochastic equation with multiplicative noise. The Fokker–Planck equation now takes the general form

$$\frac{\partial p(x,t)}{\partial t} = -\frac{\partial}{\partial x} \alpha_1(x) p(x,t) + \varepsilon \frac{\partial^2}{\partial x^2} \alpha_2(x) p(x,t) \tag{4.82}$$

Defining the transformation

$$\tilde{\alpha}(y) = \frac{\alpha_1(x)}{[\alpha_2(x)]^{1/2}} - \varepsilon [\alpha_2(x)]^{1/2} \tag{4.83}$$

Equation (4.82) can be recast in the form

$$\frac{\partial \tilde{p}(y,t)}{\partial t} = -\frac{\partial}{\partial y} \tilde{\alpha}(y) \tilde{p} + \varepsilon \frac{\partial^2}{\partial y^2} \tilde{p} \tag{4.84}$$

where

$$\tilde{p}(y,t) = p(x,t)(dx/dy) = p(x,t)[\alpha_2(x)]^{1/2}$$

The solution to Equation (4.84) can be obtained in a form similar to Equation (4.81) which when transformed into the original variables reads

$$p(x,t) = \int_{x(0)=0}^{x(t)=x} ux(t)\exp\left\{-\frac{1}{4\varepsilon\alpha_2(x)}\left[\frac{dx}{dt} - \tilde{\alpha}[\alpha_2(x)]^{1/2} - \tfrac{1}{2}[\alpha_2(x)]^{1/2}\frac{\partial}{\partial x}\tilde{\alpha}(y)\right]^{1/2}\right\} \tag{4.85}$$

The path integral formalism has been used in the literature for linear as well as nonlinear stochastic processes (Graham, 1973, 1977, 1978; Leschke and Schmutz, 1977; Kubo, Matsuo and Kitahara, 1973). The path integral techniques are often coupled with the Onsager–Machlup–Laplace (OML) approximation (Hunt and Ross, 1981; Onsager and Machlup, 1953; Machlup and Onsager, 1953; Wiegel, 1967; Langouche, Roekaerts and Tirapegui, 1979; Moreau, 1978; Lovenda, 1979) to obtain explicit relations for the conditional probability. The OML approximation, however, breaks down catastrophically for certain situations and the point has been discussed at length by Hunt, Hunt and Ross (1983).

Projection Operator Formalism

For a physical system described in terms of the Fokker–Planck, Langevin or master equation, it often turns out that some of the variables of the system vary on a characteristic time scale that is an order of magnitude higher or lower than the time scale representing the variations of the remaining variables. In other words, the system may be supposed to consist of slow variables—or the relevant variables—and fast variables—or the irrelevant variables. In situations of this type it is feasible to provide a contracted description of the system by eliminating the fast variables from it. Such situations might arise, for example, when we have the following reaction scheme:

$$A + X \rightleftharpoons X + Y$$

$$X \rightleftharpoons 2Y$$

where the forward rate constant for the second step is an order of magnitude higher than for the remaining steps. The elimination of the fast variables from the system can be easily achieved using the projector operator formalism originally developed by Nakajima (1958) and

Zwanzig (1960) and subsequently transformed into a time convolutionless formalism by Shibata, Takahashi and Hashitsume (1977) and Tokuyama and Mori (1976). It is in this latter formalism that Chaturvedi and Shibata (1979) used it for eliminating the fast variables. The detailed mathematical treatment can be understood by recourse to the original references. We highlight the main steps and present the final equations without rigorous derivations in Appendix 4.A.

Example

Obtain a contracted description (by eliminating the fast variables) of the reacting system

$$A + Y \rightleftharpoons X + Y$$

$$Y \rightleftharpoons 2X$$

using a projection operator formalism for a large forward rate constant for the second step.

Solution

For reasons of simplicity, we shall assume that all other rate constants are unity and that the rate constant k for the second forward step an order of magnitude higher than unity. The deterministic equations describing the evolution of species X and Y can be written as

$$\frac{dX}{dt} = Y - XY + 2(kY - X^2) \tag{1}$$

$$\frac{dY}{dt} = X^2 - kY \tag{2}$$

In reality, in view of the large value of k, the effective concentration of species Y will be close to zero, so that the equations can be modified to a single equation as

$$\frac{dX}{dt} = (A - X)X^2/k \tag{3}$$

We note that the last equation in reality corresponds to a contracted description of the reaction scheme which can now be identified as $A + 2X \rightleftarrows 3X$. Our major intention in this example is to show that such contraction also works at the stochastic level.

Let us now write the master equation corresponding to the reaction scheme as

$$\frac{dp(X, Y, t)}{dt} = [A(E_X^{-1} - 1)Y + (E_X - 1)XY + k(E_X^{-2}E_Y - 1)Y + (E_Y E_X^2 - 1)]p(X, Y, t) \tag{4}$$

We can identify the operator L in Equation (4) by referring to Equation (1) in Appendix 4.A. However, it is often more convenient to first convert Equation (4) into an equivalent Fokker–Planck equation (Section 4.3.1). Note that in view of the bivariate nature of the master equation, the corresponding Fokker–Planck equation also takes the bivariate form

$$\frac{\partial g(m, n, t)}{\partial t} = -\frac{\partial}{\partial m}[(a - m)n + 2(kn - m^2)]g(m, n, t) + \frac{\partial^2}{\partial m^2}(kn - m^2)g(m, n, t) - \frac{\partial}{\partial n}(m^2 - kn)g(m, n, t) + \frac{\partial^2}{\partial m\, \partial n}(kn - m^2)g(m, n, t) \tag{5}$$

using the transformation

$$p(X, Y, t) = \int dm\, dn\, e^{-m-n}\frac{m^X}{X!}\frac{n^Y}{Y!}g(m, n, t) \tag{6}$$

The above transformation is equivalent to the expansion of the function $p(X, Y, t)$ in terms of a Poisson distribution.

We now identify the operators L_0 and L_1 in Equation (5) as

$$L_0 = -\frac{\partial}{\partial n}(kn - m^2) \tag{7}$$

$$L_1 = -\frac{\partial}{\partial m}[(a - m)n + 2(kn - m^2)] + \frac{\partial^2}{\partial m^2}(kn - m^2) + \frac{\partial^2}{\partial m\, \partial n}(kn - m^2) \tag{8}$$

and introduce the projection operator $\mathscr{L}$

$$\mathscr{L} = f(n!\, m) \int dn \tag{9}$$

where f satisfies the relation $L_0 f = 0$. This fixes the function f as

$$f = \delta(kn - m^2) \tag{10}$$

We then evaluate the reduced distribution $g(m, n, t)$ by referring to Equations (6) and (8) in Appendix 4.A as

$$g(m, t) = \int dn\, g(m, n, t) \tag{11}$$

which obeys the Fokker–Planck equation

$$\frac{\partial g(m, t)}{\partial t} = \frac{1}{k}\left[-\frac{\partial}{\partial m} + 2\frac{\partial^2}{\partial m^2} - \frac{\partial^3}{\partial m^3}\right](a - m)m^2 g(m, t) \tag{12}$$

The fast varying variable is thus eliminated. We note that Equation (12) corresponds to the master equation

$$\frac{\partial p(X, t)}{\partial t} = \frac{1}{k}[A(E_X^{-1} - 1)X(X - 1) + (E_X - 1)(X - 1)]p(X, t) \tag{13}$$

when $p(X, t)$ is expanded in terms of a Poisson distribution as in Equation (6). The equation of course corresponds to the truncated reaction scheme $A + 2X \rightleftarrows 3X$.

The results indicate the equivalence of the reaction schemes both at the deterministic and stochastic levels.

The simple example is illustrative of all adiabatic elimination procedures. The basic physical assumption is that the presence of some large parameter in the system of equations forces the variable in that equation to relax to a value that can be obtained by assuming the variable to have a constant value.

The question of systematic elimination of fast variables in linear systems using the perturbation theory in the time scale ratio of the slow and fast variables has been examined in detail by Geigenmuller, Titulaer and Felderhof (1983). Systematic procedures for adiabatic elimination in stochastic systems that involves additive, multiplicative or non-additive noise have been developed by Gardiner (1984). Steyn-Ross and Gardiner (1984) have extended this method to include reaction–diffusion systems. The handbook of Gardiner also exhaustively analyses this problem for a variety of situations. A vast volume of literature exists on adiabatic elimination of fast variables (Brinkman, 1956; Wilemski, 1976; Haken, 1978; Chaturvedi and Shibata, 1979; Skinner and Woyles, 1979; Morita and Mashiyama, 1980; San Miguel and Sancho, 1980; Risken, Vollmer and Morsch, 1981; Kaneko, 1981; Titulaer, 1978, 1983; Hasegawa, Mizuno and Mabuchi, 1982; Haake, 1982; Faetti, Grigolini and Marchesoni, 1982; Grigolini *et al.*, 1983; Grigolini and Marchesoni, 1984; Marchesoni and Grigolini, 1984; Chaturvedi and Shibata, 1979). Some of these methods have recently been criticised by Marchesoni (1984), while the range of validity of a few adiabatic elimination procedures has been discussed by Festa *et al.* (1984).

Information-Theoretic Approach

To illustrate the basic notion we begin with the master equation

$$\frac{dp_n}{dt} = \sum_{m=1}^{N} W_{nm}\, p_m - W_{mn}\, p_n \tag{4.86}$$

where W_{nm} is a transition probability per unit time. Let the master Equation (4.86) possess a unique stationary probability distribution p_{st}. Thus for any solution $p(t)$ of the master equation we have $\lim_{t\to\infty} p_n(t) = p_{\text{st}}$. Equation (4.86) can also be written in terms of the expectation values of a set of linearly independent functions f_n^μ, $\mu = 1, 2, \ldots q$ as

$$\frac{d}{dt}\langle f_n^\mu \rangle_p = \sum_{n,m=1}^{N} f_n^\mu (W_{nm}\, p_m - W_{mn}\, p_n) \tag{4.87}$$

where the expectation value is defined as

$$\langle . \rangle_P = \sum_{n=1}^{N} (\,.\,) p_n(t) \tag{4.88}$$

We now use the following approximation:

$$p_n(t) = p_n^*(t) = p_{\text{st}} \exp\left[-\lambda_0(t) - \sum_{\mu=1}^{q} \lambda_\mu(t) f_n^\mu \right] \tag{4.89}$$

where it is assumed that at initial time $p_n(t)$ and $p_n^*(t)$ give the same expectation values of the function f_n^μ, $\mu = 1, \dots q$. The probability distribution $p_n^*(t)$ is referred to as information gain solution (Kociszewski, 1981). The first term in the exponent refers to the normalisation factor.

The closed set of equations describing the evolution of the expectation values of f_n^μ, $\mu = 1, \dots q$ is obtained when Equation (4.89) is used in Equation (4.87). We thus have

$$\frac{d}{dt}\langle f^\mu(t)\rangle_{p^*} = \sum_{n,m=1}^{N} f_n^\mu(W_{nm}\, p_m^* - W_{mn}\, p_n^*) \tag{4.90}$$

Equations (4.90) and (4.88) together can be used to determine the functions λ_μ, $\mu = 1, \dots q$ which in turn can be used in Equation (4.89) to obtain $p_n^*(t)$.

Perturbation Method

The perturbation method utilises the practical reality, viz. the smallness of the noise, and provides the solution linearised around the macroscopic equation in the form of an expansion series. To illustrate the approach we shall consider a simple example of a linear stochastic differential equation with additive noise, the exact solution for which is known. The equation can be written as

$$dx = -jx\, dt + D\, dw \tag{4.91}$$

or equivalently in terms of the Fokker–Planck equation as

$$\frac{\partial p}{\partial t} = \frac{\partial}{\partial x}(jxp) + \tfrac{1}{2}D^2 \frac{\partial^2}{\partial x^2} p \tag{4.92}$$

where D refers to the noise intensity and is generally a small parameter. In the limit of $D \to 0$, Equation (4.91) reduces to the macroscopic equation, while Equation (4.92) gets converted from a second-order to a first-order equation showing singular behaviour. The exact solution to Equation (4.91) can be written easily as

$$x = x_0 + Dx_1 \tag{4.93}$$

where x_0 refers to the simple macroscopic solution. Equation (4.92) can also be solved to obtain the moments

$$\langle x \rangle = A_1 \exp(-jt), \qquad \sigma^2[x(t)] = \frac{D^2}{2j}[1 - \exp(-2jt)] \tag{4.94}$$

The solution to the Fokker–Planck Equation (4.92) being Gaussian, the conditional probability can be written as

$$p(x,t) = \frac{1}{D}\left[\frac{j}{\pi(1-e^{-2jt})}\right]^{1/2} \exp\left\{-\frac{j}{D^2}\frac{(x-\langle x \rangle)^2}{(1-e^{-2jt})}\right\} \tag{4.95}$$

which, for the limiting case of $D \to 0$, reduces to the deterministic trajectory

$$p(x,t) = \delta[(x - \langle x \rangle)] \tag{4.96}$$

The difficulty with this conditional probability is that it cannot be expanded easily in power series in D. For this purpose we need to redefine the variable x as

$$y = \frac{x - \langle x \rangle}{D} \qquad \text{or} \qquad x = \langle x \rangle + Dy \tag{4.97}$$

The probability density for y can be written in terms of that for x:

$$p(y,t) = p(x,t)\, dx/dy = \left[\frac{j}{\pi(1-e^{-2jt})}\right]^{1/2} \exp\left[-\frac{y^2 j}{1-e^{-2jt}}\right] \tag{4.98}$$

Note that this transformed equation does not depend on D and in the limiting case of $D \to 0$ is therefore not singular.

The simple example considered here suggests that for a general stochastic equation

$$dx = f(x)\, dt + Dg(x)\, dw \tag{4.99}$$

we can write the solution by comparison with Equation (4.93) as

$$x(t) = x_0(t) + Dx_1(t) + D^2 x_2(t) + \cdots \tag{4.100}$$

where x_0 refers to the macroscopic solution. The equivalent Fokker–Planck description of Equation (4.99), viz.

$$\frac{\partial p}{\partial t} = -\frac{\partial}{\partial x} f(x)p + \tfrac{1}{2}D^2 \frac{\partial^2}{\partial x^2} g^2(x)p \tag{4.101}$$

needs to be scaled using the transformation given by Equation (4.97) which can then be solved in terms of the perturbation expression to give

$$p(y,t) = p_0(y,t) + Dp_1(y,t) + D^2p_2(y,t) + \cdots \tag{4.102}$$

To obtain the explicit equations for p_0, p_1, p_2, etc. we substitute Equation (4.97) in (4.101):

$$\frac{\partial p(y,t)}{\partial t} = -\frac{\partial}{\partial y}\left\{\frac{f(x+Dy)-f(x)}{D}p(y,t)\right\} + \frac{1}{2}\frac{\partial^2}{\partial y^2}\left\{q(x(t)+Dy)p(y,t)\right\} \tag{4.103}$$

The coefficients of f and g in this equation can be expanded in powers of D as

$$\begin{aligned} f(x(t)+Dy) &= \sum_{i=0}^{\infty} A_i D^i y^i \\ g(x(t)+Dy) &= \sum_{i=0}^{\infty} B_i D^i y^i \end{aligned} \tag{4.104}$$

We also have the relation $p(y,t) = \sum_{i=0}^{\infty} D^i p_i$. Substituting Equation (4.104) in (4.103) and equating coefficients of like powers of D gives

$$\frac{\partial p_0}{\partial t} = -A_1(t)\frac{\partial}{\partial y}(yp_0) + \tfrac{1}{2}B_0(t)\frac{\partial^2 p_0}{\partial y^2} \tag{4.105}$$

for the zeroth approximation, and for the general i^{th} order approximation

$$\frac{\partial p_i}{\partial t} = -\frac{\partial}{\partial y}\left[\sum_{n=0}^{i} y^{i-n+1}A_{i-n+1}(t)p_n\right] + \frac{1}{2}\frac{\partial^2}{\partial y^2}\left[\sum_{n=0}^{i} y^{i-n}B_{i-n}(t)p_n\right] \tag{4.106}$$

Note that only Equation (4.105) in this hierarchy is the Fokker–Planck equation. The boundary conditions for the equations, however, are difficult and at times can pose some problems. One can obtain the stationary and transient moments and autocorrelation functions easily.

Example 4.4 *Critical slowing down in autocatalytic reaction*

Tambe, Kulkarni and Doraiswamy (1985d) have considered the case of an autocatalytic reaction $X_1 + X_2 \rightleftarrows 2X_1$ which is known to exhibit the phenomenon of critical slowing down (see Example 4.1). The example has earlier been worked out using the Markovian master equation formalism. In the present case we shall rework this example using alternative techniques such as the path integral method, the transformation technique and the numerical methods. The governing stochastic equation or its equivalent representation in terms of the Fokker–Planck equation, for the special case of a reaction in CSTR with no input concentration of the autocatalytic species and for a large input concentration of X_2, can be written following Tambe, Kulkarni and Doraiswamy (1985d) as

$$\frac{\partial p}{\partial t} = -\frac{\partial}{\partial x_1}[K(x_1)p(x,t)] + \frac{\partial^2}{\partial x_1^2}[D(x_1)p(x,t)] \tag{1}$$

where

$$K(x_1) = -ax_1 - x_1^2 + \frac{ax_{20}x_1 + x_1^3}{b + x_1} + \frac{\varepsilon x_1 b}{(b + x_1)^3} \tag{2}$$

$$D(x_1) = x_1^2/(b + x_1)^2 \tag{3}$$

and $a = F/Vk_2$, $b = F/Vk_1$ and ε represents the intensity of noise.

We shall now obtain the solution to Equation (1) using different techniques.

(i) *Path Integral Method*

The Fokker–Planck equation can be written in more compact form as

$$\frac{\partial p}{\partial t} = -\mathscr{L}(\hat{x}_1, \hat{y}_1)p \tag{4}$$

where $x_1 \to \hat{x}_1$, $\hat{y}_1 \to i\,\partial/\partial x_1$ and the operator $\mathscr{L}$ can be defined as

$$\mathscr{L} = i\hat{y}_1 K(\hat{x}_1) + \tfrac{1}{2}\hat{y}_1^2 D(\hat{x}_1) \tag{5}$$

The probability function p can now be evaluated as a path integral in terms of the original variable x_1:

$$p(x_1, t) = \int_{x_1(0)=0}^{x_1(t)=x_1} x_1(t) \times \exp\left\{-\frac{(b + x_1)^2}{4\varepsilon x_1^2}\left[\frac{dx_1}{dt} - \left(\frac{ax_{20}x_1 + x_1^3}{b + x_1}\right.\right.\right.$$

$$\left.\left. + \frac{\varepsilon b x_1}{(b + x_1)^2} - ax_1 - x_1^2 - \frac{\varepsilon x_1 b}{(b + x_1)^2}\right)\right]^2 - \frac{x_1}{2(b + x_1)} \times$$

$$\frac{d}{dx_1}\left[\frac{b+x_1}{x_1}\left(\frac{ax_{20}x_1+x_1^3}{b+x_1}+\frac{\varepsilon bx_1}{(b+x_1)^3}-ax_1-x_1^2-\frac{\varepsilon bx_1}{(b+x_1)^2}\right)\right]\Bigg\}dx_1 \tag{6}$$

Equation (6) for small values of ε leads to further simplification. Also, if we denote y_m as the most probable path, then the fluctuations or the variance $\langle(x_1 - y_m)^2\rangle = \varepsilon\sigma$ can be obtained from the probability distribution $(x_1 - y_m)$. Denoting $(x_1 - y_m) = Z$ for small ε, Equation (6) can be written as

$$p(Z,t) = \int_{Z(0)=0}^{Z(t)=Z} Z(t)\exp\quad -\frac{\left(\dfrac{dZ}{dt} - K'(y_m(t))Z\right)^2 [b + x_1 y_m(t)]^2}{4\varepsilon x_1^2 y_m(t)}\,dZ \tag{7}$$

where K' refers to the differentiation of the drift term K in Equation (2) with respect to x_1 (or y_m). The most probable path $y_m(t)$ can be obtained by maximising the integral:

$$\frac{dy_m}{dt} = Ky_m(t) \tag{8}$$

and the variance σ is given by

$$\frac{d\sigma}{dt} = 2K'y_m(t)\sigma + 2D(y_m(t)) \tag{9}$$

with the solution

$$\sigma(t) = \sigma_0\left[\frac{Ky_m(t)}{Ky_m(0)}\right]^2 + K^2(y_m(t))\int_{y_m(0)}^{y_m(t)} \frac{D(y_m)}{K(y_m)^3}\,dy_m \tag{10}$$

(ii) *Solution Using Scaling Technique*

The general scaling theory, discussed elaborately by Suzuki (1981), in essence provides an asymptotic evaluation of transient fluctuations. For this purpose the time scale is divided into initial, second and final regions. In the initial regime the random force and initial fluctuations are important, while in the second regime the nonlinearity of the process dominates the evolution. It is in this second regime that the macroscopic order (or macroscopic fluctuations) sets in. Thus the Gaussian treatment, while adequate for the initial and final regimes, represents a poor approximation for the intermediate regime, and scaling relations for time, strength of fluctuations, nonlinearity of the system, initial extent of fluctuations, etc. are necessary.

The stochastically equivalent Langevin equation corresponding to the Fokker–Planck Equation (1) for small values of ε and $b \gg x_1$ can be rewritten as

$$\frac{dx_1}{dt} = \left(\frac{ax_{20} - ab}{b}\right)x_1 - \frac{a+b}{b}x_1^2 + x_1\eta_1(t) \tag{11}$$

where

$$\eta_1(t) = \frac{1}{b}\eta(t) \tag{12}$$

Following Suzuki (1981) we define the time-dependent nonlinear transformation

$$\bar{T} = F^{-1}[\exp(-\gamma t)F(x_1)] \tag{13}$$

where the function $F(x_1)$ is defined as

$$F(x_1) = \exp \int^{x_1} \frac{\dfrac{d}{dx_1}\left[\left(\dfrac{ax_{20} - ab}{b}\right)x_1 - \dfrac{a+b}{b}x_1^2\right]_{x_1=0}}{\left[\dfrac{(ax_{20} - ab)x_1}{b} - \dfrac{a+b}{b}x_1^2\right]} dx_1 \tag{14}$$

with

$$\gamma = \left[\frac{(ax_{20} - ab)}{b}\right] > 0 \qquad \text{(i.e. } x_{20} > b\text{)} \tag{15}$$

Using the transformation given by Equation (13) in Equation (11) and expanding the multiplicative term $\eta_1(t)$ in terms of $\bar{T}$ we obtain in the intermediate time regime

$$\frac{d\bar{T}}{dt} = \eta_1(t)\bar{T} \tag{16}$$

The solution of Equation (16) in terms of the original variable x_1 can be readily obtained as

$$[x_1(t)]_{sc} = \left\{x_1(0)\exp\int_0^t \eta_1(t')\,dt'\right\}\left\{\exp\left(\frac{ax_{20}-ab}{b}\right)t\right\}$$
$$\times\left[1+\frac{(a+b)}{ax_{20}-ab}\left\{\exp\left(\frac{ax_{20}-ab}{b}\right)t-1\right\}\right.$$
$$\left.\times\left\{x_1(0)\exp\int_0^t \eta_1(t')\,dt\right\}\right]^{-1} \tag{17}$$

Performing the indicated integration and taking the average over the random force $\eta_1(t)$ easily leads to

$$[\langle x_1(t)\rangle]_{sc} = \frac{ax_{20}-ab}{(a+b)}\left[1-\exp\left(\frac{(ax_{20}-ab)}{b}\right)t\right]^{-1}$$
$$\times\left[1-\frac{ax_{20}-ab}{(a+b)x_1(0)}\exp\left(-\frac{ax_{20}-ab-b\varepsilon}{b}t\right)\right] \tag{18}$$

where further terms

$$O\left[\exp\left(\frac{ax_{20}-ab-b\varepsilon}{b}t\right)\right]$$

are neglected. The equation clearly indicates critical slowing down in the average $\langle x_1\rangle_{sc}$.

(iii) *Solution Using Numerical Method*

The fluctuation equation corresponding to the Fokker–Planck Equation (1) can be alternatively written in the form

$$\frac{dx_1}{dt} = f(x_1) + g(x_1)\xi(t) \tag{19}$$

The solution to this problem is unique if $\xi(t)$ is a Gaussian delta correlated random process with zero mean and unit intensity:

$$\langle\xi(t)\rangle = 0, \qquad \langle\xi(t_1)\,.\,\xi(t_2)\rangle = \delta(t_1-t_2) \tag{20}$$

The functions $f(x_1)$ and $g(x_1)$ can be easily identified as

$$f(x_1) = -ax_1 - x_1^2 + \frac{ax_{20}x_1 + x_1^3}{b+x_1} - \frac{bx_1}{(b+x_1)^3} \tag{21}$$

$$g(x_k) = x_1/b + x_1 \tag{22}$$

For ease of computation the time scale $(0 - t)$ is divided into n equal intervals and the values of the dependent variable (x_j) are assigned according to the formula

$$x_1(t_{j+1}) - x_1(t_j) = f(x_1)_j \frac{t}{n} + g(x_1)_j u \frac{t}{n} \tag{23}$$

Equation (23) has been solved to compute a large number of trajectories which are averaged to obtain the mean $\langle x \rangle$, variance $\langle x^2 \rangle$ and intensity of fluctuations $[(x^4)/\langle x^2 \rangle^2 - 1]$ during the transient solution. Once the stationary distribution for these quantities is obtained, the value of k_f (or equivalently that of a and b) is suddenly changed to a new value. The transients that emerge are sketched in Figures 4.4 and 4.5.

Figure 4.4 shows the situation where the parameters selected correspond to the case of critical slowing down, while Figure 4.5 shows the transients when the system exhibiting the phenomenon of critical slowing down is disturbed. As evident from Figure 4.5, the relaxation towards the new equilibrium value now proceeds faster. The curve in the figure describing the intensity of fluctuations, however, shows sharp maxima along the time axis. The phenomenon typically resembles the tunneling process known in the contemporary literature.

4.4 Application to Bistable Systems

In the analyses presented thus far, which include both the single (Section 3.2.3) and multivariable (Section 4.2) cases, we considered systems, the macroscopic description of which indicated the existence of a single unique stationary state that is stable. The application of the expansion method to such situations leads to the construction of a mesostate that corresponds to the macrostate of the macroscopic equation. This mesostate is sharply peaked around the macrostate with a width of the order of $\Omega^{1/2}$. In the present section we shall analyse the situation when the macroscopic description of the system indicates the existence of more than one macrostate. Several cases of practical interest, involving either single or multiple variables, subscribe to this type of situation.

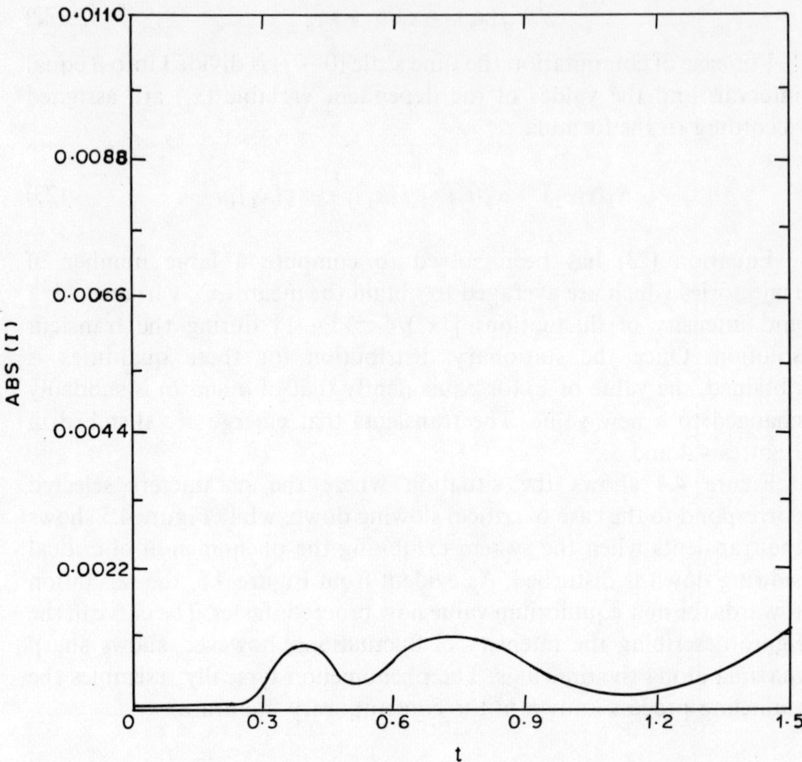

FIGURE 4.4 Time evolution of intensity (I) for $k_f = 1$ (critical slowing down case).

To illustrate this point we shall begin by considering a simple example of a chemically reacting system, the macroscopic equation of which is given by

$$\frac{dx}{dt} = f(x) \tag{4.107}$$

It is assumed that the function $f(x)$ is nonlinear and the equation for certain sets of parameter values indicates the existence of three stationary macrostates, the end ones of which (x_1, x_3) are stable while the middle one (x_2) is unstable. The typical form of the rate function is shown in Figure 4.6.

In view of the fact that the states x_1 and x_3 are locally stable, i.e. the

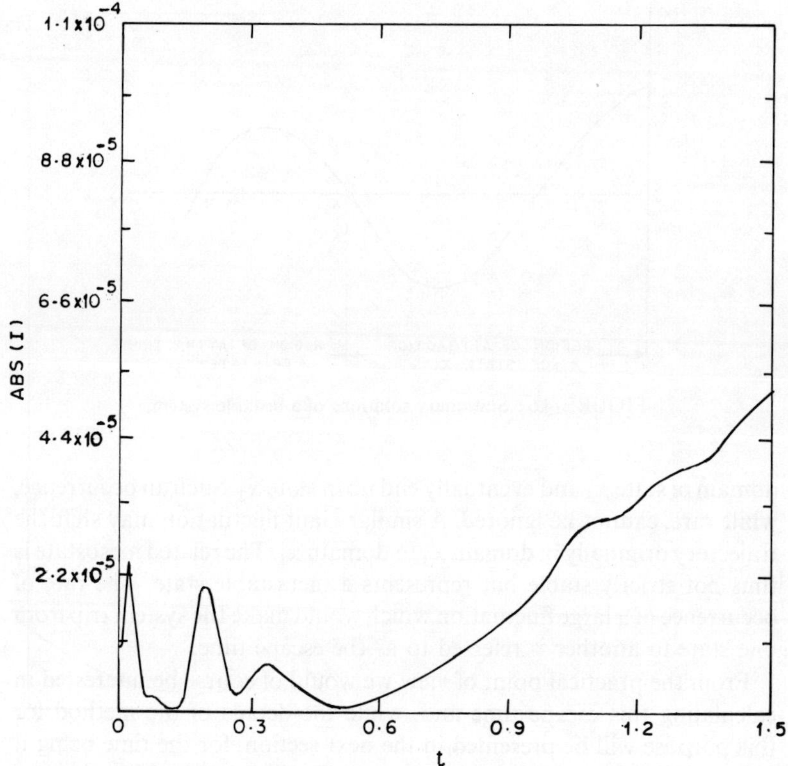

FIGURE 4.5 Time evolution of intensity (I) for $k_f = 0.8$.

condition of $\alpha_1^1(x) < 0$ is met for these states, any local perturbation or fluctuations would die away in the neighbourhood of these states. There thus exists a region around these states, within which if we start, the trajectory would finally end up in one of these states. The regions of attraction around the states are also indicated in Figure 4.6. In view of the smallness of fluctuations, we could apply the method of expansion to this type of situation and generate the mesostate corresponding to the macrostate. The argument indicates that the trajectory starting in the region of attraction of states x_1 or x_3 would eventually end up in state x_1 or x_3. The argument, of course, ignores the possibility of occurrence of a large fluctuation during the actual transient that can abruptly cause the trajectory originally in the domain of state x_1 to move over to the

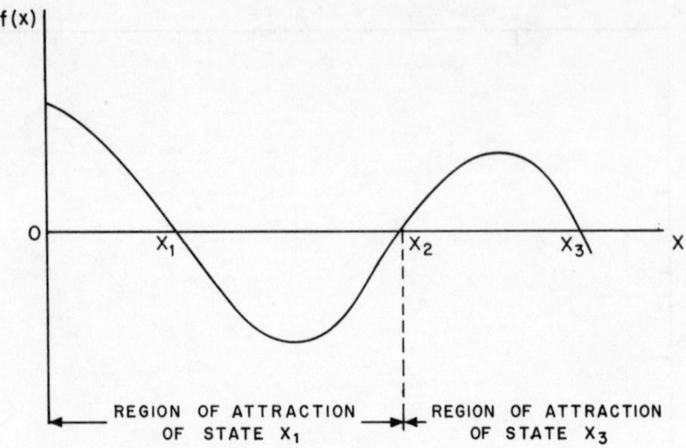

FIGURE 4.6 Stationary solutions of a bistable system.

domain of state x_3 and eventually end up in state x_3. Such an occurrence, while rare, cannot be ignored. A similar giant fluctuation may shift the trajectory originally in domain x_3 to domain x_1. The related mesostate is thus not strictly stable but represents a metastable state. The rate of occurrence of a large fluctuation which would make the system trip from one state to another is referred to as the escape time.

From the practical point of view we would of course be interested in calculating this escape time and, while the details of the method for this purpose will be presented in the next section, for the time being it is intuitively clear to us that the escape time depends on how close we start to the middle unstable state (x_2). It would also depend on the actual magnitude of the probability distribution peak at the unstable point x_2. In view of the nature of x_2 there is reason to expect that the initial fluctuations might shift the trajectory from one region to another. There is therefore a non-negligible probability that a system starting in the region of attraction for state x_1 but closer to the unstable state x_2 would not follow a macroscopic path leading to state x_1. In fact, for an initial condition starting near or at the unstable point, the problem of calculating the probability that the system would evolve either to state x_1 or x_3 becomes important. The calculation of this splitting probability will also be presented in Section 4.4.2.

We note from the discussion presented above that the rate of occurrence of a giant fluctuation depends in some measure on the magnitude of the probability distribution peak around the middle

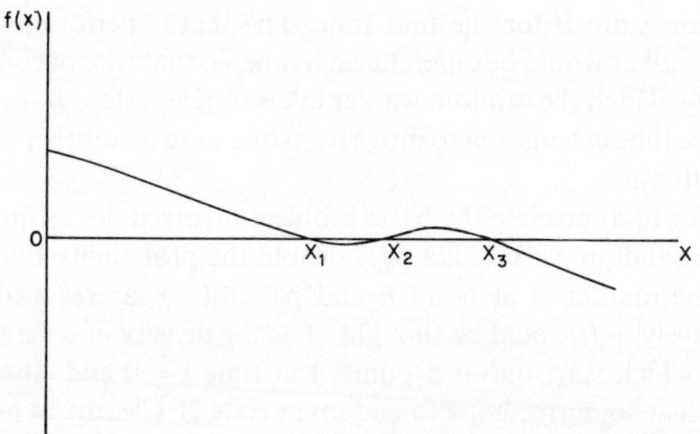

FIGURE 4.7 Stationary solutions of a bistable system near critical point.

unstable state. Clearly this magnitude can be changed by altering the parameters of the system. In other words, the typical form of the rate function in Figure 4.6 can be altered from a well-defined bistable situation to a less clear and obscure shape as in Figure 4.7. It is of interest in a situation of this type, where the stabilities and instabilities of the states are not well pronounced, to calculate the escape time and splitting probability. The typical case corresponding to the critical region is discussed in Section 4.4.3.

The main difference between the two types of situations shown in Figures 4.6 and 4.7 lies in the time scales. For cases such as in Figure 4.7, the two time scales—one on which the transient state approaches the stationary value and the other on which the giant fluctuation occurs—are widely different. For the latter situation the two time scales lose this distinction.

4.4.1 The First Passage Problem: The Case of the Random Walker

The discussion in the previous section brings into focus the fact that the occurrence of a large disturbance or fluctuation during the evolution of a trajectory from a given initial state may lead to a situation where the system trips from one stable state to another. This specific problem may be cast in terms of a more general formulation where we assume a random walker starting at a particular point A at zero time. After making random trips to adjacent points, some time later it reaches the destination state B for the first time. The repeat performance of the random walker would be different each time, so that we can conceive of a mean time which the random walker takes to reach state B. In principle the escape time calculation to move from one state to another is identical to this situation.

In order to appreciate the basic problem involved, let us quantify the case of a random walker. Let $p_n(t)$ denote the probability that at some time t the marker is at point n and has not so far reached state B. Alternatively, $p_n(t)$ could be thought of as the density of several random walkers which start out at a point A at time $t = 0$ and, after making independent sojourns, hope to end up in state B. Clearly, at any further time t, not all random walkers would reach state B. Those that have reached are counted out from the original collection, and evolution of the density p_n thus follows the master equation of the random walk, viz.

$$\frac{dp_n}{dt} = p_{n+1}(t) - 2p_n(t) + p_{n-1}(t), \qquad n < B - 2 \tag{4.108}$$

$$\frac{dp_{n-1}}{dt} = p_{n-2} - 2p_{n-1} \tag{4.109}$$

The set of Equations (4.108) and (4.109) can be solved subject to initial condition $p_n(0) = \delta_{n,A}$ and the number of walkers that have reached the destination state B at time t is given by $\hat{p}$ where

$$\frac{d\hat{p}}{dt} = p_{B-1} \tag{4.110}$$

The first passage problem as formulated above can be recognised as a problem with absorbing boundary. The equivalence can be generalised and, for any one-step process with arbitrary birth and death terms, we can write equivalent forms of Equations (4.108) and (4.109) as

$$\frac{dp_n}{dt} = (E - 1)r_n p_n + (E^{-1} - 1)g_n p_n, \qquad n < B - 2 \tag{4.111}$$

$$\frac{dp_{n-1}}{dt} = -r_{B-1}p_{B-1} + g_{B-2}p_{B-2} - g_{B-1}p_{B-1}, \qquad p_n(0) = \delta_{n,A} \tag{4.112}$$

with the first passage distribution given by

$$\frac{d\hat{p}}{dt} = g_{B-1}p_{B-1} \tag{4.113}$$

If we represent the number of walkers that reach state B between times t and $t + \Delta t$ by $f_A\,dt$, then $f_A\,dt$ is given by Equation (4.113). The total probability of reaching state B is then given by an integral

$$\pi = \int_0^\infty f_A(t)\,dt \tag{4.114}$$

and the mean passage time by

$$t_{BA} = \int_0^\infty t f_A(t)\,dt \tag{4.115}$$

Alternative ways such as the use of the renewal equation have been discussed by van Kampen (1981).

(A) *Mean Passage Time for Discrete Processes Described by a Master Equation: Exact, Asymptotic and Approximate Relations*

The general concept of the mean passage time was presented above by considering the example of a random walker. The present section considers the exact, asymptotic and approximate relations for the mean passage time for discrete processes that are described by the master equation. The general theory is applicable to unimodal and bimodal systems.

The earlier attempt to calculate the mean first passage time for a system described by the master equation was due to Montroll and Shuler (1958) who considered the case of dissociation of a diatomic molecule. Subsequently, Weiss (1966) developed a general formulation and in recent years a large number of publications reporting the mean passage time for unimolecular decomposition (Procaccia, Mukamel and Ross, 1978) and for chemical reactions with fluctuation-induced transition from one state to another (Matheson, Walls and Gardiner, 1975; Oppenheim, Shuler and Weiss, 1977; Gillespie, 1979) have appeared. These analyses involve cumbersome calculation procedures, and generalisation of a procedure from one problem to another is difficult. In recent years, additionally, two different procedures for calculating the mean passage time for unit-step one-dimensional random walk have appeared (Gillespie, 1979; Seshadri, West and Lindenberg, 1980). The computational scheme of Gillespie is easily adaptable to numerical computation on digital machines, while the analysis of Seshadri, West and Lindenberg (1980) is more suitable for analytical calculations. Gillespie (1981) has subsequently compared these two approaches and confirmed that the two lead to identical results. Approximate relations to quickly compute the mean passage time for unimodal and bimodal systems have also been proposed.

The analysis of Seshadri, West and Lindenberg (1980a) begins with the master equation

$$\frac{\partial}{\partial t} p(n, t \mid n_0) = W_{n,n+1} p(n+1, t \mid n_0) + W_{n,n-1} p(n-1, t \mid n_0) - W_{n+1,n} + W_{n-1,n} p(n, t \mid n_0) \tag{4.116}$$

where $p(n, t \mid n_0)$ represents the probability that the random variable $N(t)$ assumes a value of n at time t given that $N(0) = n_0$ and $W_{n,m}$ represents the transition rate to move from state m to state n. The random variable $N(t)$ assumes the integer values $0, 1, \ldots B$. We now define another probability $f(n, t \mid n_0)$ representing the probability that $N(t) = n$ and that the system during the time $(0 - t)$ has not reached a threshold value, say M. If we define $T(M|n_0)$ as the time at which the random variable assumes the value $N(t) = M$ for the first time, then

$$T(M|n_0) = \mathrm{Min}[\tau \mid N(t) = M \mid N(0) = n_0] \tag{4.117}$$

The cumulative distribution function $F(M, t \mid n_0)$ representing the probability that the time spent by the system to reach state M exceeds the value t is written as

$$\begin{aligned} F(M, t \mid n_0) &= \mathrm{Prob}\{T(M|n_0) > t\} \\ &= f(n, t \mid n_0) \end{aligned} \tag{4.118}$$

where summation over all the states below or above state M (depending on whether the initial state n_0 is below or above M) is implied. The distribution function can be used to calculate the moments of the first passage time distribution to state M. The general r^{th} moment can be obtained as

$$T_r(M|n_0) = r \int_0^\infty dt\, t^{r-1} F(M, t \mid n_0) \tag{4.119}$$

We now return to the master equation where, in view of the range of $N(t)$ defined, we can equate $W_{-1,0}$ to $W_{0,-1}$ to $W_{B,B+1}$ to zero. Also, if the initial condition $N(0) = n_0$ is chosen such that it lies below or above state M, then correspondingly $W_{M-1,M}$ or $W_{M+1,M}$ should also be zero. The probability $f(n, t \mid n_0)$ should also satisfy the master equation which can be abbreviated to

$$\frac{\partial f}{\partial t} = \mathscr{L}\mathbf{f}, \qquad \mathbf{f}(0) = I = \text{unit matrix} \tag{4.120}$$

where $\mathbf{f}$ is the $M \times M$ matrix whose $(n+1, n_0+1)$ element is $f(n, t \mid n_0)$ and $\mathscr{L}$ the $M \times M$ matrix of the transition probability per unit time. Let us now denote $\bar{\mathbf{f}}$ as the Laplace transform of $\mathbf{f}$; the transformed Equation (4.120) for $s = 0$ can be written as

$$\begin{aligned} W_{1,0}\bar{f}_{0,n_0} &= W_{0,1}\bar{f}_{1,n_0} + \delta_{0,n_0} \\ (W_{0,1} + W_{2,1})\bar{f}_{1,n_0} &= W_{1,0}\bar{f}_{0,n_0} + W_{1,2}\bar{f}_{2,n_0} + \delta_{1,n_0} \\ &\cdots\cdots\cdots\cdots\cdots\cdots \\ &\cdots\cdots\cdots\cdots\cdots\cdots \\ (W_{M-2,M-1} + W_{M,M-1})\bar{f}_{M-1,n_0} &= W_{M-1,M-2}\bar{f}_{M-2,n_0} + \delta_{M-1,n_0} \end{aligned} \tag{4.121}$$

where $\bar{f}_{n,n_0}$ is an abbreviation used for $\bar{f}_{n,n_0} = \bar{f}(n, 0 \mid n_0)$. The set of Equations (4.121) can be solved to obtain the recursive relation

$$\bar{f}_{n,n_0} = \phi_n \bar{f}_{0,n_0} - \xi_{n,n_0} \tag{4.122}$$

where

$$\phi_0 = 1, \qquad \phi_n = \frac{W_{1,0}W_{2,1}W_{3,2}\cdots W_{n,n-1}}{W_{0,1}W_{1,2}W_{2,3}\cdots W_{n-1,n}}, \qquad 1 \leqslant n \leqslant M-1 \tag{4.123}$$

and

$$\xi_{n,n_0} = 0, \qquad \text{if} \quad n \leqslant n_0$$

$$\phi_n = \frac{1}{W_{n-1,n}}\left(1 + \frac{W_{n,n-1}}{W_{n-2,n-1}} + \cdots + \frac{W_{n,n-1}W_{n-1,n-2}\cdots W_{n_0+2,n_0+1}}{W_{n-2,n-1}W_{n-3,n-2}\cdots W_{n_0,n_0+1}}\right) \tag{4.124}$$

if $n_0 < n \leqslant M-1$.

The general Equation (4.122) can be substituted in the last identity in Equation (4.121) and rearranged to obtain

$$\bar{f}_{0,n_0} = \frac{\xi_{M-1,n_0}}{\phi_{M-1}} + \frac{1}{\phi_{M-1}W_{M,M-1}} \tag{4.125}$$

In order to calculate the mean passage time, we shall return to Equation (4.119) and write it in a Laplace transformed form as

$$T_1(M|n_0) = \sum f(n, 0 \mid n_0) \tag{4.126}$$

where use of Equation (4.122) leads to

$$T_1(M|n_0) = \left(\frac{1 + W_{M,M-1}\xi_{M-1,n_0}}{W_{M,M-1}\phi_{M-1}}\right) \sum_{i=0}^{M-1} \phi_i - \sum_{i=0}^{M-1} \xi_{i,n_0} \tag{4.127}$$

Equation (4.127) represents an exact relation which can be further simplified and written in terms of the equilibrium distribution as

$$T_1(M|n_0) = \sum_{n=n_0+1}^{M} \frac{\sum_{l=0}^{n-1} p_s(l)}{p_s(n-1)W_{n,n-1}} \qquad \text{if} \quad n_0 < M \tag{4.128}$$

$$= \sum_{n=M}^{n_0-1} \frac{\sum_{l=n+1}^{B} p_s(l)}{p_s(n+1)W_{n,n+1}} \qquad \text{if} \quad n_0 > M \tag{4.129}$$

Here p_s refers to the steady state solution of the master equation. In certain cases when the stationary solution of the master equation does not possess a proper form as, for example, when $W_{0,1} = 0$ (i.e. when the null state is absorbing implying that $p_s(l) = \delta_{l,0}$), clearly Equation (4.129) can no longer be used. The analytical expressions are, however, extremely useful when $p_s(l)$ has a nontrivial form.

An alternative algorithm that also involves a straightforward analysis of the number of transitions between the adjacent states has been derived by Gillespie (1979) and can be employed. The algorithm makes use of a quantity $t_{n_0,M}(x)$ that implies the average total time spent by the system in a state x during the random walk from state n_0 to state M and is given by

(i) $n_0 < M$

$$t_{n_0,M}(M-1) = \frac{1}{W_{M,M-1}} \tag{4.130}$$

$$t_{n_0,M}(x) = \frac{1}{W_{x+1,x}}\{\theta(x+1-n_0) + W_{x,x+1}t_{n_0,M}(x+1)\} \tag{4.131}$$

where $\theta(z) = 1$ if $z > 0$ and $\theta(z) = 0$ if $z \leqslant 0$.

The mean passage time is then obtained as

$$T(M|n_0) = \sum_{x=0}^{M-1} t_{n_0,M}(x) \qquad (n_0 < M) \tag{4.132}$$

(ii) $n_0 > M$

$$t_{n_0,M}(M+1) = \frac{1}{W_{M,M+1}} \tag{4.133}$$

$$t_{n_0,M}(x) = \frac{1}{W_{x-1,x}}[\theta(n_0 + 1 - x) + W_{x,x-1}t_{n_0,M}(x-1)] \tag{4.134}$$

$$x = M+2, M+3, \ldots$$

and

$$T(M|n_0) = \sum_{x=M+1}^{\infty} t_{n_0,M}(x), \qquad n_0 > M \tag{4.135}$$

A more elaborate analysis by Gillespie (1980) also presents a procedure for calculating the variance of the first passage time. The equivalence of Equations (4.130)–(4.132) or Equations (4.133)–(4.135) for calculating the mean first passage time with Equation (4.128) or Equation (4.129) derived by Seshadri, West and Lindenberg (1980) has been shown by Gillespie (1981). The trick involves the iteration of Equation (4.131) or (4.134) and use of detailed balancing to express the transition probability in terms of the stationary probability distribution. An interim result of this calculation is an expression for the mean time spent in state x during the random walk from state n_0 to state M:

$$t_{n_0,M}(x) = p_s(x) \sum_{n=\mathrm{Max}(x,n_0)}^{M-1} \frac{1}{W_{n+1,n}\,p_s(n)}$$

$$(n_0 < M,\ 0 \leqslant x \leqslant M-1) \tag{4.136}$$

or

$$t_{n_0,M}(x) = p_s(x) \sum_{n=M+1}^{\min(x,n_0)} \frac{1}{W_{n-1,n}\,p_s(n)}$$

$$(n_0 > M,\ x \geqslant M+1) \tag{4.137}$$

The analytical form of these expressions is extremely useful in calculating the mean number of transitions from any state x to its adjacent state during the random walk of the system from state n_0 to M.

This mean number is simply obtained as

$$\begin{pmatrix}\text{mean number of}\\ \text{transitions from}\\ \text{state } x \text{ to } x+1\end{pmatrix} = W_{x+1,x}t_{n_0,M}(x) \tag{4.138}$$

$$\begin{pmatrix}\text{mean number of}\\ \text{transitions from}\\ \text{state } x \text{ to } x-1\end{pmatrix} = W_{x-1,x}t_{n_0,M}(x) \tag{4.139}$$

Where appropriate an expression for $t_{n_0,M}(x)$ should be used. Using Equation (4.136) or (4.137) it is also possible to calculate the mean of the total number of visits to state x during the random walk from n_0 to M:

$$\begin{pmatrix}\text{total number of visits}\\ \text{to state } x \text{ during random}\\ \text{walk from } n_0 \text{ to } M\end{pmatrix} = [W_{x+1,x} + W_{x-1,x}]t_{n_0,M}(x) \tag{4.140}$$

A clear understanding of the physical significance of Equation (4.136) or (4.137) also affords clues for simplifying the calculation of mean first passage times for certain cases. We shall return to these equations later in the section on approximate relations for the mean passage time.

The terms appearing under the summation sign in Equations (4.136) and (4.137) can be identified as physical quantities that can be assigned some precise meaning. To appreciate this point let us analyse these terms. We first note that $[p_s(x)W_{x+1,x}\,dt]$ represents the probability that the system will make a transition from state x to $x+1$ in the time interval dt after it has evolved for a sufficiently long period of time (t_∞). This is equivalent to suggesting that the time interval after t_∞ to mark the first occurrence of the transition from state x to $x+1$ is an exponentially distributed random variable with $[p_s(x)W_{x+1,x}]$ as a decay constant. The mean of the exponential distribution, as can be calculated using the prescription presented in Chapter 1, is simply the inverse of the decay constant. We then have

$$T_\infty^+(x) = [p_s(x)W_{x+1,x}]^{-1} \tag{4.141}$$

Similarly we can define

$$T_\infty^-(x) = [p_s(x)W_{x-1,x}]^{-1} \tag{4.142}$$

The two equations are related to each other through the detailed balance as $T_\infty^+(x) = T_\infty^-(x-1)$. Using Equation (4.141) in (4.136) we have the mean first passage time as

$$T(M|n_0) = \sum_{x=n_0}^{x} T_\infty^+(x) \sum_{n=0}^{x} p_s(n) \qquad \text{if} \quad n_0 < M$$

$$= \sum_{x=M+1}^{n_0} T_\infty^-(x) \sum_{n=x}^{\infty} p_s(n) \qquad \text{if} \quad n_0 > M \tag{4.143}$$

Similar relations can be written for other quantities of interest such as the mean time spent in state x, the mean number of transitions to states $(x + 1)$ and $(x - 1)$ from state x, etc.

Equation (4.143) for the mean passage time and other similar equations for other quantities of interest do not involve the transition probabilities per unit time explicitly. These equations were derived first by Gillespie (1982) who also noted the fundamental nature of the newly defined functions $T_\infty^+(x)$ and $T_\infty^-(x)$. Like the equilibrium distribution $p_s(x)$, they can be directly calculated from a knowledge of the transition probabilities per unit time. Their specific relation to the transition probabilities per unit time and the fact that they are complementary to the stationary distribution $p_s(x)$ are brought out clearly in Appendix 4.B.

Equation (4.143), simple as it is for the calculation of the mean first passage time, can be further approximated for specific systems if its stationary distribution is known. These approximations will be discussed in the section on approximate relations for the mean passage time along with other approximations.

Asymptotic Results for Mean Passage Time

The exact equations, Equation (4.128) or (4.129) for the calculation of the mean passage time, while quite simple, do not often converge rapidly. It is possible to approximate these relations in the asymptotic regime if the physical characteristics of the system are known. Thus, we shall consider two examples where the equilibrium distribution $p_s(n)$ expressed in terms of its potential $V(n)$ as

$$p_s(n) = p(0) \exp[-V(n)] \tag{4.144}$$

has the forms as shown in Figure 4.8. In the first case the potential exhibits a single minimum, while in the second two minima exist. Several physical examples of interest pertain to either of these situations. In the

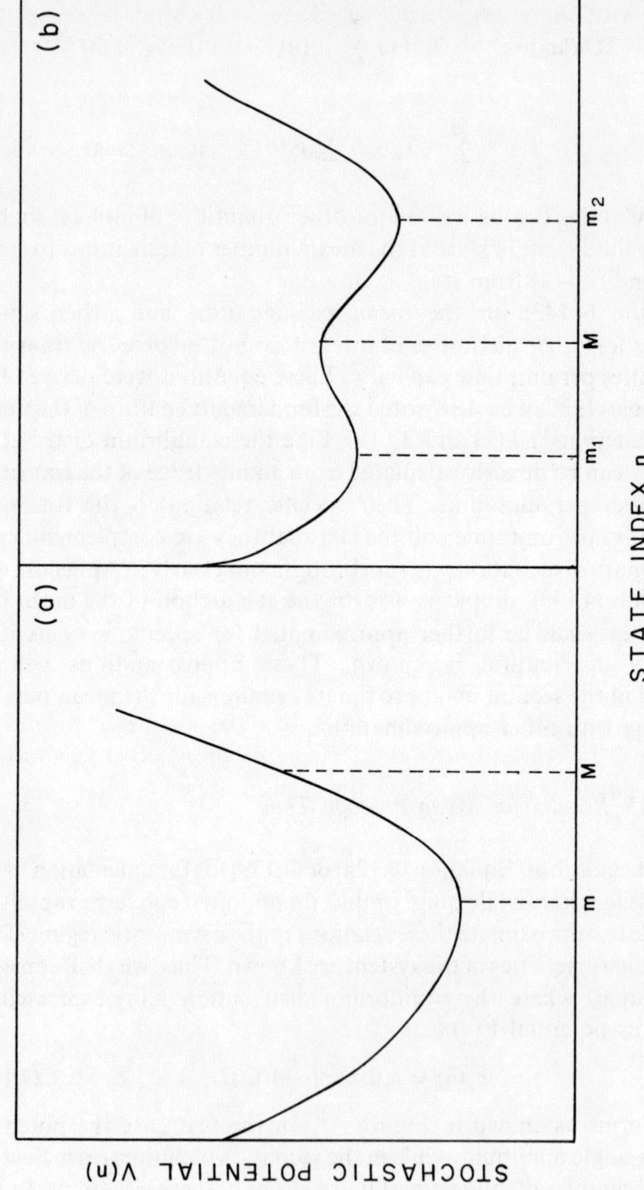

FIGURE 4.8 Stochastic potential with a single and double minimum.

first example for $n_0 < M$, we note that $p(M) \ll p(m)$ (see Figure 4.8). The main contribution to Equation (4.128) then comes from the region near to state M. Denoting the summand in Equation (4.128) as t_n, we can rewrite it as

$$T_1(M|n_0) = \sum_{n=n_0+1}^{M} t_n \frac{[1 - t_{n-1} \mid t_n]}{[1 - t_{n-1} \mid t_n]} \tag{4.145}$$

Using the discrete analog of the rule of integration by parts, this equation can be rewritten in the form

$$T_1(M|n_0) = \frac{t_M}{1 - t_{M-1} \mid t_M}\left[1 - \frac{(t_{M-1} \mid t_M)^2 - (t_{M-2} \mid t_M)}{(1 - t_{M-2} \mid t_{M-1})^2} + \cdots\right] \tag{4.146}$$

where the second and subsequent terms for most physical situations are considerably less than unity and can be ignored. The ratio terms in the denominator can be expressed in terms of transition probabilities using the detailed balance as

$$\frac{t_{n-1}}{t_n} = \frac{W_{n,n-1}}{W_{n-2,n-1}} \frac{\sum_{l=0}^{n-2} p(l)}{\sum_{l=0}^{n-1} p(l)} \tag{4.147}$$

which, for n nearer to M, measures the ratio of the transition rate upward to downward from a given state. Depending on whether this ratio is $\ll 1$ or is equal to unity, we have the asymptotic relations as

$$T_1(M|n_0) \simeq t_m$$

or (4.148)

$$\simeq \frac{1}{1 - t_{M-1} \mid t_M} \qquad (=1)$$

A similar analysis for the case of $n_0 > M$ leads to the asymptotic relation

$$T_1(M|n_0) = \frac{t_M}{1 + t_{M+1} \mid t_M} \tag{4.149}$$

The asymptotic expressions have been used by Seshadri, West and Lindenberg (1980, 1980a) to calculate the mean passage time for specific systems exhibiting single and double peak minima of the stochastic potential.

Approximate Relations for Mean Passage Time

In this section we shall present some approximate relations for calculating the mean passage time. For reasons of generality we again consider unimodal and bimodal systems of the type sketched in Figure 4.9. As shown in the figure we are interested in calculating the average time required by the system to move from a state K to state L. It is clear

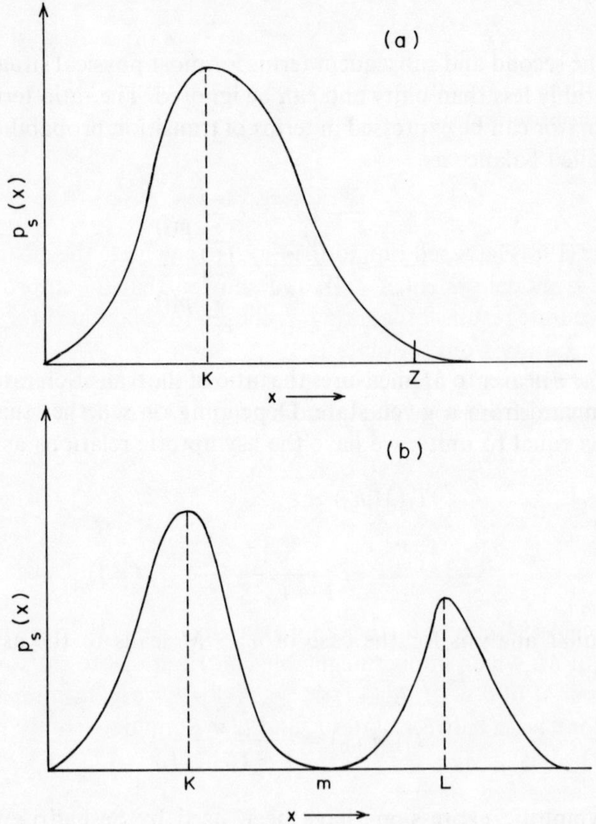

FIGURE 4.9 Stationary solutions for unimodal and bimodal systems.

that the system would spend most of its time experiencing small fluctuations around state K until a large fluctuation carries it to state L. This is just the mean passage time T_{KL} and can of course be calculated using Equation (4.132) or any of the expressions presented in the earlier sections. In fact we can even approximate Equation (4.132) for this case as

$$T_{KL} = \sum_{\substack{x \in K \\ x \leqslant L}} t_{KL}(x) \tag{4.150}$$

since a large portion of time is going to be spent in states that are under the K peak but well below L. Similarly Equation (4.136) can be approximated as

$$t_{KL}(x) = p_s(x) \sum_{n=K}^{L-1} \frac{1}{W_{n+1,n}\, p_s(n)} \tag{4.151}$$

Substitution of Equation (4.151) into Equation (4.150) leads to the desired approximation

$$T_{KL} = \sum_{n=K}^{L-1} \frac{1}{W_{n+1,n}\, p_s(n)} \qquad (L \gg K) \tag{4.152}$$

Gillespie (1981) has used Equation (4.152) to calculate the mean passage time for a model system $X \rightleftarrows Y$ and showed that the approximation yields accurate results if the state L is chosen to be about three standard deviations away from the state K.

Similar reasoning applied to the case of bimodal systems leads to approximate equations corresponding to (4.132) and (4.136):

$$T_{KL} = \sum_{\substack{x \in |K| \\ x \ll m}} t_{KL}(x) \tag{4.153}$$

$$t_{KL}(x) = p_s(x) \sum_{n=\max(x,K)}^{n-1} \frac{1}{W_{n+1,n}\, p_s(n)} \tag{4.154}$$

Equation (4.154) can be further approximated due to the deep minimum in $p_s(n)$ at M, which implies negligible contributions to the summation due to $n \ll M$ or $n \gg M$. Also $1/W_{n+1,n}$ is slowly varying compared with $1/p_s(n)$ for n lying between states K and L. We could rearrange Equation (4.154) then as

$$t_{KL}(x) = \frac{p_s(x)}{W_{M+1,M}} \sum_{n=K}^{L} \frac{1}{p_s(n)} \tag{4.155}$$

Substituting this equation for t_{KL} in Equation (4.153) gives

$$T_{KL} = \frac{a_K}{W_{M+1,M}} \sum_{n=K}^{L} \frac{1}{p_s(n)} \alpha a_K \tag{4.156}$$

where a_K refers to the area under the K peak $[\sum_{x=0}^{M} p_s(x)]$.

Similar analysis gives

$$T_{LK} = \frac{a_L}{W_{M-1,M}} \sum_{n=K}^{L} \frac{1}{p_s(n)} \alpha a_L \tag{4.157}$$

Equations (4.156) and (4.157) indicate that the transition time from state K to L or vice versa is approximately proportional to the area under the appropriate quasi-stationary state. It follows from these equations that the ratio of the mean transition time for K–L to that for L–K transitions is proportional to the ratio of the areas under the K and L peaks.

We shall now return to Equation (4.143) to see how we could approximate this relation. For specifying we consider the example of a bimodal system possessing the stationary states K and L that are well separated by an unstable state M (see Figure 4.9). Clearly, $K < M < L$ and our interest lies in calculating the first passage time from state K to L. The mean passage time from state K to L is given by Equation (4.132) as

$$T_{KL} = \sum_{x=0}^{L-1} t_{KL}(x) \tag{4.158}$$

where Equation (4.136) for t_{KL} can be written as

$$t_{KL}(x) = p_s(x) \sum_{n=\max(x,K)}^{L-1} T_{\infty}^{+}(n), \qquad x < L \tag{4.159}$$

We note that a large portion of this transit time will be spent in the states that lie under the peak at K in $p_s(x)$. Denoting these states by $\{K\}$, we can approximate Equations (4.158) and (4.159) by

$$T_{KL} = \sum_{x \in \{K\}} t_{KL}(x) \tag{4.160}$$

$$t_{KL}(x) = p_s(x) \sum_{x \in \{M\}} T_{\infty}^{+}(n), \qquad x \in K \tag{4.161}$$

where $\{M\}$ denotes the states under the peak at M in $T_{\infty}^{+}(x)$. This approximation becomes clear if we examine the $T_{\infty}^{+}(x)$–x diagram

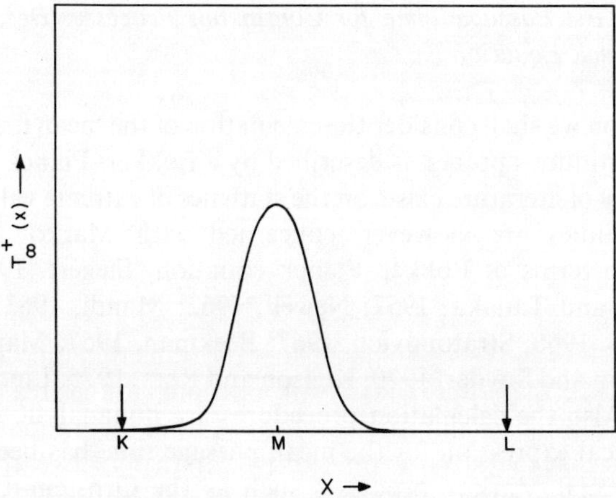

FIGURE 4.10 Typical plot showing $T_\infty^+(x)$ variations for a bimodal system.

(Figure 4.10) which indicates that $T_\infty^+(x)$ takes its largest value near M and negligibly small values at K and L. Substituting Equation (4.161) in (4.160) gives

$$T_{KL} = \sum_{x \in \{K\}} p_s(x) \sum_{n \in \{M\}} T_\infty^+(n) \tag{4.162}$$

Similarly, for transition from state L to K, we can write the approximate relation

$$T_{LK} = \sum_{x \in \{L\}} p_s(x) \sum_{n \in \{M\}} T_\infty^+(n) \tag{4.163}$$

The second summation in Equation (4.163) should have been $T_\infty^-(n)$; however, as we have noted earlier, and particularly at M, the two differ insignificantly. The relations (4.162) and (4.163) afford a simpler definition of the mean passage time, which can be recognised as simply the product of the areas under the initial stable steady state peak in the $p_s(x)$ curve times the area under the intermediate unstable steady state peak in the T_∞^+ curve. The mean passage time being proportional to the area under the initial stable steady state peak has been noted earlier in Equation (4.156). The present result clarifies the constant of proportionality as the area under the intervening peak between the stable states.

(*B*) *Mean First Passage Time for Continuous Processes Described by Fokker–Planck Equation*

In this section we shall consider the calculation of the mean first passage time for continuous processes described by a Fokker–Planck equation. A large body of literature exists on the statistics of extreme values. Most of these studies are, however, concerned with Markov processes described in terms of Fokker–Planck equation (Siegert, 1951, 1953; Maruyama and Tanaka, 1957; Newell, 1962; Mandl, 1963; Keilson, 1964; Weiss, 1966; Stratonovich, 1967; Beekman, 1967; Mandl, 1968; Stone, Belkin and Snyder, 1970; Keilson and Ross, 1975; Lindenberg *et al.*, 1975). Also the calculation procedures are nonanalytic in nature. The analytical expression for the mean passage time has been derived only for special Markov processes such as the Ornstein–Uhlenbeck process (Newell, 1962; Keilson and Ross, 1975; Lindenberg *et al.*, 1975; Helstrom, 1959). For the asymptotic region, i.e. for large extents of time, Lindenberg and Seshadri (1979) have developed an analytical method which seems especially useful for practical calculations. In the present section therefore we shall outline the main features of their method.

We begin with the general Fokker–Planck equation

$$\frac{\partial p(x,t\mid x_0)}{\partial t}=\frac{\partial}{\partial x}[-\alpha_1(x)p(x,t\mid x_0)]+\frac{1}{2}\frac{\partial^2}{\partial x^2}[\alpha_2(x)p(x,t\mid x_0)] \tag{4.164}$$

where the several terms appearing in the equation have already been defined. One could start with an equivalent description corresponding to Equation (4.164) in the Langevin form as

$$\frac{dX}{dt}=\alpha_1(X)-\tfrac{1}{4}\alpha_2(X)+\sqrt{\alpha_2(X)}F(t) \tag{4.165}$$

where X refers to the random variable and $F(t)$ to the zero centred Gaussian distributed random variable with $\langle F(t)Ft'\rangle=\delta(t-t')$.

We now define a probability density $f(x,t\mid x_0)$ which denotes the probability that $X(t)$ lies within the range $(x-x+dx)$ during the time $(0-t)$ without ever having crossed a certain preset threshold value of

$X = \pm\chi$. The cumulative distribution function is then defined as

$$F(\chi, t \mid x_0) = \int_{-\chi}^{\chi} dx\, f(\chi, t \mid x_0) \tag{4.166}$$

If $Y(t)$ represents the highest value which the variable X takes during $(0 - t)$, then we have

$$Y(t) = \text{Max}\{|X(\tau)|\}, \qquad 0 \leqslant \tau \leqslant t \tag{4.167}$$

and

$$F(\chi, t \mid x_0) = \text{Prob}\{Y(t) < \chi \mid X(0) = x_0\} \tag{4.168}$$

or alternatively

$$F(\chi, t \mid x_0) = \text{Prob}\{T(\chi) > t \mid X(0) = x_0\}$$

where $T(\chi)$ is the time when the random variable first reaches $(\pm x)$ and is given by

$$T(\chi) = \text{Min}\{\tau \mid |X(\tau)| = \chi\} \tag{4.169}$$

The cumulative distribution function can be used to calculate the moments of the extremal value $Y(t)$ and the mean passage time $T(x)$. The general equations for the moments are

$$Y_n(t \mid x_0) = n \int_0^{\infty} d\chi\, \chi^{n-1}[1 - F(\chi, t \mid x_0)] \tag{4.170}$$

and

$$T_n(\chi \mid x_0) = n \int_0^{\infty} dt\, t^{n-1} F(\chi, t \mid x_0) \tag{4.171}$$

Let us now return to the Fokker–Planck Equation (4.164) which is also satisfied by the probability density $f(x, t \mid x_0)$

$$\frac{\partial f(x, t \mid x_0)}{\partial t} = \frac{\partial}{\partial x}[-\alpha_1(x) f(x, t \mid x_0)] + \frac{1}{2}\left[\frac{\partial^2}{\partial x^2}\alpha_2(x) f(x, t \mid x_0)\right] \tag{4.172}$$

along with the conditions

$$f(\pm\chi, t \mid x_0) = 0 \tag{4.173}$$

$$f(x, 0 \mid x_0) = \delta(x - x_0) \tag{4.174}$$

The solution to this equation can be written in terms of the eigenfunction expression as

$$f(x, t \mid x_0) = \sum_{k=0}^{\infty} p_s(x) \frac{U_k(x)U_k(x_0)}{N_k} e^{-\lambda_n t} \tag{4.175}$$

Here $p_s(x)$ is the stationary solution of Equation (4.164) and the eigenfunctions $U_k(x)$ and the eigenvalues λ_n satisfy the Sturm–Liouville equation

$$\frac{d}{dx}\left[\psi(x)\frac{dU_k(x)}{dx}\right] + \lambda_k p_s(x)U_k(x) = 0 \tag{4.176}$$

where

$$\psi(x) = \tfrac{1}{2}\alpha_2(x)p_s(x)$$

Both $\psi(x)$ and $p_s(x)$ are positive functions with finite number of poles. Obtaining a solution to Equation (4.176) is in general a difficult task. However, for rare events such as the occurrence of a giant fluctuation that takes the system from one quasi-stationary state to another, Lindenberg and Seshadri (1979) argued that it is adequate to know the lowest eigenvalue. For sufficiently large x, the cumulative distribution function is then obtained as

$$F(\chi, t) = \exp[-\lambda_0(\chi)t] \tag{4.177}$$

and the moments of first passage time and extremum value as

$$T_n(\chi) = n!/[\lambda_0(\chi)]^n \tag{4.178}$$

$$Y_n(t) = n\int_0^{\infty} d\chi\, \chi^{n-1}[1 - \exp(-\lambda_0(\chi)t] \tag{4.179}$$

The lowest eigenvalue λ_0 as required in the above equations can be independently evaluated. For this purpose we start with the backward form of the Fokker–Planck equation that is satisfied by the probability density $f(x, t \mid x_0)$ in Equation (4.172). Thus we have

$$\frac{\partial f(x, t \mid x_0)}{\partial t} = -\frac{\partial}{\partial x_0}\alpha_1(x_0)f(x, t \mid x_0) + \frac{1}{2}\frac{\partial^2}{\partial x^2}\alpha_2(x_0)f(x, t \mid x_0) \tag{4.180}$$

This Fokker–Planck equation can be converted into a differential-difference equation form. This involves multiplying it by nt^{n-1} and integrating with respect to x and t over the ranges $(-\chi$ to $\chi)$ and (0 to ∞). The resulting equation for the first passage moments is

$$\tfrac{1}{2}\alpha_2(x_0)\frac{\partial^2 T_n(\chi\,|\,x_0)}{\partial x_0^2} + \alpha_1(x_0)\frac{\partial T_n(\chi\,|\,x_0)}{\partial x_0} = nT_{n-1}(\chi\,|\,x_0) \quad (4.181)$$

with

$$T_0(\chi\,|\,x_0) = 1, \qquad \tau_n(X\,|\,\pm\chi) = 0$$

Integrating the equation for $n = 1$ gives

$$T_1(\chi\,|\,x_0) = \int_{x_0}^{\chi} dq \frac{\left[\int_0^q p_s(y)\,dy\right]}{\tfrac{1}{2}\alpha_2(q)p_s(q)} \quad (4.182)$$

For large χ, the integrals in Equation (4.182) can be easily integrated to obtain

$$\tau_1(\chi\,|\,x_0) = \frac{1}{\lambda_0} \qquad (\chi \text{ large}) \quad (4.183)$$

The result can now be used in Equations (4.178) and (4.179) to obtain analytical forms for the moments of the mean first passage time and the extremum values.

The theory of Lindenberg and Seshadri (1979) as discussed above has been applied to a wide variety of situations, which include the constant and variable diffusion terms $[\alpha_2(x)]$ in the Fokker–Planck equation for a variety of nonlinear drift terms $[\alpha_1(x)]$. The expressions for the mean first passage time and the mean extremum values for several of these situations have been presented by Lindenberg and Seshadri (1979). The results indicate that as the dependence of the drift term $[\alpha_1(x)]$ on x increases, it takes more time for the system to reach a given threshold value. Also the mean passage time is very sensitive to the nature of the expression $\alpha_1(x)$. Knowing $T_1(\chi\,|\,x_0)$ and $T_2(\chi\,|\,x_0)$ for a given $\alpha_1(x), \alpha_2(x)$, we can compute the dispersion $\mathscr{D}$ defined as

$$\frac{[T_1(\chi\,|\,x_0) - T_2^2(\chi\,|\,x_0)]^{1/2}}{T_1(\chi\,|\,x_0)}$$

$\mathscr{D}$ evaluated for the various nonzero forms for $\alpha_1(x)$ indicates that for $X \to \infty$ the dispersion approaches unity. First passage times are therefore a sensitive and broadly distributed measure of the structure of the system. Additionally, the extremal value calculations also afford a very sensitive and sharp measure of the structure of the system.

4.4.2 Splitting Probability

In the previous sections we were concerned with the first passage problem from one state to another, both of which represented the stable steady states of a bistable system. In the present section, we continue with this type of system but consider the first passage problem from the unstable state that lies in between the two quasi-stationary states.

For the sake of illustration we suppose that, to begin with, the system is at some point m between K and L (see Figures 4.6, 4.9). We are specifically concerned with the case when the point m corresponds to the unstable point M, our intention being to know the first time the system reaches either state K or L. We are also interested in knowing the probability that the system at M would get caught up in the region of attraction of state K or L and eventually evolve to either of these states.

The first passage time to move from state M to state K can be obtained referring to Equation (4.129) as

$$T_1(K|M) = \sum_{n=K}^{M-1} \frac{\sum_{l=n+1}^{\Omega} p_s(l)}{p_s(n+1)W_{n+1,n}} \tag{4.184}$$

which also includes the transition time from the unstable state M to state L and back from state L to state K. As we noted in the previous section these contributions are substantial. In obtaining the relaxation from an unstable state to a metastable state K or L, we are normally not interested in this entire time period but would like to know the time for transition straight from the unstable state M to state K or L. In such instances we could assume an artificial reflecting boundary at the unstable state M so that transitions from state M to $M + 1$ are not allowed. We now calculate the mean passage time from M to K in presence of this reflecting boundary. The presence of the reflecting boundary at M affects the total stationary probability which can be bounded as

$$\sum_{l=n+1}^{M} p_s(l) \leqslant (M - n - 1)p(n + 1) \tag{4.185}$$

Substitution of Equation (4.185) into (4.184) yields the desired bounds on the passage time:

$$T_1(K|M) \leqslant \sum_{n=K}^{M-1} \frac{(M-n-1)}{W_{n,n+1}} \tag{4.186}$$

We now turn to the question of the probability that the system starting at state M, that is contained between states K and L, will reach state L before it reaches K. For simplicity of notation, let $W_{\pm}(x)$ denote the transition probability per unit time to move from state x to $(x \pm 1)$. The quantity $W_{\pm}(x)/[W_{+}(x) + W_{-}(x)]$ then represents the probability that the system starting at state x will move to state $(x \pm 1)$. The probability of reaching state L can then be written as

$$\pi_L(x) = \left[\frac{W_-(x)}{W_+(x) + W_-(x)}\right]\pi_L(x-1) + \left[\frac{W_+(x)}{W_+(x) + W_-(x)}\right]\pi_L(x+1)$$
$$K+1 \leqslant x \leqslant L-1 \tag{4.187}$$

provided $\pi_L(K) = 0$, $\pi_L(L) = 1$.

Rearranging Equation (4.187) we have

$$\pi_L(x+1) - \pi_L(x) = \frac{W_-(x)}{W_+(x)}[\pi_L(x) - \pi_L(x-1)] \tag{4.188}$$

which is a recursion relation and can be iterated to obtain

$$\pi_L(x+1) - \pi_L(x) = \pi_L(K+1) \prod_{l=K+1}^{x} \frac{W_-(l)}{W_+(l)}$$
$$K+1 \leqslant x \leqslant L-1 \tag{4.189}$$

where we have used the condition $\pi_L(K) = 0$. We now sum Equation (4.189) from $x = n$ to $x = L-1$ to obtain

$$1 - \pi_L(n) = \pi_L(K+1) \sum_{x=n}^{L-1} \prod_{l=K+1}^{x} \frac{W_-(l)}{W_+(l)}$$
$$K+1 \leqslant n \leqslant L-1 \tag{4.190}$$

In Equation (4.190), if we set $n = K+1$, the equation can be solved to obtain $\pi_L(K+1)$ which when substituted back in Equation (4.190) yields

$$\pi_L(n) = \left[1 + \sum_{x=K+1}^{n-1} \prod_{l=K+1}^{x} \frac{W_-(l)}{W_+(l)}\right] \times \left[1 + \sum_{x=K+1}^{L-1} \prod_{l=K+1}^{x} \frac{W_-(l)}{W_+(l)}\right]^{-1} \qquad K+1 \leqslant n \leqslant L-1 \quad (4.191)$$

If state K is nonabsorbing, then we can multiply Equation (4.191) by the ratio $W_-(K)/W_+(L)$ to obtain a simpler relation:

$$\pi_L(n) = \sum_{x=K}^{x-1} \prod_{l=K}^{x} \frac{W_-(l)}{W_+(l)} \bigg/ \sum_{x=K}^{L-1} \prod_{l=K}^{x} \frac{W_-(l)}{W_+(l)} \qquad K+1 \leqslant n \leqslant L-1 \quad (4.192)$$

If no states are absorbing ($p_s(x) > 0$, $x \geqslant 0$), then [using Equation (3), Appendix 4.B]

$$\prod_{l=K+1}^{x} \frac{W_-(l)}{W_+(l)} = \frac{W_+(K)p_s(K)}{W_-(x)p_s(x)}, \qquad x \geqslant K \quad (4.193)$$

When substituted in Equation (4.191) this yields

$$\pi_L(x) = \sum_{n=K}^{x-1} [W_+(n)p_s(n)]^{-1} \bigg/ \sum_{n=K}^{L-1} [W_+(n)p_s(n)]^{-1} \qquad K < x < L \quad (4.194)$$

(Obviously the probability that the system starting at some point $x \in (K, L)$ will reach K before L is $\pi_K(x) = 1 - \pi_L(x)$.) Equation (4.194) can be simplified if we invoke the definition of $T_\infty^+(x)$ [Equation (4.141)]:

$$\pi_L(x) = \sum_{n=K}^{x-1} T_\infty^+(n) \bigg/ \sum_{n=K}^{L-1} T_\infty^+(n), \qquad K < x < L \quad (4.195)$$

As noted earlier the function $T_\infty^+(x)$ can be easily calculated from a knowledge of W's; we can therefore calculate $\pi_L(x)$ easily and for a well-defined bistable system the typical shape of the $\pi_L(x)$–x plot is indicated in Figure 4.11. We notice from the figure that for x closer to the steady state K the values of $\pi_L(x)$ lie close to zero, rising sharply as $x = M$ is approached, eventually levelling off to unity as the point $x = M$ is crossed. Of course, the behaviour would be reversed for $\pi_K(x)$. To

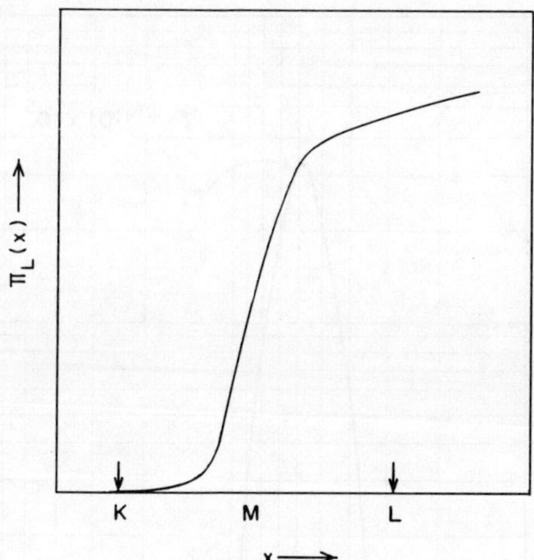

FIGURE 4.11 Plot of probability that the system will reach state L before it reaches K.

calculate the slope of the curve in Figure 4.11, we take

$$\pi_L(x+1) - \pi_L(x) = T_\infty^+(n) \Big/ \sum_{n=K}^{L-1} T_\infty^+(n), \qquad K < x < L \quad (4.196)$$

which assumes a maximum value at $x = M$ since $T_\infty^+(n)$ is lowest at this point (see Figure 4.12). The range of x values over which the curve rises sharply can be approximated as the inverse of this slope. We therefore have the effective width of transition as

$$\Delta_M = \sum_{n=K}^{L-1} T_\infty^+(n) \Big/ T_\infty^+(x) = \sum_{n \in M} T_\infty^+(n) \Big/ T_\infty^+(M) \qquad (4.197)$$

Equation (4.197) has a simple geometric interpretation in terms of the $T_\infty^+(n)$ plot where Δ_M is merely the effective width of the M peak.

Also, inserting Equation (4.197) in the equations for the passage time [Equations (4.162) and (4.163)], we obtain

$$T_{KL} = T_\infty^+(M)\Delta_M \sum_{x \in M} p_s(x) \qquad (4.198)$$

$$T_{LK} = T_\infty^+(M)\Delta_M \sum_{x \in L} p_s(x) \qquad (4.199)$$

These equations again afford a simple geometric interpretation of the

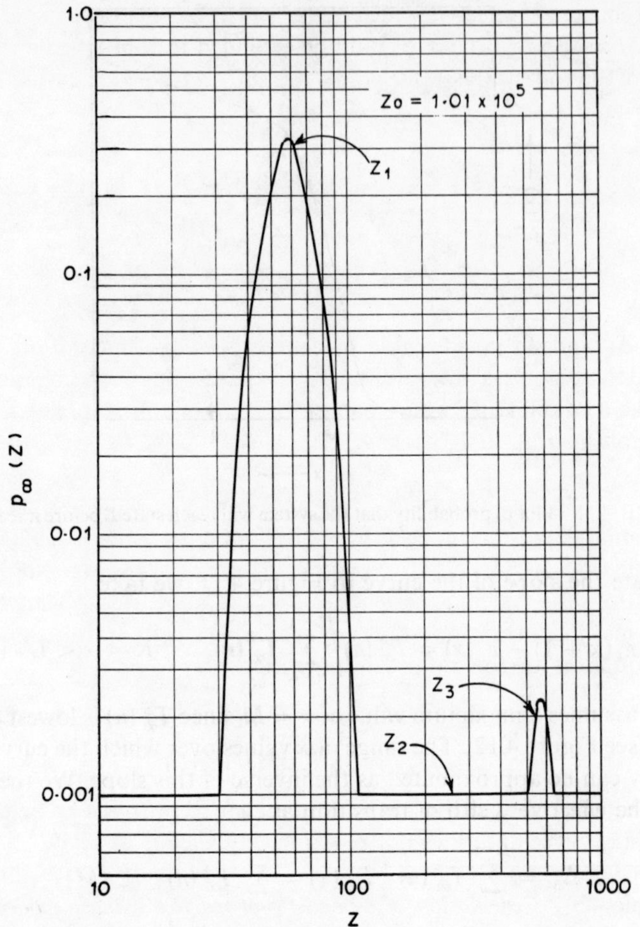

FIGURE 4.12 Plot of probability density versus Z.

mean first passage time from one quasi-stationary state to another. More importantly, all the quantities therein can be estimated exclusively from the $T_\infty^+(x)$ and $p_s(x)$ curves. $T_\infty^+(M)$ represents the mean waiting time for the transition from state x to $(x \pm 1)$, and $p_s(x)$ represents the width of the transition region; both these quantities can be estimated once we have the $T_\infty^+(x)$–x curve. $\sum_{x \in M} p_s(x)$ or $\sum_{x \in L} p_s(x)$ represents the area under the initial steady state peak and is known once the $p_s(x)$–x curve is available.

The concept of effective width of the transition region at the unstable point M in the $T_\infty^+(n)$ curve could be extended to think of the effective width of the zone at the stable points K and L in the $p_s(x)$ curve. Thus we can write

$$\Delta_K = \sum_{x \in K} p_s(x) \Big/ p_s(K)$$

and (4.200)

$$\Delta_L = \sum_{x \in L} p_s(x) \Big/ p_s(L)$$

Using Δ_K and Δ_L it is feasible to define the mean fluctuation period around these states. Thus, defining T_{KK} as the mean time the system takes to make a round trip from K back to K, and of course precluding the transition from K to L, we have

$$T_{KK} = \Delta_K / W_{K+1,K} = \Delta_K / W_{K-1,K} \tag{4.201}$$

A similar equation can be written for T_{LL}. Of particular significance is the ratio T_{KL}/T_{KK} or T_{LK}/T_{LL}, which measures the average number of fluctuations the system experiences about the stable steady state K or L before making a transition to the other stable steady state L or K. The final equation for this ratio can be written as

$$\begin{aligned} T_{KL}/T_{KK} &= \Delta_M T_\infty^+(M)/T_\infty^+(K) \\ T_{LK}/T_{LL} &= \Delta_M T_\infty^+(M)/T_\infty^+(L) \end{aligned} \tag{4.202}$$

Note that the last quantity can be estimated merely from a knowledge of the $T_\infty^+(x)$ curve and does not require a knowledge of $p_s(x)$.

Example 4.5 *Calculation of stochastic properties using a model system exhibiting bistability on a catalyst surface*

Let us consider a simple model that exhibits the phenomenon of bistability on a catalyst surface (Bykov and Yablonskii, 1981). The reaction scheme is represented as

$$\begin{aligned} Z &\rightleftharpoons AZ \\ AZ + 2Z &\rightleftharpoons 3Z \\ Z &\rightleftharpoons Q \end{aligned} \tag{1}$$

where Z represents the number of available sites and the product Q

TABLE 4.3
Steady state solutions for various values of Z_0

$Z_0 \times 10^{-5}$	Z_1	Z_2	Z_3
0.99	57	278	549
1.00	58	266	569
1.01	59	255	588
1.02	61	246	605
1.03	62	237	622
1.04	63	228	638
1.05	64	220	653
1.06	66	213	668
1.07	67	206	683

Z_1 and Z_3 are the points at which maxima occur for $p_\infty(Z)$ curve.
Z_2 is the minimal point for $p_\infty(Z)$ curve.

desorbs as fast as it is formed. The macroscopic equation corresponding to the reaction scheme may be written as

$$\begin{aligned}\frac{dZ}{dt} &= -k_1 Z + k_{-1}(Z_0 - Z) - k_{-2}Z^3 + k_2 Z^2(Z_0 - Z) - k_3 Z \\ &= f(Z) \qquad (2)\end{aligned}$$

where Z_0 represents the total number of available sites on the surface and k's the various rate constants. The stationary solutions of Equation (2) for a set of parameter values coincide with those given in Table 4.3.

The stochastic properties for this case will now be evaluated (Tambe and Kulkarni, 1986). The master equation describing the evolution of the probability density can be formulated for the reaction scheme as

$$\begin{aligned}\frac{\partial p(Z,t)}{\partial t} &= (k_1 + k_3)Zp(Z,t) + k_{-1}(Z_0 - Z)p(Z,t) \\ &\quad + k_{-2}Z(Z-1)(Z-2)p(Z,t) \\ &\quad + k_2(Z_0 - Z)Z(Z-1)p(Z,t) \qquad (3)\end{aligned}$$

The transition probability $W_\pm(Z)$ defining the probability that the system in state Z at time t will step to state $(Z \pm 1)$ in the next infinitesimal time interval $(t - t + dt)$ can then be identified as

$$W_+(Z) = k_{-1}(Z_0 - Z) + k_2(Z_0 - Z)Z(Z-1) \qquad (4)$$

$$W_-(Z) = (k_1 + k_3)Z + k_{-2}Z(Z-1)(Z-2) \qquad (5)$$

Let $p_\infty(Z)$ be the fully relaxed solution of the master Equation (3) of the process. This solution also satisfied the detailed balance recursion relation

$$p_\infty(Z) = \frac{W_+(Z-1)}{W_-(Z)} p_\infty(Z-1), \qquad Z \geqslant 1 \tag{6}$$

which can be iterated over the domain of Z to obtain

$$p_\infty(Z) = \frac{\prod_{n=1}^{Z} \frac{W_+(Z-1)}{W_-(Z)}}{1 + \sum_{Z=1}^{\infty} \prod_{n=1}^{Z} \frac{W_+(Z-1)}{W_-(Z)}} \tag{7}$$

Knowing the transition probabilities per unit time, we can construct the stationary solution $p_\infty(Z)$ of the process.

Gillespie (1982) defines another function $T_\infty^+(Z)$ that measures the mean time spent in a state Z before the system moves to state $(Z \pm 1)$. The formal definition can be given as $T^+(Z) = [W_+(Z)p(Z)]^{-1}$ where $[p_\infty(Z)W_+(Z)\,dt]$ represents the probability that the system will make a transition from Z to $Z \pm 1$ in time dt. From a definition analogous to that for $p_\infty(Z)$, we can obtain

$$T_\infty^+(Z) = \frac{1}{W_+(0)}\left[1 + \sum_{Z=1}^{\infty} \prod_{n=1}^{Z} \frac{W_+(Z-1)}{W_-(Z)}\right] \frac{W_-(Z)}{W_+(Z)}, \qquad Z \geqslant 1 \tag{8}$$

The equations for $p_\infty(Z)$ and $T_\infty^+(Z)$ are the basic relations in the use of the theory proposed by Gillespie (1982) and can be easily evaluated once the functions $W_\pm(Z)$ are known. All other stochastic properties of the system such as the mean passage time to move from one state to another, the mean time spent in any state Z during this passage time, and the mean number of transitions $Z \rightarrow Z \pm 1$ executed during the passage time, can be calculated once the $p_\infty(Z)$, $T_\infty^+(Z)$ functions are known. Additionally, we can calculate the probability that the system in state Z, that lies between the states Z_1 and Z_3, would first reach state Z_1 before reaching Z_3. The relations for several of these properties are easily derived and are presented in Table 4.3.

The deterministic Equation (2) has been solved for a set of parameter values $k_{-1} = 0.15 \times 10^{-2}$, $(k_1 + k_3) = 3.333$, $k_{-2} = 0.0166 \times 10^{-3}$, $k_2 = 0.15 \times 10^{-6}$. The equation possesses three roots for values of Z_0 lying in the region of 0.95×10^5–1.07×10^5. The transition probabilities

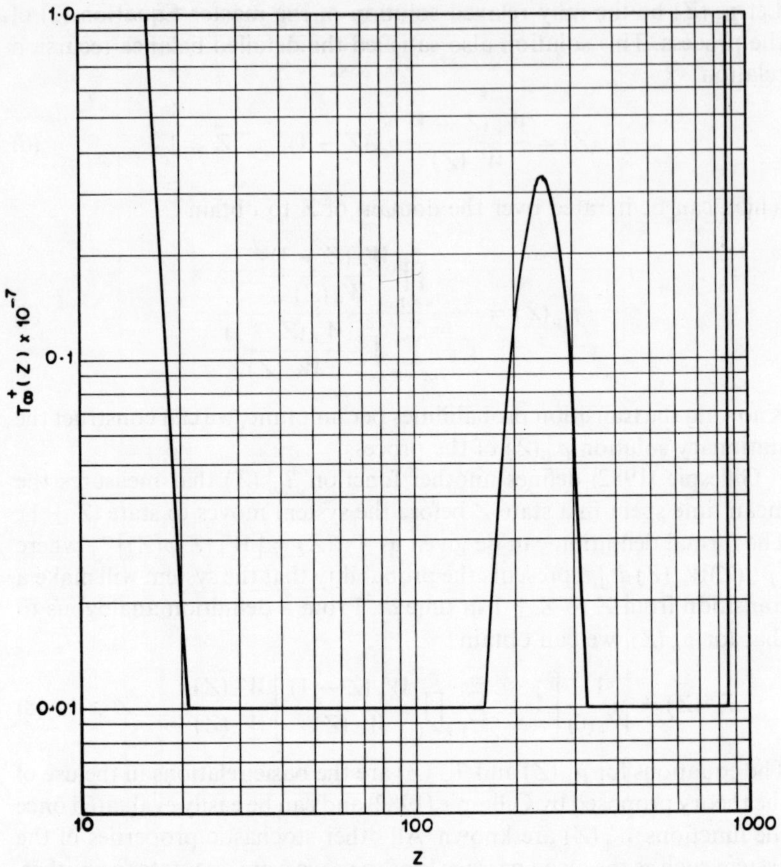

FIGURE 4.13 Plot of $T_{\infty}^{+}(Z)$ versus Z ($Z_0 = 1.01 \times 10^5$).

per unit time can be calculated using Equations (4) and (5), which in turn are used in Equations (6) and (7) to obtain the $p_{\infty}(Z)$ and $T_{\infty}^{+}(Z)$ variations. Typical curves for $Z_0 = 1.01 \times 10^5$ are shown in Figures 4.12 and 4.13. It can be easily verified that the peaks in the $p_{\infty}(Z)$ curve correspond to the stable steady state solutions of the deterministic equation, while the minima correspond to the unstable state. In the $T_{\infty}^{+}(Z)$ curve, the stable steady states appear as minima while a peak corresponds to the unstable solution. Similar curves for various values of Z_0 can be prepared and Table 4.3 lists the three stationary solutions at

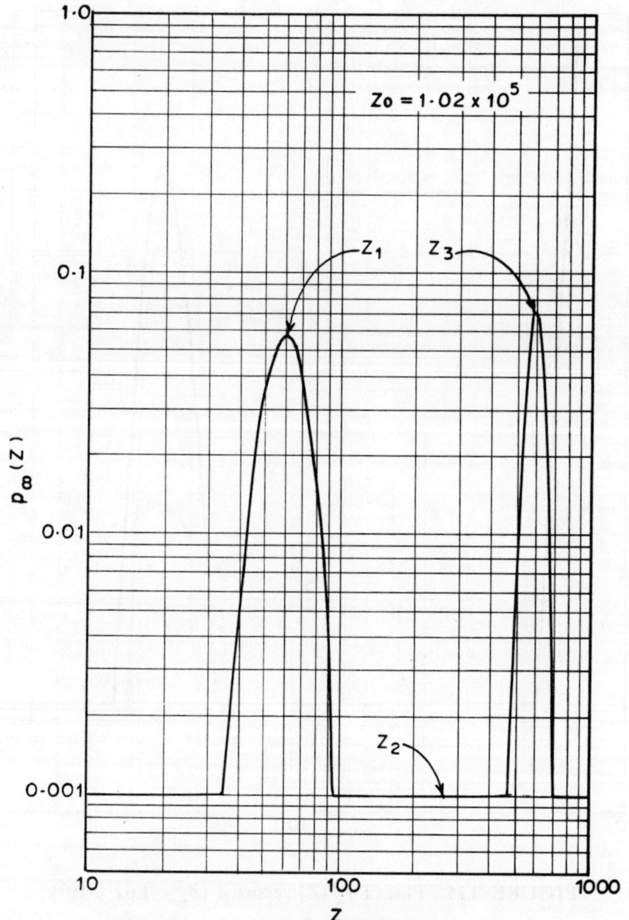

FIGURE 4.14 Plot of probability density versus Z.

which the maxima and minima occur in the $p_{\infty}(Z)$ curve. The values of course tally with the deterministic solutions.

It is evident from the solutions given in the table that increasing the number of sites on the catalyst surface (i.e. Z_0) has the influence of bringing the states Z_1 and Z_2 closer while Z_2 and Z_3 move apart. As indicated in Figures 4.14 and 4.15, with an increase in Z_0 the probability peak around Z_1 continuously decreases, while that around Z_3

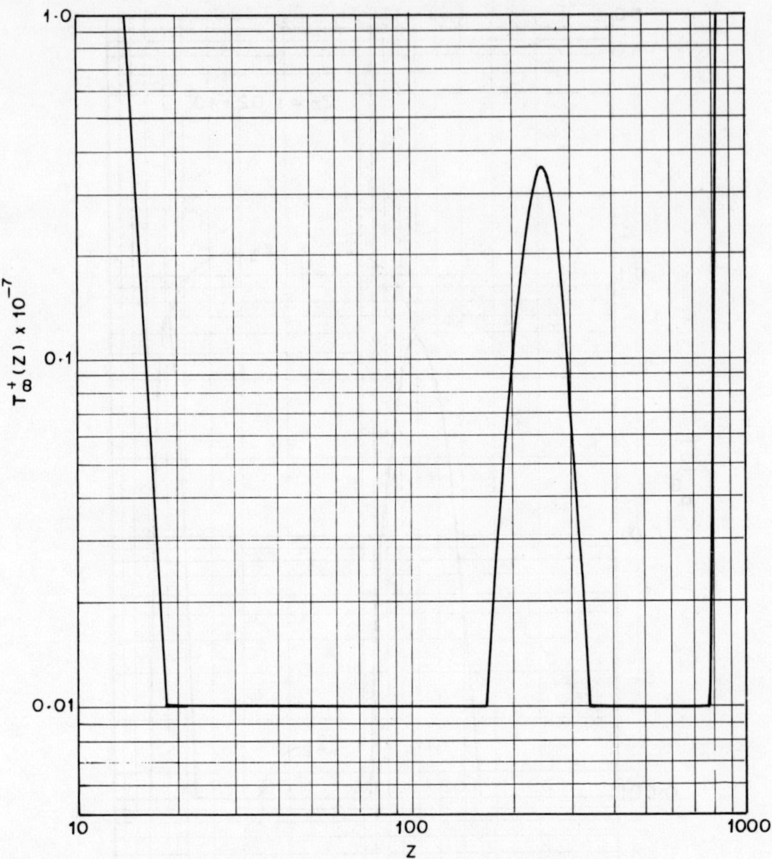

FIGURE 4.15 Plot of $T_{\infty}^{+}(Z)$ versus Z ($Z_0 = 1.02 \times 10^5$).

continuously increases. At some intermediate value of Z_0 (around $Z_0 = 1.015 \times 10^5$), the two peaks are equal in their heights and the peak around Z_3 exceeds that around Z_1 for values of Z_0 higher than this value. As regards the $T_{\infty}^{+}(Z)$ curve, the maximum of the peak goes through a minimum with increasing Z_0.

We have noted earlier that the stable steady states Z_1 and Z_3 of the macroscopic equation appear as metastable state in the probabilistic approach. The states Z_1 and Z_2 in Figure 4.12 are therefore metastable and the occurrence of a sufficiently large fluctuation can cause the system

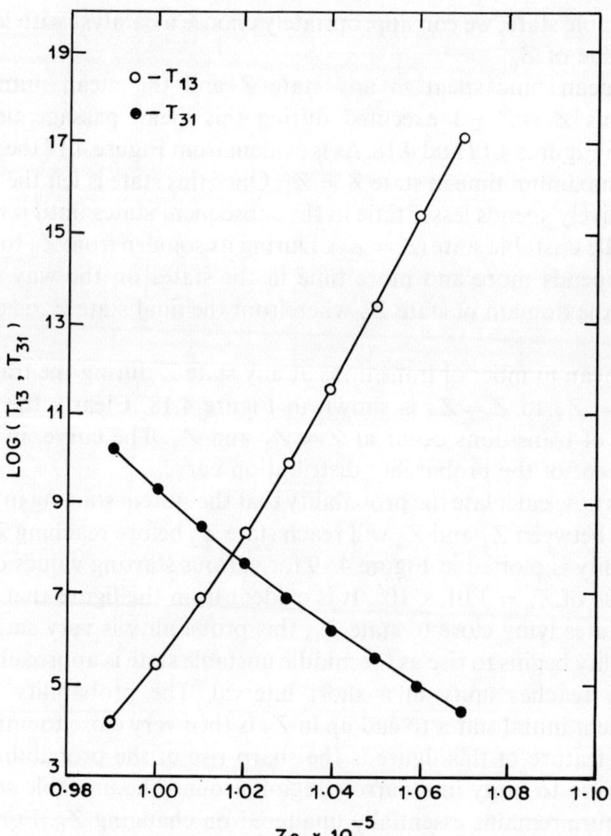

FIGURE 4.16 Mean passage time to move from state Z_1 to Z_3 (T_{13}) and from state Z_3 to Z_1 (T_{31}) for different values of Z_0.

operating at $Z = Z_1$ to move over to $Z = Z_3$. The mean time that is required before this transition can take place can be evaluated using the appropriate relations presented earlier. This mean passage time has been calculated for the case under consideration when Z_0 on the catalyst surface varies. The results are plotted in Figure 4.16 where the times required to move from Z_1 to Z_3 (T_{13}) and Z_3 to Z_1 (T_{31}) are indicated. As may be noted from the figure, T_{13} shows a monotonically increasing tendency with Z_0 while T_{31} decreases. Depending on whether Z_1 or Z_3 is the desirable state, we can appropriately choose a catalyst with large or small value of Z_0.

The mean time spent in any state Z and the mean number of transitions $Z \rightarrow Z \pm 1$ executed during this mean passage time are shown in Figures 4.17 and 4.18. As is evident from Figure 4.17 the system spends maximum time in state $Z = Z_1$. Once this state is left the system progressively spends lesser time in the subsequent states until it reaches the middle unstable state ($Z = Z_2$). During its sojourn from Z_2 to Z_3 the system spends more and more time in the states on the way until it reaches the domain of state Z_3 wherefrom the final state Z_3 is quickly reached.

The mean number of transitions at any state Z during the transition from $Z = Z_1$ to $Z = Z_3$ is shown in Figure 4.18. Clearly the largest number of transitions occur at $Z = Z_1$ and Z_3. The curve, in fact, is reminiscent of the probability distribution curve.

Let us now calculate the probability that the system starting in state Z that lies between Z_1 and Z_3 will reach state Z_3 before reaching Z_1. The probability is plotted in Figure 4.19 for various starting values of Z for one value of $Z_0 = 1.01 \times 10^5$. It is evident from the figure that for the initial states lying close to state Z_1, this probability is very small. The probability begins to rise as the middle unstable state is approached and virtually reaches unity in a short interval. The probability for the subsequent initial states to end up in Z_3 is then very close to unity. The notable feature of this figure is the sharp rise of the probability from almost zero to unity in a narrow region around the unstable state Z_2. The feature remains essentially unaltered on changing Z_0. For higher values of Z_0 than those shown here the curve shifts to the left, while for lower values it moves to the right.

Of special interest is the situation where we begin at $Z = Z_2$. The logical question is whether the system would eventually evolve to state Z_1 or Z_3. For this purpose we can prepare plots of the type shown in Figure 4.19 to obtain the probability to move to state Z_3 if we begin at Z_2 for several values of Z_0. The probability of moving to state Z_3 if we begin at the unstable state for various values of Z_0 is plotted in Figure 4.20. Clearly evident from this figure is the fact that this probability is always larger than 0.5 within the range of Z_0, implying that in this range the system would always evolve to state Z_3. The probability of going to state Z_1 is of course one minus the probability of moving to state Z_3.

This result has an important implication, especially if we take a look

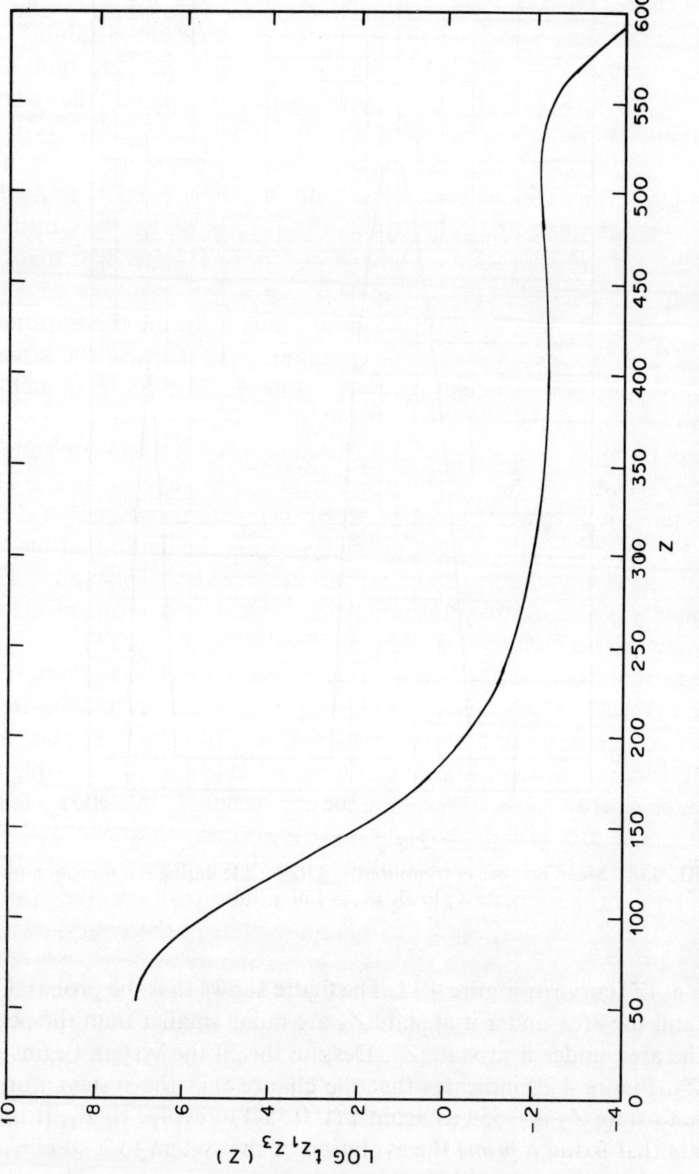

FIGURE 4.17 Mean time $[t_{Z_1 Z_3}(Z)]$ spent in state Z during passage from state Z_1 to Z_3

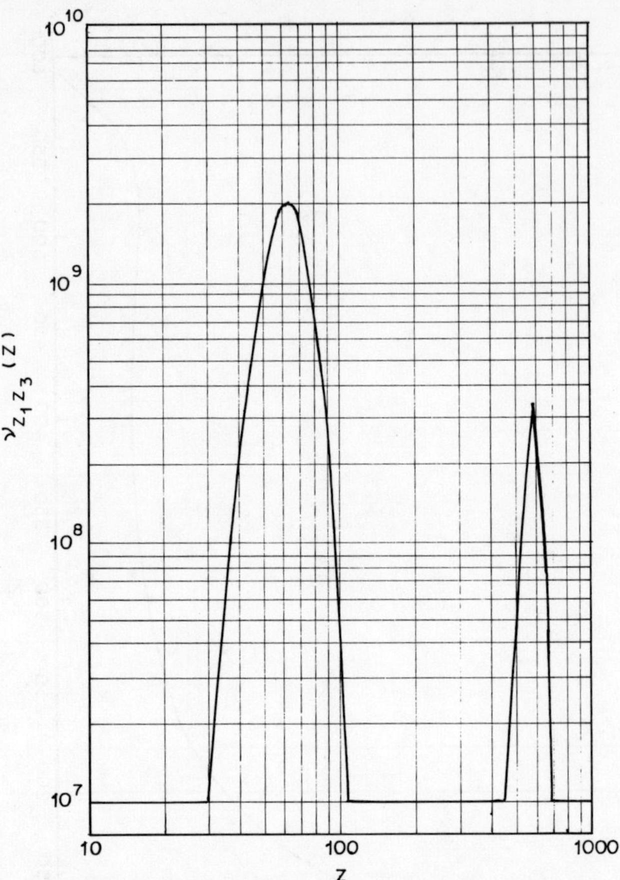

FIGURE 4.18 Mean number of transitions $\nu_{Z_1 Z_3}(Z)$ made during the transition from state Z_1 to Z_3 ($Z_0 = 1.01 \times 10^5$).

at the $p_\infty(Z)$ curve in Figure 4.12. The figure shows that the probability peak and the area under it at state Z_3 are much smaller than the peak and the area under it at state Z_1. Despite this, if the system begins at $Z = Z_2$, Figure 4.20 indicates that the chance that the system would evolve to state Z_3 is 0.554 as against (1–0.554) to evolve to Z_1. It thus appears that fixing *a priori* the evolution of the system to a state with larger probability peak could be wrong. In fact, as can be noted from the

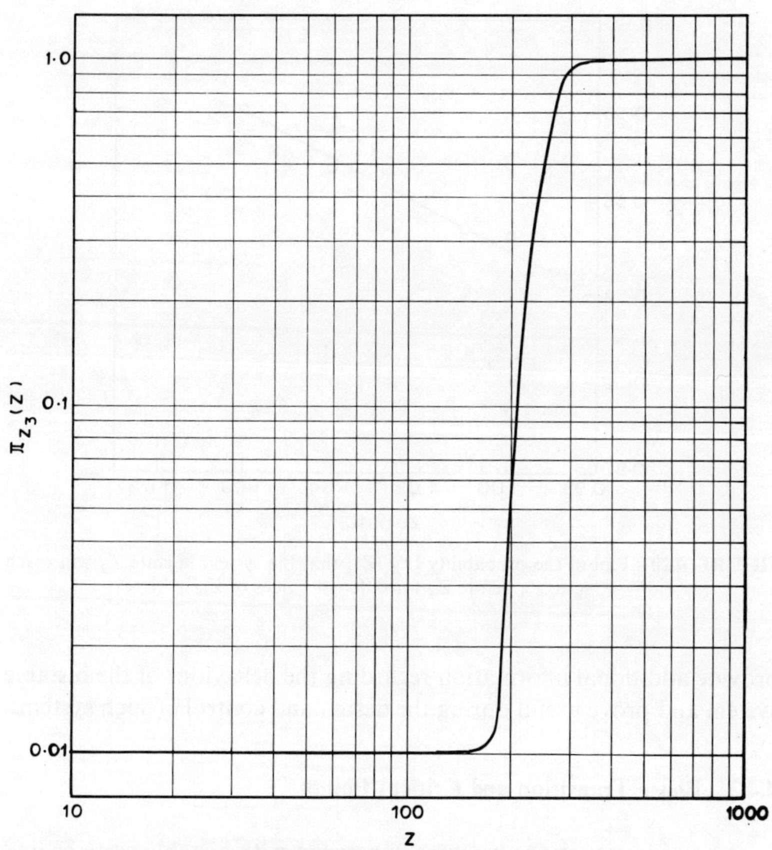

FIGURE 4.19 Plot of the probability $\pi_{Z_3}(Z)$ that the system starting in state $Z\varepsilon(Z_1, Z_3)$ will reach state Z_3 before it reaches state Z_1 ($Z_0 = 1.01 \times 10^5$).

equation, the splitting probability has nothing to do with $p_\infty(Z)$ and its features are governed solely by $T_\infty(Z)$.

The present example calculates the stochastic properties of a bistable system by considering the case of a simple chemical reaction on the catalyst surface. The properties such as mean time required to move from one state to another, the mean time spent in different states and the transitions $Z \rightarrow Z \pm 1$ made at any state Z during the transition, the probability of reaching one state before reaching another state, etc.

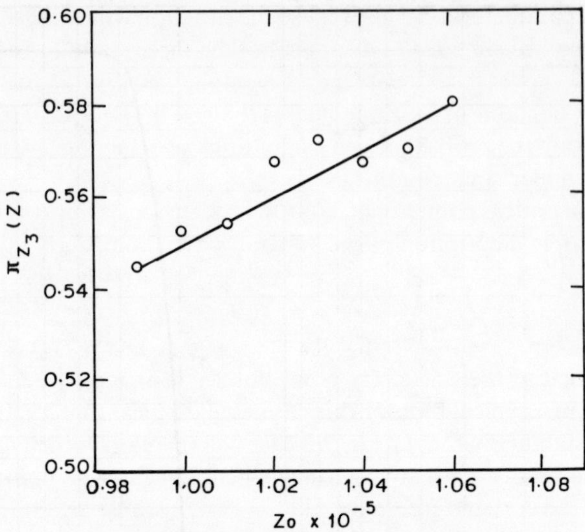

FIGURE 4.20 Plot of the probability $[\pi_{Z_3}(Z)]$ that the system in state Z_2 will reach state Z_3 before Z_1 for different values of Z_0.

provide additional information regarding the behaviour of the bistable system and prove useful during the design and control of such systems.

4.4.3 Phase Transition and Critical Points

In the previous sections we were concerned with bistable systems and analysed the problems of the transit time required by the system to move from one state to another in presence of fluctuations. Several variants of the problem, such as the passage time from one quasi-stationary state to another or from one quasi-stationary state to the middle unstable state or vice versa, were examined. In the present section we shall be concerned with another variant of the bistable system. In the beginning of Section 4.4 it was remarked that a well-defined bistable system represents a clear-cut separation between two time scales. The time scale on which the system approaches a quasi-stationary state starting within its region of attraction is much smaller compared with the time scale on which a giant fluctuation occurs. We also noted that these time scales can be altered, by altering the operational parameters of the system so that for a set of parameters one may realise the situation where the distinction between the two time scales disappears. The three states in the bistable system now move closer together and eventually merge into a single state which may be stable or unstable. This is the case of the critical region and our interest in the problem lies in evaluating the behaviour of the system operating under such conditions.

The term critical point applied to such a state seems appropriate—for a macroscopic description of the system written compactly in the form

$$\frac{d\phi}{dt} = \alpha_{1,0}(\phi) \tag{4.203}$$

indicates that at the stationary point both $\alpha'_{1,0}(\phi)$ and $\alpha''_{1,0}(\phi)$ are zero. Reference to a typical plot for this case would reveal that $\alpha_{1,0}(\phi)$ is also zero. The stability of the state in this case will therefore be decided by the next higher order derivative term. Depending on whether $\alpha'''_{1,0}(\phi)$ is greater or lower than zero, we would have an unstable or stable state. In the case of a critical point we would have a stable steady state.

In Section 4.2 we considered a case where $\alpha_{1,0}(\phi)$ takes a value equal to zero, and it was shown that the conventional expansion method needs modification. In situations of this type it was further shown that the first term in the expansion of the master equation is a nonlinear Fokker–Planck equation. The present case differs from that case in that the next two higher order derivatives are also zero. The fluctuations are thus characterised by a power different from what has been used in the conventional method.

To recapitulate, the conventional method presupposes a Gaussian distribution of fluctuations around the deterministic state $x = [\Omega\phi + \Omega^{1/2}y]$. In instances when $\alpha_{1,0}(\phi)$ becomes zero, we argued that the fluctuations would grow to the size of the system; x must therefore scale in inverse proportion to the size of the system. In the present case we assume an arbitrary power and define

$$x = \Omega\phi + \Omega^{\alpha}y \tag{4.204}$$

where α represents an arbitrary parameter with the only restriction of being less than unity. It has been suggested, on the basis of matching of terms in the expansion, that α should take a value of 3/4 (van Kampen, 1981). Using this relation and following the expansion procedure (see Appendix 3.A) the resulting dominant term can be written as a nonlinear

Fokker–Planck equation

$$\frac{\partial \pi}{\partial \tau} = \Omega^{-1/2}\left[\tfrac{1}{6}|\alpha'''_{1,0}(\phi)|\frac{\partial}{\partial y}\, y^3\pi + \tfrac{1}{2}\alpha_{2,0}(\phi)\frac{\partial^2 \pi}{\partial y^2}\right] \quad (4.205)$$

which can be solved using the methods discussed in Section 4.3.3. We note from this equation that the π distribution is no more Gaussian. Also the fluctuations at the critical point now show a higher dependency on the system size (3/4 as against 1/2 in the conventional method).

The equation describing the evolution of the system such as Equation (4.205) has been obtained on the assumption of vanishing first and second derivatives, i.e. at the critical point. For any other point where the conditions are not met, this equation therefore does not describe the behaviour of the system. Thus, for a point lying just above or below the critical point, this method would not give the right results and for such cases we must revert to the original method. The use of van Kampen's expansion method therefore envisages different transformations for different situations. The use of asymptotic representation of the master equation in the form of a nonlinear Fokker–Planck equation avoids this difficulty. The subject matter has already been discussed in Section 4.3.2, and we shall apply it to the case of a reacting system operating at the critical point.

A model reacting system that exhibits the phenomenon of bistability and critical point behaviour for certain ranges of parameter values is the Schlogl reaction scheme

$$A + 2X \rightleftharpoons 3X$$
$$X \rightleftharpoons B$$

The system has been analysed earlier from the deterministic viewpoint and operating conditions for the existence of multistable states and the critical point have been well established (Schlogl, 1972; Nicolis and Turner, 1977). Assuming that the concentrations of species A and B are controlled by an external agency so as to remain constant, the macroscopic equation describing the evolution of the species X can be written as

$$\frac{dx}{dt} = -k_2x^3 + k_1Ax^2 - k_3x + k_4B \quad (4.206)$$

where x denotes the macroscopically observed number of particles of species X, and the rate constants k_1 and k_2 involve the volume of the system so as to have the appropriate units to make Equation (4.206) dimensionally consistent. At steady state Equation (4.206), after appropriate scaling, can be written as

$$x^3 - \alpha x^2 + \gamma x - \beta = 0 \tag{4.207}$$

where

$$\alpha = k_3 A/k_2, \qquad \beta = k_4 B/k_2 \qquad \text{and} \qquad \gamma = k_3/k_2$$

The cubic Equation (4.207) is known to possess triple roots:

$$\alpha = (3\gamma)^{1/2}, \qquad \beta = \alpha^3/27 = \alpha\gamma/9 \tag{4.208}$$

Operating the system corresponding to the condition (4.208) therefore would ensure operation at the critical point. To make sure that this point is away from the equilibrium point, we take detailed balancing of the two steps in the reaction scheme, which at equilibrium indicates that

$$\alpha_{eq} x^2 = x^3 \qquad \text{and} \qquad \beta_{eq} = \gamma_{eq} x \tag{4.209}$$

Eliminating x from these equations leads to $\beta_{eq} = \alpha_{eq}\gamma_{eq}$, which contradicts the second condition in Equation (4.208). In operating the system at points corresponding to condition (4.208) we therefore have a system that is operated far from the equilibrium state, and our interest lies in evaluating the behavioural features of the system around this point. To study this, we redefine the parameters in Equation (4.206) as

$$k_2 A^2 = 1, \qquad k_1 A^2 = 3, \qquad k_4 = 1, \qquad k_3 = 3 + \delta$$

and

$$B = A(1 + \delta') \tag{4.210}$$

which can be written as

$$x^3 - 3Ax^2 + (3 + \delta)A^2 x + (1 + \delta')x^3 = 0 \tag{4.211}$$

and admits a triple root $x = A$ when both δ and δ' take values equal to zero. In order to study the behaviour near the triple root we set $x = A(1 + y)$, which converts Equation (4.211) into a form

$$y^3 + \delta(y + 1) = \delta' \tag{4.212}$$

that resembles the van der Waals equation near the critical point. Equation (4.212) possesses one real solution $y = 0$ for $\delta = \delta' > 0$. For the situation when $\delta' < 0$, the equation, however, yields the additional roots,

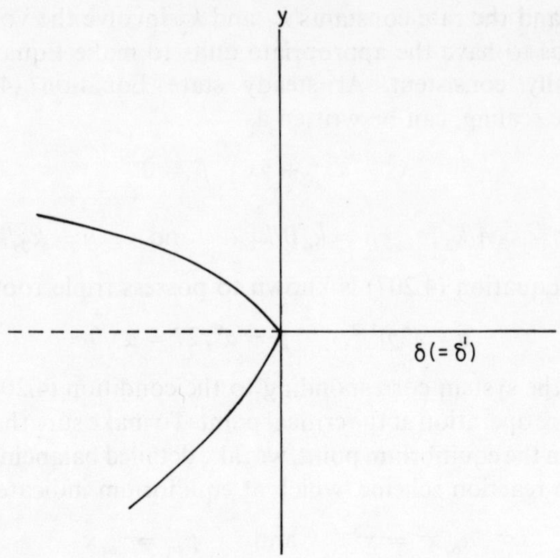

FIGURE 4.21 Bifurcation plot for the Schlogl's model

$y = \pm\sqrt{\delta}$, indicating multiplicity. The typical bifurcation plot is shown in Figure 4.21 where the stability of the branches, as indicated by linear stability analysis, is also indicated by broken (unstable) and continuous (stable) lines.

The question that concerns us here is the way in which the behaviour near this point is influenced by the presence of fluctuations. Clearly we need to formulate the master equation corresponding to this situation which, for the reaction scheme analysed, can be written as

$$\frac{\partial p(X,t)}{\partial t} = (E-1)[X(X-1)(X-2) + (3+\delta)A^2X]p(X,t)$$

$$+ (E^{-1}-1)[3AX(X-1) + (1+\delta')A^3]p(X,t) \tag{4.213}$$

or, using the generating function $[\psi(s) = \sum_X s^X p(X,t)]$, as

$$s^2\left(\frac{d^3\psi}{ds^3} - 3A\frac{d^2\psi}{ds^2}\right) + (3+\partial)A^2\frac{d\psi}{ds} - (1+\delta')A^3\psi = 0 \tag{4.214}$$

This affords an easy way to compute the moments of the probability distribution (see Chapter 1). The solution to this third-order differential equation has been obtained by Nicolis and Turner (1977) using the method of Laplace transforms. The solution expressed in terms of the hypergeometric series and Barne's type of integral takes the formal form

$$\psi(s) = \text{constant} \int_C dt\, f(t) \exp[V\alpha g(t, s)] \tag{4.215}$$

where

$$f(t) = t^{-1/2}\xi^2(t^2 + \lambda^2)/[\lambda^2(t^2 + \xi^2)] + O(V)V^{-1} \tag{4.216}$$

and

$$\begin{aligned} g(t, s) = t - t\ln t + t\ln 3 + (t - i\xi)\ln(t - i\xi) \\ + (t + i\xi)\ln(t + i\xi) - (t - i\lambda)\ln(t - i\lambda) \\ - (t + i\lambda)\ln(t + i\lambda) + t\ln s \end{aligned} \tag{4.217}$$

The parameters ξ and λ appearing in these equations are defined as

$$\begin{aligned} \xi^2 &= (1 + \delta')/3 - 1/(4\alpha^2 V^2) \\ \lambda^2 &= 3 + \delta - 1/(4\alpha^2 V^2) \end{aligned} \tag{4.218}$$

The asymptotic behaviour of the variance $\langle(\delta X)^2\rangle$ can be calculated using Equation (4.215) for large values of V at, below or above the critical point. Nicolis and Turner (1977) suggest a modified method of steepest descent for this purpose, leading to

$$\langle(\delta X)^2\rangle = \begin{cases} (4/\delta + 1)\alpha V + O(1), & \delta' = \delta > 0 \\ 4(\alpha V)^{3/2}\Gamma\tfrac{3}{4}/\Gamma\tfrac{1}{4} + O(V), & \delta' = \delta = 0 \\ (-2/\delta - 3/\sqrt{-\delta} + 1)\alpha V + O(1), & -1 < \delta' = \delta < 0 \end{cases} \tag{4.219}$$

At the critical point ($\delta = \delta' = 0$) the results diverge as $\langle(\delta X)^2\rangle$ is not of the order of α.

Equation (4.219) represents the exact results derived from the master equation. As is known, except for some model systems, the exact solution of the master equation is generally difficult. It is necessary then to devise alternative methods such as systematic approximation schemes to obtain the desired results. One such scheme employed by Nicolis and Turner (1977) uses the singular perturbation analysis where the original master equation, Equation (4.213) or (4.214), is converted into an alternative form

$$\varepsilon^2 q'' + 3\varepsilon q'(q - 1) + q^3 - 3q^2 + \frac{3 + \delta}{s^2} q - \frac{1 + \delta}{s^2} = 0 \tag{4.220}$$

using the cumulant generating function $F(s) = e^{\alpha f(s)}$. The various terms appearing in Equation (4.220) have been defined as

$$q = \frac{df}{ds} = f', \qquad \delta = \delta' \qquad \text{and} \qquad \varepsilon = 1/\alpha \tag{4.221}$$

Clearly, for large values of α, Equation (4.220) defines a singular perturbation problem, and setting $\varepsilon = 0$ we can obtain the outer solution that yields exactly the same information regarding the variance as given by Equation (4.219). The outer solution, of course, breaks down at the critical point ($\delta = 0$) and we need to construct an inner solution valid in the near vicinity of ($\varepsilon = 0$, $\delta = 0$). The method uses a straightforward application of singular perturbation analysis and gives identical results, and can be used in cases where the exact solution of the master equation presents difficulties.

The analysis indicates that the variance of X as $\delta \to 0$ diverges as inverse of δ in agreement with mean field theories. However, at the bifurcation point the variance behaves in a qualitatively different way. In the region of multiple states ($\delta < 0$) the variance assumes finite values, implying that the fluctuations are as important as averages. The phenomenological equation therefore differs from the equation for the averages.

We remarked in Section 4.3.2 that the general criticism regarding the inconsistency of the Fokker–Planck equation as obtained by the Kramers–Moyal expansion of the master equation can be overcome by formulating a Fokker–Planck equation employing alternative schemes. Two alternative ways of formulating the Fokker–Planck equation were discussed in Sections 3.2.3 and 4.3.2. We shall employ these Fokker–Planck equation formulations to treat the present case. The exact results being already known, the application would provide a test of the applicability of these formulations to such situations and, if found acceptable, should also confirm the use of these equations as alternative ways for solving unknown situations.

We refer to the master Equation (4.213) describing the evolution of the system under study, and using the methodology presented in Section 4.3.2 write the corresponding Fokker–Planck equation as

$$\frac{\partial p(x,t)}{\partial t} = \frac{\partial}{\partial x}[x^3 - 3x^2/V - 3\alpha x^2 + 3\alpha x/V + (3+\delta)\alpha^2 x - (1+\delta')\alpha^3]p(x,t$$

$$+\frac{1}{2V}\frac{\partial^2}{\partial x^2}[x^3 + 3\alpha x^2 + (3+\delta)\alpha^3 x + (1+\delta')\alpha^2]p(x,t) \tag{4.222}$$

The stationary solution to this equation can be obtained as

$$p_s(x) = N\tilde{f}(x)\exp[V\tilde{g}(x)] \tag{4.223}$$

where N refers to the normalisation constant and $\tilde{f}$ and $\tilde{g}$ are defined as

$$\tilde{f}(x) = [f_1(x)]^{-1}\exp 2\int\frac{(3x^2 - 3\alpha x)\,dx}{f_1(x)}$$

$$f_1(x) = [x^3 + 3\alpha x^2 + (3+\delta)\alpha^2 x + (1+\delta')\alpha^3] \tag{4.224}$$

$$\tilde{g}(x) = 2\int\frac{-x^3 + 3\alpha x^2 - (3+\delta)\alpha^2 x + (1+\delta')\alpha^3}{f_1(x)}\,dx$$

Using the stationary solution we can calculate the variance $\langle(\delta X)^2\rangle$ for large V for the same situation as in the master equation formalism. Using again the method of steepest descent we obtain exactly results as given by Equation (4.219). The simple example therefore illustrates the soundness of the Fokker–Planck approximation as envisaged in Section 4.3.2 and suggests its use as a right alternative in cases where the solution to the original master equation poses problems.

An alternative form of the Fokker–Planck equation has been derived by Grabert, Hanggi and Oppenheim (1983) on the basis of the nonlinear transport theory and the formulation is discussed in Section 3.2.3. We can formulate this Fokker–Planck equation corresponding to the master Equation (4.213) by referring to the prescriptions presented in that section. Hanggi *et al.* (1984) have employed such a formulation to test its use in describing the behaviour near the phase transition point in the Schlogl reaction scheme. The results indicate the usefulness of this approach as an alternative equivalent description of the phenomenon in terms of the master equation.

Several studies on the behaviour of systems at or near the phase transition points have been reported so far. Most of these studies consider the basic models of Schlogl (1972) that give the chemical analogue for first- and second-order phase transitions. Some studies relating to the behaviour near the critical point have already been mentioned. Some others such as those of Grossmann and Schranner (1978) use a master equation formalism that is solved using the projector operator technique discussed in Appendix 4.A, to obtain dynamical correlations of the particle number $\langle x \rangle$. The mean particle number $\langle x \rangle$ and the variance are plotted in Figure 4.22 for variuous parameters of the system. It is seen that for Schlogl's model I for $(k_2k_3/k_1k_4) =$ 1, 10, 1000, the mean particle number and other higher cumulants are simply given by (k_1p_A/k_2). Also, for small values of (k_1p_A/k_4) [lying between 0 and 1] we have $\langle x \rangle \simeq (k_3p_A/k_4)$, while for $(k_1p_A/k_4) > 1$ we have $\langle x \rangle = (k_1p_A/k_2 + k_4k_2)$. In the intermediate regime the mean particle number changes appreciably with the ratio (k_1p_A/k_2), and memory effects are expected to be important.

For the second-order model of Schlogl (first-order transition) Figure 4.23 shows the mean particle number and variance as functions of the system parameter. It is seen that for $(k_1k_4/k_2k_3) = 1$ chemical equilibrium exists and the mean particle number and all higher cumulants are simply given by $(k_3k_4/k_1k_2)(k_1^3k_3/k_2k_4^3)p_A$. The phase transition occurs around $(k_1^3k_3/k_2k_4^3)p_A = 1$, and for $(k_1k_4/k_2k_3) = 3$ a critical point is reached. For smaller values of $(k_1^3k_3/k_2k_4^3)p_A$ the mean particle number is given by

$$\langle x \rangle = \frac{k_1^3k_3}{k_2k_4^3}p_A\left[\frac{k_3k_4}{k_1k_2}\right]^{1/4} \Bigg/ \left[\frac{k_1k_4}{k_2k_3}\right]^{1/2} \tag{4.225}$$

while for larger values by

$$\langle x \rangle = \frac{k_1^3k_3}{k_2k_4^3}p_A\left[\frac{k_3k_4}{k_1k_2}\right]^{1/4}\left[\frac{k_1k_4}{k_2k_3}\right]^{1/2} \tag{4.226}$$

For intermediate values of $[k_1^3k_3/k_2k_4^3]p_A$ the mean particle number changes linearly and the slope increases with increase in $[k_3k_4/k_1k_2]^{1/4}$ and shows a discontinuity similar to first-order transitions for very high values of $[k_3k_4/k_1k_2]^{1/4}$.

Schlogl's models have also been analysed by Horsthemke and Brenig (1977) and more recently by Hanggi *et al.* (1984) using alternative forms of the Fokker–Planck equation.

Studies relating to reaction schemes other than Schlogl's models have also been analysed to confirm the behaviour at the phase transition points. Thus Schnakenberg (1980) considered a two-component case

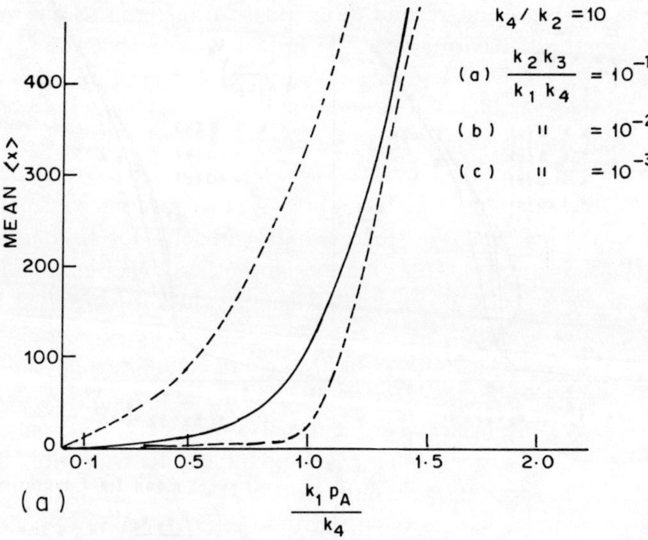

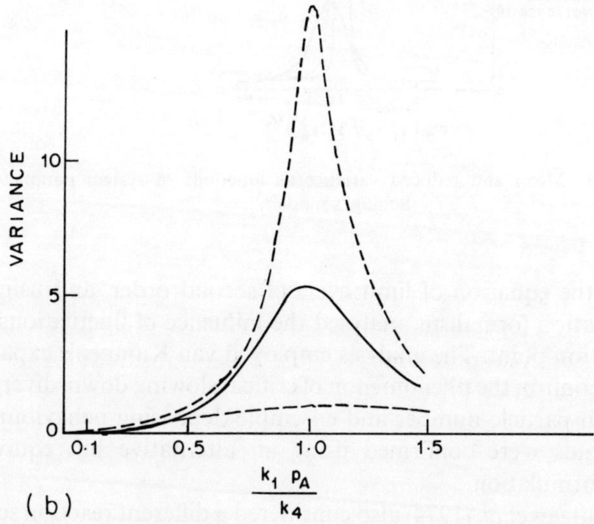

FIGURE 4.22 Mean and reduced variance as functions of system parameters for Schlogl's model

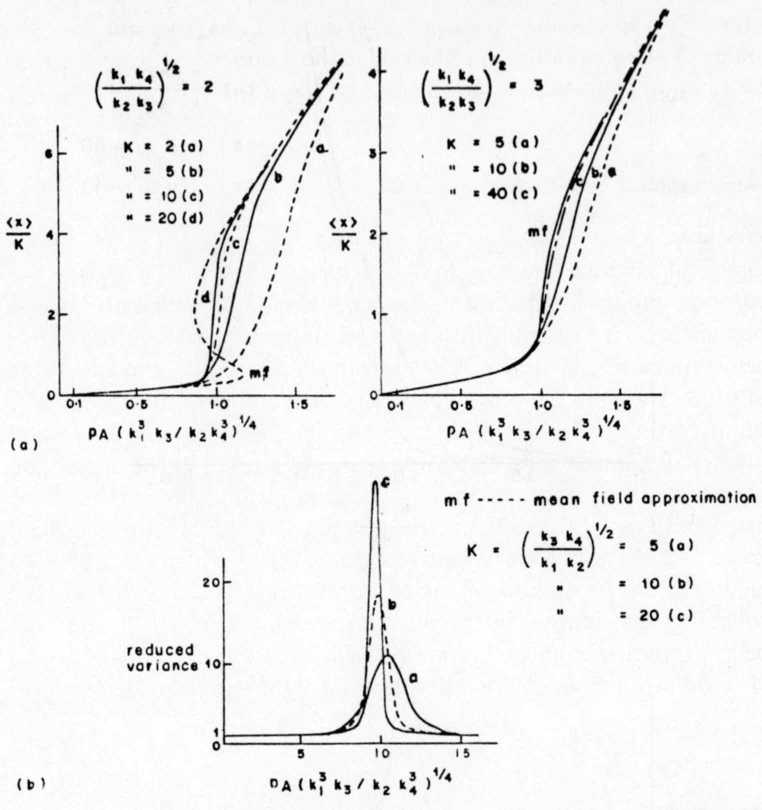

FIGURE 4.23 Mean and reduced variance as functions of system parameters for Schlogl's model 2.

leading to the equation of limit cycle of second order, and using the master equation formalism analysed the influence of fluctuations near the bifurcation point. The analysis employed van Kampen's expansion method to confirm the phenomenon of critical slowing down, divergence or the mean particle number and ensemble dephasing behaviour. The general trends were confirmed using an alternative but equivalent Langevin formulation.

Earlier Nitzan *et al.* (1974) also considered a different reaction scheme obeying the macroscopic steady state equation $\mu x + \lambda - x^3 = 0$. The scheme exhibits both hard and soft transitions for appropriate parameter values. The analysis employing the master equation and Fokker–Planck equation obtained using the Kramers–Moyal technique results in a divergence of the variance at the critical point.

4.4.4 General Comments

Before we close the discussion on bistable systems, it would be worthwhile to draw attention to the fact that such systems can serve as a prototype model for describing several other and apparently diverse phenomena. To illustrate this point we refer to Kramers' theory of chemical kinetics (Kramers, 1940) where the reaction is modelled as a diffusion process in the phase space. The reactants are considered to be noninteracting particles subjected to the external force derived from a potential and to the Brownian force arising as a result of the interaction of the particles with the surrounding medium. It is assumed that all the particles are initially confined to the potential well and their subsequent escape over the potential barrier represents the chemical reaction. The reaction rate as per this model can be calculated from a knowledge of the probability distribution in the phase space as a function of time that obeys a typical nonlinear Fokker–Planck equation.

For diffusion in an external potential $\phi(x)$ the equation has the form

$$\frac{\partial p(x, t)}{\partial t} = \frac{\partial}{\partial x_1} \phi'(x)p + \tfrac{1}{2}\alpha_2 \frac{\partial^2 p}{\partial x^2}, \qquad \alpha_2 = k_B T \tag{4.227}$$

When the potential $\phi(x)$ represents a potential with two wells as shown in Figure 4.8b, the analogy with the discussion on bistable systems presented thus far becomes extremely clear. The equation also arises in the description of several other phenomena such as superfluorescence, laser activity, spinodal decomposition and also in some of the problems in electronics and biology.

The simplest bistable system containing a one-dimensional potential of two wells separated by an unstable maximum thus serves as a prototype model for description of several apparently diverse phenomena. The model is ideal for studying systems initially poised near the unstable point and the general problem has been extensively investigated in this spirit. For a system initially in an unstable state its further passage is governed solely by the stochastic forces leading to a rapid growth in the fluctuations of the observable quantity. In the intermediate regime the fluctuations and the macroscopic observables tend to macroscopic order, while in the final time regime where the system approaches the stable stationary state the fluctuations shrink to a finite microscopic level.

A complete solution of the equation covering all the three time regimes is possible only for a few forms of the potential $\phi(x)$. In most other cases, approximations suitable to each time regime are employed to obtain the time development of $p(x, t)$. For more recent developments in this direction we shall merely refer to the literature (Weiss, 1982; Suzuki, 1979, 1983; Weaver, 1982; Hongler and Zheng, 1982; Banerjee, Bhattacharjee and Mani, 1984).

4.5 Application to Unstable Situations (Limit Cycle Behaviour)

In Section 4.4 we discussed the general methodology for the stochastic analysis of reacting systems characterised by complexity in the form of multistationarity. In the present section we shall consider a situation, the macroscopic equations of which indicate the existence of an instability. We shall specifically be concerned with situations where a unique stationary state that is unstable prevails. An important difference between the cases treated in the earlier section and the cases to be considered now is that the phenomenon of multistationarity can be exhibited by systems described by a single variable, while for limit cycles to appear consideration of multivariable systems becomes essential. In Section 4.2 we noted that the general method of expansion can be applied to multivariable cases provided the solution is unique and stable. In the event such a solution becomes unstable we would logically like to enquire whether the method is suitable.

The contention is best illustrated by considering a specific example of reaction on a catalyst surface where components A and B are adsorbed and then interact with each other in presence of surface heterogeneity effects. Mechanically, the system can be described as

$$A(g) + [Z] \underset{k_{-1}}{\overset{k_1}{\rightleftharpoons}} [AZ]$$

$$B(g) + [Z] \underset{k_{-2}}{\overset{k_2}{\rightleftharpoons}} [BZ]$$

$$[AZ] + [BZ] \xrightarrow{k_3} 2[Z] + \text{Products}$$

and leads to the following nondimensional mass balance equations under isothermal conditions:

$$\frac{dx}{dy} = a_1(1 - x - y) - b_1 x - xy\, e^{-\mu y} \tag{4.228}$$

$$\frac{dy}{dt} = a_2(1 - x - y) - b_2 y - xy\, e^{-\mu y} \tag{4.229}$$

where the coefficients a_1, b_1, a_2 and b_2 involve the rate constants of the appropriate steps and are defined as

$$a_1 = k_1[A(g)]/k_3^0, \qquad b_1 = k_{-1}/k_3^0$$

$$x = [AZ]/L, \qquad y = [BZ]/L \tag{4.230}$$

$$a_2 = k_2[B(g)]/k_3^0, \qquad b_2 = k_{-2}/k_3^0, \qquad L = [Z] + [AZ] + [BZ]$$

and

$$k_3 = k_3^0 \exp[-\mu y] \tag{4.231}$$

The set of Equations (4.228) and (4.229) have been analysed by Pikios and Luss (1977) and a criterion for the occurrence of sustained oscillations has been derived.

The present development starts with this simple model, and Figure 4.24 sketches the variation in the steady state solution (x_s) as a function of the system parameters. Two parameters a_1 and μ are considered; however, one could choose any other parameter such as a_2, b_1, b_2, etc. The important point to note here is that, on progressive decrease in μ, the steady state solution suddenly jumps at a critical value ($\mu = \mu_c = 12.538$) where the stability of the solution also undergoes a change. Further, the solution being unique all through the parameter range, this figure indicates that a critical value of μ exists for defined values of other parameters where a unique stable solution changes over to a unique unstable solution. This soft mode instability occurs at a critical value of μ and the phenomenon typically resembles second-order phase transition. A similar case was investigated in the previous section. Our intention in the present section has been to see the effects of fluctuations on the behaviour of the system operating near the critical point ($\mu_c < \mu = 14$) in the region where periodic solution is obtained.

We employ the master equation approach where the situation is viewed as consisting of a population of species x and y, which have a certain probability of getting consumed either through their own reaction or due to interaction with each other. The joint probability distribution $P(x, y, t)$ is written following Tambe, Kulkarni and Doraiswamy (1985a):

$$\begin{aligned}\frac{dP(x, y, t)}{dt} &= \frac{a'_1}{L}(E_x^{-1} - 1)P + \frac{a'_1 + b'_1}{L}(E_x - 1)xP \\ &\quad + \frac{a'_1}{L}(E_x - 1)yP + (E_x - 1)xy\, e^{-\mu y}P \\ &\quad + \frac{a'_2}{L}(E_y^{-1} - 1)P + \frac{a'_2}{L}(E_y - 1)xP \\ &\quad + \frac{a'_2 + b'_2}{L}(E_y - 1)yP + (E_y - 1)xy\, e^{-\mu y}P\end{aligned} \tag{4.232}$$

where E_x and E_y refer to the difference operators acting on x and y, respectively.

The original master equation in its exact form is difficult to solve, and it is necessary to have an approximation scheme in the form of a power series expansion in some physical parameter. The parameter L refers to the total number of sites on the surface and therefore in some sense measures the size of the system. The expansion method of van Kampen (1976) can then be used to solve Equation (4.232). For this purpose we define the transformations

$$x = L\phi(t) + L^{1/2}n \qquad \text{and} \qquad y = L\psi(t) + L^{1/2}m \tag{4.233}$$

where n and m now refer to the new variables and $\phi(t)$ and $\psi(t)$ can be appropriately chosen. The first part of these equations refers to the macroscopic part, while the second refers to the fluctuating parts of x and y. According to transformations (4.233) the probability distribution of x and y now becomes the probability distribution of n and m which can be differentiated to obtain

$$\frac{\partial P(x, y, t)}{\partial t} = \frac{\partial \pi}{\partial t} - L^{1/2}\frac{\partial \phi}{\partial t}\frac{\partial \pi}{\partial x} - L^{1/2}\frac{\partial \psi}{\partial t}\frac{\partial \pi}{\partial y} \tag{4.234}$$

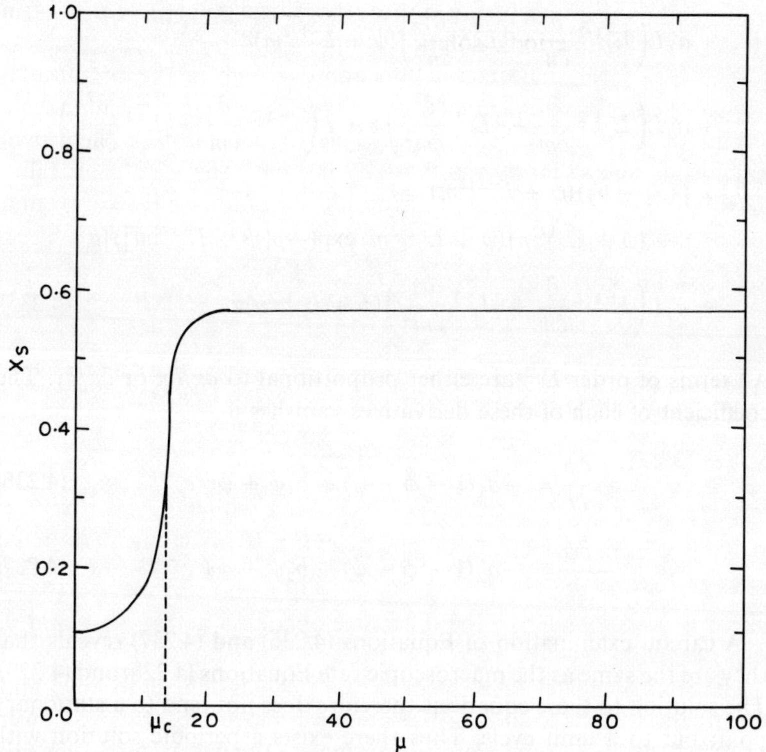

FIGURE 4.24 Evolution of steady state solution (X_s) as a function of parameter μ ($a_1 = 0.0084$, $a_2 = 0.0084$, $b_1 = 0.0007$, $b_2 = 0.0011$).

Replacing Equation (4.234) for $\partial P/\partial t$ in Equation (4.232) and making use of Equation (4.233) leads to the new distribution $\partial \pi/\partial t$:

$$\frac{\partial \pi}{\partial t} - L^{1/2}\frac{\partial \phi}{\partial t}\frac{\partial \pi}{\partial x} - L^{1/2}\frac{\partial \psi}{\partial t}\frac{\partial \pi}{\partial y}$$

$$= a_1' L\left(-L^{-1/2}\frac{\partial}{\partial n} + \tfrac{1}{2}L^{-1}\frac{\partial^2}{\partial n^2}\right)\pi + L\left(L^{-1/2}\frac{\partial}{\partial n} + \tfrac{1}{2}L^{-1}\frac{\partial^2}{\partial n^2}\right)$$

$$\times \left[(a_1' + b_1')(\phi + L^{-1/2}n)\right.$$

$$\left. + (\phi + L^{-1/2}n)(\psi + L^{-1/2}m)\exp(-\mu[\psi + L^{-1/2}m])\right]\pi$$

$$+ a_1' L\left(L^{-1/2} \frac{\partial}{\partial n} + \tfrac{1}{2} L^{-1} \frac{\partial^2}{\partial n^2} \right)(\psi + L^{-1/2} m)\pi$$

$$+ a_2' L\left(L^{-1/2} \frac{\partial}{\partial m} + \tfrac{1}{2} L^{-1} \frac{\partial^2}{\partial m^2} \right)\pi + L\left(L^{-1/2} \frac{\partial}{\partial m} + \tfrac{1}{2} L^{-1} \frac{\partial^2}{\partial m^2} \right)$$

$$\times [(a_2' + b_2')(\psi + L^{-1/2} m)$$

$$+ (\phi + L^{-1/2} n)(\psi + L^{-1/2} m) \exp(-\mu[\psi + L^{-1/2} m])]\pi$$

$$+ a_2' L\left(L^{-1/2} \frac{\partial}{\partial m} + \tfrac{1}{2} L^{-1} \frac{\partial^2}{\partial m^2} \right)(\phi + L^{-1/2} n)\pi \tag{4.235}$$

All terms of order $L^{1/2}$ are either proportional to $\partial\pi/\partial x$ or $\partial\pi/\partial y$. The coefficient of each of these derivatives vanishes if

$$-\frac{\partial \phi}{\partial t} = -a_1'(1 - \phi - \psi) + b_1'\phi + \phi\psi\, e^{-\mu\psi} \tag{4.236}$$

$$-\frac{\partial \psi}{\partial t} = -a_2'(1 - \phi - \psi) + b_2'\psi + \phi\psi\, e^{-\mu\psi} \tag{4.237}$$

A careful examination of Equations (4.236) and (4.237) reveals that they are the same as the macroscopic rate Equations (4.228) and (4.229). The solution to these equations therefore does not tend to a stationary point but to a limit cycle. Thus there exists a periodic solution with period T for the system. The terms of order L^0 in the expansion of Equation (4.235) give the evolution of fluctuations as

$$\frac{\partial \pi}{\partial t} = \frac{\partial}{\partial n}\{(a_1' + b_1')n + n\psi + m\phi\, e^{-\mu\psi} + a_1' m\}\pi$$

$$+ \frac{1}{2}\frac{\partial^2}{\partial n^2}\{a_1' + (a_1' + b_1')\phi + \phi\psi\, e^{-\mu\psi} + a_1'\psi\}\pi$$

$$+ \frac{\partial}{\partial m}\{(a_2' + b_2')m + n\psi + m\phi\, e^{-\mu\psi} + a_2' n\}\pi$$

$$+ \frac{1}{2}\frac{\partial^2}{\partial m^2}\{a_2' + (a_2' + b_2')\psi + \phi\psi\, e^{-\mu\psi} + a_2'\phi\}\pi \tag{4.238}$$

The equation for fluctuations contains the second derivative terms which tend to spread out the distribution π. In case the macroscopic

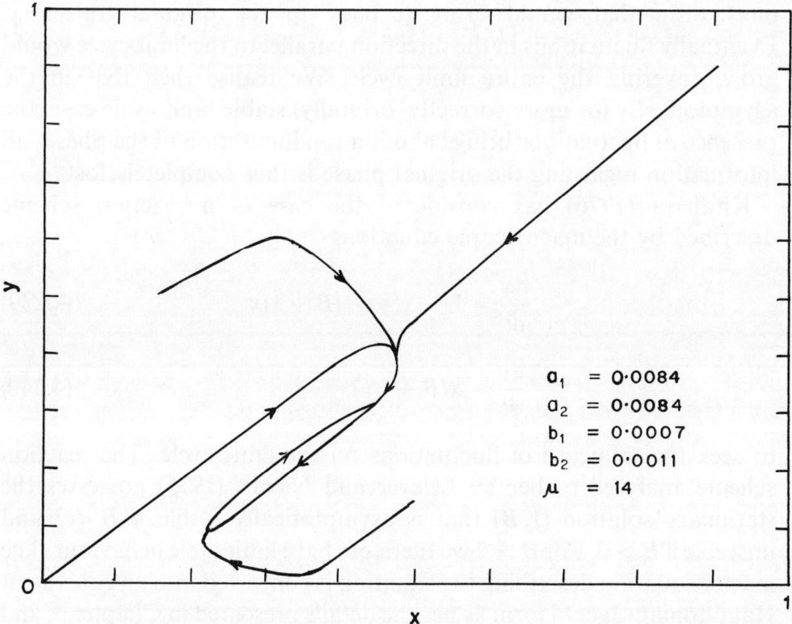

FIGURE 4.25 Phase plane plot for different initial conditions.

solution is asymptotically stable the first-order derivatives in the equation annul this effect, thus restoring the stability. In our present example, where limit cycle exists, the solutions are asymptotically stable—but not in the usual sense. To explain this point let us take the phase–plane plot of the macroscopic equation shown in Figure 4.25. Even if we were to start at any arbitrary initial point, the trajectory would eventually end up in the orbit defining the limit cycle. This implies that all the solutions are asymptotically stable with the same time period and amplitude of oscillations. However, depending on the initial point, between two solutions a phase difference exists and this phase difference would never die out.

Thus, if the fluctuations described by Equation (4.238) were such as to shift the trajectory perpendicular to the direction of the limit cycle, the asymptotic stability of the solution would ensure that the fluctuations would die out. However, if the fluctuations were to shift the trajectory in the direction parallel and along the limit cycle, there exists no mechanism that would bring it back to the original trajectory. Eventually fluctuations in the direction parallel to the limit cycle would grow, covering the entire limit cycle. We realise then that in the asymptotically (or more correctly, orbitally) stable limit cycle case, the presence of fluctuations brings about a randomisation of the phase; all information regarding the original phase is thus completely lost.

Kitahara (1976) has considered the case of a reaction scheme described by the macroscopic equations

$$\frac{dx}{dt} = 1 + x^2y - (B + 1)x \tag{4.239}$$

$$\frac{dy}{dt} = x(B - xy) \tag{4.240}$$

to seek the influence of fluctuations on the limit cycle. The reaction scheme analysed earlier by Lefever and Nicolis (1975) possesses the stationary solution $(1, B)$ that is asymptotically stable if $B < 2$ and unstable if $B > 2$. For $B > 2$ we therefore have limit cycle behaviour. The master equation describing the situation is converted into an equivalent Hamiltonian–Jacobi form as per the details presented in Chapter 3, and the resulting equations solved to confirm the randomisation of phase in such systems.

Other techniques discussed in this and the previous chapters can also be used, although no explicit use of these methods for this situation seems to be available in the literature.

First Passage Problems

In the present section, we shall consider the mean first passage time and mean external values of processes that describe the linear or nonlinear oscillating systems. The problem has already been discussed in Section 4.4.1, where however the treatment was necessarily restricted to single variable cases. The application of theory presented there to cases exhibiting limit cycle behaviour requires its extension to multivariable cases.

The simplest macroscopic description of an oscillating system requires the consideration of two first-order differential equations which may be combined into a single variable second-order differential equation. A typical form of such an equation would be

$$\frac{d^2x}{dt^2} + \mu F[x, dx/dt] + G(x) = 0 \tag{4.241}$$

Our interest in the problem lies in evaluating the mean passage time and the extremal values of the system when it is disturbed externally by, say, a white noise. Equation (4.241) with the suitable noise term appended to the rhs can be written as

$$\frac{dx}{dt} = q, \qquad \frac{dq}{dt} = -\mu F(x, q) - G(x) + f(t) \tag{4.242}$$

or, in terms of the Fokker–Planck equation, as

$$\frac{\partial p(x, q, t)}{\partial t} = \left[-q\frac{\partial}{\partial x} + \frac{\partial}{\partial q}[\mu F(x, q) + G(x)] + \varepsilon \frac{\partial^2}{\partial q^2} \right] p(x, q, t) \tag{4.243}$$

where 2ε is the white noise spectral density. The solution to this bivariate form often poses considerable problems. It is possible, however, to give an approximate description of the equation in terms of a single variable. For this purpose, we define

$$E = \frac{1}{2}\left(\frac{dx}{dt}\right)^2 + u(x)$$

$$u(x) = \int G(x)\, dx \tag{4.244}$$

and write Equation (4.243) as

$$\frac{\partial p(E, t)}{\partial t} = \left[\frac{\partial}{\partial E}\left\{ \mu \frac{\psi(E)}{\xi(E)} - \varepsilon \right\} + \varepsilon \frac{\partial^2}{\partial E^2} \frac{\phi(E)}{\xi(E)} \right] p(E, t) \tag{4.245}$$

where

$$\xi(E) = \frac{1}{2}\int_{E > u(x)} \frac{dx}{E - u(x)}$$

$$\phi(E) = \int_{E > u(x)} dx \sqrt{E - u(x)} \tag{4.246}$$

$$(E) = \frac{1}{\sqrt{2}} \int_{E > u(x)} F(x, \sqrt{2(E - u)})\, dx$$

Equation (4.245) is now of the form that has been treated in Section 4.4.1.

The equation is more convenient when the $F(x, dx, dt)$ term in Equation (4.243) is linear while the $G(x)$ term takes a nonlinear form. Seshadri, West and Lindenberg (1980) have used this equation for varying nonlinearities of the function $G(x)$ to obtain the mean first passage time to a large threshold value.

In case the damping force $F(x, dx, dt)$ is nonlinear, it is convenient to define an alternative transformation such as

$$A = (x^2 + (dx/dt)^2/\omega_0^2)^{1/2} \tag{4.247}$$

where the restoring force is taken to be harmonic as $\omega_0^2 x$ and rewrite the Fokker–Planck equation (4.243) as

$$\frac{\partial p(A,t)}{\partial t} = \left\{\frac{\partial}{\partial A}\left[\mu\frac{N(A)}{\omega_0} - \frac{\varepsilon}{2A\omega_0^2}\right] + \frac{\varepsilon}{2\omega_0^2}\frac{\partial^2}{\partial A^2}\right\} p(A,t)$$

where

$$N(A) = -\frac{1}{2\pi}\int_0^{2\pi} F(A\cos\theta, \omega_0 A\sin\theta)\sin\theta\, d\theta \tag{4.248}$$

Equation (4.248) has also been used by Seshadri, West and Lindenberg (1980) to obtain the mean passage time and moments.

Equation (4.247) or (4.248) corresponds to the approximate single variable formulation of the two variable system. The methods deployed in Section 4.4.1 can then be used to obtain the desired information. We shall illustrate the use of this technique by the following example of oscillating chemical systems.

We consider the case of a simple chemical oscillator described by the reaction sequence

$$A + S \rightleftharpoons AS$$

$$AS + B \rightleftharpoons ABS$$

$$ABS + 2S \rightleftharpoons 3S + \text{Product}$$

This simple kinetic scheme analysed earlier by Takoudis, Schmidt and Aris (1981) can be written in dimensionless form as

$$\frac{d\phi_1}{dt} = \alpha_1(1 - \phi_1 - \phi_2) - \gamma_1\phi_1 - \alpha_2\phi_1 + \gamma_2\phi_2 \tag{4.249}$$

$$\frac{d\phi_2}{dt} = \alpha_2\phi_1 - \gamma_2\phi_2 - \phi_2(1 - \phi_1 - \phi_2)^2 \tag{4.250}$$

where α, γ and the ϕ's are dimensionless parameters appropriately defined. It is possible to eliminate ϕ_2 in terms of ϕ_1 from Equations (4.249) and (4.250) to obtain a single equation in the form of Equation (4.241):

$$\frac{d^2\phi_1}{dt^2} + F(\phi_1, \dot{\phi}_1) + G(\phi_1) = 0 \tag{4.251}$$

where the functions F and G can be identified as

$$F(\phi_1, \dot{\phi}_1) = -A_1\dot{\phi}_1^3 - \dot{\phi}_1^2(A_2\phi_1 + A_3) - \dot{\phi}_1(A_4\phi_1^2 + A_5\phi_1 + A_6) \tag{4.252}$$

$$G(\phi_1) = A_7\phi_1^3 + A_8\phi_1^2 + A_9\phi_1 + A_{10}$$

and A_1 to A_{10} are constants the details of which have been presented by Tambe, Kulkarni and Doraiswamy (1985e).

The functions $F(\phi_1, \dot{\phi}_1)$ and $G(\phi_1)$ can be identified with the damping and restoring forces, respectively. Further, consider a disturbance $f(t)$, which can be taken as a stationary Gaussian white noise, added to the rhs of Equation (4.241). The Fokker–Planck equation corresponding to the system given by these functions can be written following Tambe, Kulkarni and Doraiswamy (1985e):

$$\frac{\partial P}{\partial t} = -\frac{\partial}{\partial E}[A(E)P] + \frac{1}{2}\frac{\partial^2}{\partial E^2}[B(E)P] \tag{4.253}$$

where the parameters $A(E)$ and $B(E)$, according to the Stratonovich prescription, are obtained as

$$A(E) = \sqrt{2(E-u)}\, F[\phi_1, \sqrt{2(E-u)}] - \varepsilon$$

and (4.254)

$$B(E) = 4(E-u)\varepsilon$$

To illustrate the use of Equation (4.253) let us further simplify the forms of the functions F and G. Thus, for instance, when the reverse reaction in step 2 of the reaction mechanism is taken to be an order of magnitude larger than the other rate processes, the functions F and G in Equation (4.252) simplify to

$$F(\phi_1, \dot{\phi}_1) = -A_6\dot{\phi}_1$$

$$G(\phi_1) = A_9\phi_1 + A_{10} \tag{4.255}$$

Employing these functions in Equations (4.254) leads to

$$A(E) = -2A_6(E-u) - \varepsilon = \psi - \varepsilon, \qquad B(E) = \frac{2\psi\varepsilon}{A_6} \tag{4.256}$$

In view of these definitions it is convenient to recast Equation (4.253) as

$$\frac{\partial \pi(\psi, t)}{\partial t} = \frac{\partial}{\partial \psi}[-m_1(\psi)\pi(\psi, t)] + \frac{1}{2}\frac{\partial^2}{\partial \psi^2}[m_2(\psi)\pi(\psi, t)] \tag{4.257}$$

where $\pi(\psi, t)\, d\psi = P(E, t)\, dE$ and

$$m_1(\psi) = 1 - \frac{\psi}{\varepsilon}, \qquad m_2(\psi) = \frac{2\psi}{A_6} \tag{4.258}$$

Equation (4.257) is of the proper form and the stationary probability distribution can be obtained as

$$\pi(\psi, \infty) = \frac{A_6 K}{2\psi} \exp\left(A_6 \ln \psi - \frac{A_6 \psi}{\varepsilon} \right\} \tag{4.259}$$

where the constant K can be evaluated as

$$K = 2(A_6 - 1)!\left(\frac{A_6}{\varepsilon}\right)^{A_6} \tag{4.260}$$

The mean passage time to a threshold value of $\psi = \psi_c$ subject to the condition of $\psi = \psi_0$ initially is given as

$$t_m = \int \frac{1}{K} \frac{e^{(A_6\psi/\varepsilon)}}{\psi^{A_6}}\, d\psi$$

or

$$t_m = \frac{1}{2(A_6 - 1)!\,[A_6/\varepsilon]^{A_6}} \left[\frac{1}{A_6 - 1} \left(-\frac{e^{(A_6\psi/\varepsilon)}}{\psi^{A_6 - 1}} + \frac{A_6}{D} \int \frac{e^{(A_6\psi/\varepsilon)}}{\psi^{A_6 - 1}}\, d\psi \right) \right]_{\psi_0}^{\psi_c} \tag{4.261}$$

Equation (4.261) requires a knowledge of ψ_c, which is related to E and ϕ_1 according to Equation (4.256). To obtain an explicit relation between E and ϕ_1, Equation (4.244) can be integrated to give

$$E = -F(\phi_1, \dot{\phi}_1)\, d\phi_1 = -\phi_1[\sqrt{2(E-u)}] \tag{4.262}$$

Equation (4.262) together with (4.256) can then be used to calculate the threshold value of the parameter ψ_c, which can be used in Equation (4.261) to calculate the mean passage time for reaching that threshold value. The equation is analytic and can be easily solved.

Appendix 4.A

Elimination of Fast Variables: Projector Operation Formalism

To illustrate the main steps in the application of the projector operation formalism, we begin with a general equation

$$\frac{\partial p}{\partial t} = Lp(t) \tag{1}$$

where L represents a linear operator. The equation represents a typical form of a Liouville equation, a Fokker–Planck equation, a master equation of the birth and death type or a set of stochastic differential equations, depending on the nature of L and p.

The method envisages identification of the operator L in the form

$$L = L_0 + qL_1 \tag{2}$$

where q represents a formal expansion parameter. Further, defining

$$\tilde{p}(t) = \exp(-L_0 t)p(t) \tag{3}$$

$$\tilde{L}_1(t) = \exp(-L_0 t)L_1 \exp(L_0 t) \tag{4}$$

and using them in Equation (1) we have

$$\frac{\partial \tilde{p}}{\partial t} = q\tilde{L}_1(t)\tilde{p}(t) \tag{5}$$

The time convolutionless projection operator formalism is obtained by introducing a projector operator $\mathscr{L}$ in Equation (5) yielding the following equation for $\mathscr{L}p(t)$:

$$\frac{\partial \mathscr{L}\tilde{p}(t)}{\partial t} = K(t)\mathscr{L}\tilde{p}(t) \tag{6}$$

where the relation $(1 - \mathscr{L})p(0) = 0$ is assumed and $K(t)$ is a complex function that is usually difficult to evaluate completely. However, for most practical cases we can expand it in powers of q. Expansion of $K(t)$ to second order in q gives

$$K(t) = q\mathscr{L}_1(t) + q^2 \int_0^t dt_1[\mathscr{L}\tilde{L}_1(t)\tilde{L}_1(t_1) - \mathscr{L}\tilde{L}_1(t)\mathscr{L}\tilde{L}_1(t_1)] \quad (7)$$

Equation (6) along with (7) then provides the contracted description.

Appendix 4.B

Functional Relation Between T_∞^+, T_∞^- and the Transition Probability Rates

We consider a random walk governed by the transition probability rate $W_+(x)$ that defines the probability that the system in state x at time t will move to state $(x + 1)$ in the next small interval Δt. Similarly we can define $W_-(x)$. Written more explicitly, they imply $W_+(x) = W_{x+1,x}$ and $W_-(x) = W_{x-1,x}$. We also define $P_\infty(x)$ as the probability that the system will be in state x after a sufficiently long time t_∞. The function $P_\infty(x)$ is of course the stationary solution of the governing master equation for the random walk process. The function $P_\infty(x)W_+(x)\,dt$ represents the probability that the system will make the first transition from state x to $x + 1$ in any interval dt after t_∞. The first occurrence of such a transition is an exponentially distributed function with the mean given by

$$T_\infty^+(x) = [W_+(x)P_\infty(x)]^{-1}, \qquad T_\infty^-(x) = [W_-(x)P_\infty(x)]^{-1} \quad (1)$$

The fundamental role of these functions has been noted in Equation (4.141).

Let us now write the detailed balance recursion relation satisfied by $P_\infty(x)$:

$$P_\infty(x)/P_\infty(x-1) = W_+(x-1)/W_-(x) \quad (2)$$

which can be iterated to obtain

$$P_\infty(x) = P_\infty(0) \prod_{n=1}^{x} \frac{W_+(n-1)}{W_-(n)}, \qquad x \geqslant 1 \quad (3)$$

where $P_\infty(0)$ can be determined from the normalisation condition $\sum_x P_\infty(x) = 1$ as

$$P_\infty(0) = \left[1 + \sum_{x=1}^{\infty} \prod_{n=1}^{x} \frac{W_+(n-1)}{W_-(n)}\right]^{-1} \tag{4}$$

Referring to Equations (1) and (2), we note that $T_\infty^+(x) = [1 - T_\infty^-(x)]$. Knowledge of one is therefore sufficient. We now multiply Equation (2) by $W_+(x)$ to obtain

$$W_+(x)P_\infty(x) = \frac{W_+(x-1)W_+(x)P_\infty(x-1)}{W_-(x)}, \qquad x \geqslant 1 \tag{5}$$

This equation can be rearranged using Equation (1) to yield the recursion relation for $T_\infty^+(x)$ in terms of $W_\pm(x)$ as

$$T_\infty^+(x)W_+(x) = W_-(x)T_\infty^+(x-1), \qquad x \geqslant 1 \tag{6}$$

or, in more expanded form after iteration, as

$$T_\infty^+(x) = T_\infty^+(0) \prod_{n=1}^{x} \frac{W_-(n)}{W_+(n)}, \qquad x \geqslant 1 \tag{7}$$

The quantity $T_\infty^+(0)$ can be found from Equations (1) and (4) as

$$T_\infty^+(0) = \frac{1}{W_+(0)}\left[1 + \sum_{x=1}^{\infty} \prod_{n=1}^{x} \frac{W_+(n-1)}{W_-(n)}\right] \tag{8}$$

We note from these derivations that Equations (2) and (6) and Equations (8) and (4) are identical in structure. Also, $T_\infty^+(x)$ can be calculated from a knowledge of $W_\pm(x)$ as easily as $p_\infty(x)$.

REFERENCES

Arnold, L., *Stochastische Differentialgleichungen Münchenweins*, Oldenburg, 1973.
Banerjee, K., Bhattacharjee, J. K. and Mani, H. S. *Phys. Rev. A* **30**, 1118, 1984.
Bedeaux, D. *Phys. Lett.* **62A**, 10, 1977.
Beekman, J. A., *Trans. Am. Math. Soc.*, **126**, 29, 1967.
Bharoocha-Reid, A. T., *Random Integral Equations*, Academic Press, New York, 1972.
Brinkman, H. C., *Physica* **22**, 29, 1956.
Bykov, V. I. and Yablonskii, G. S., *Reac. Kinet. Catal. Lett.* **16**, 377, 1981.
Caroli, B., Caroli, C. and Roulet, B. J. *Stat. Phys.* **21**, 415, 1979.
Chaturvedi, S. S. and Shibata, F. *Z. Phys. B* **35**, 297, 1979.
Chaturvedi, S. *Z. Phys. B* **51**, 271 (1983.
Czajkowski, G. *Z. Phys. B* **43**, 87, 1981.
Dashen, R., *J. Math. Phys.* **20**, 84, 1979.
Faetti, S., Grigolini, P. and Marchesoni, F. *Z. Phys. B* **47**, 353, 1982.

Festa, C., Fronzoni, L., Grigolini, P. and Marchesoni, F. *Phys. Letts. A* **102**, 1984, 1984.
Feynman, R. P. and Hibbs, A. R., *Quantum Mechanics and Path Integrals*, McGraw-Hill, New York, 1965.
Frank, U. F., *Angew. Chem. Int. Ed. Engl.* **17**, 1, 1978.
Gardiner, C. W., *A Handbook of Stochastic Methods for Physics, Chemistry and Natural Sciences*, Springer-Verlag, Berlin, 1983.
Gardiner, C. W. and Chaturvedi, S., *J. Stat. Phys.* **17**, 429, 1977.
Gardiner, C. W., *Phys. Review A* **29**, 2814, 1984.
Geigenmuller, U., Titulaer, U. M. and Felderhof, B. U., *Physica* **119A**, 41, **120A**, 635, 1983.
Gichmann, I. I. and Skorochod, A. W., *Stochastische Differentialgleichungen*, Akademie-Verlag, Berlin, 1971.
Gillespie, D. T., *J. Chem. Phys.* **76**, 3762, 1982.
Gillespie, D. T., *J. Chem. Phys.* **74**, 5295, 1981.
Gillespie, D. T., *Physica A* **95A**, 69, 1979.
Gillespie, D. T., *Physica A* **101A**, 535, 1980.
Grabert, H., Hanggi, P. and Oppenheim, I., *Physica A* **117**, 300, 1983.
Gortz, R., *J. Phys. A* **9**, 1089, 1976.
Gortz, R. and Walls, D. F., *Z. Phys. B* **25**, 423, 1976.
Graham, R., *Springer Tracts in Modern Physics*, Springer-Verlag, Heidelberg, **66**, 1, 1973.
Graham, R., *Z. Phys. B* **26**, 281, 1397, 1977.
Graham, R., in *Stochastic Processes in Non-Equilibrium Systems*, Eds. L. Garrido, P. Segler and P. Shophard, Springer-Verlag, New York, 1978.
Grechannikov, A. V. and Yablonskii, G. S., *React. Kinet. Catal. Lett.* **19**, 321, 1982.
Grigolini, F. and Marchesoni, F. *Adv. Chem. Phys.* Eds. M. W. Evans, F. Grigolini, and G. Pastori-Parravicini, 1984.
Grigolini, P., Grosso, G., Pastori-Parravicini, G. and Sparpoglione, M., *Phys. Rev. B* **27**, 7342, 1983.
Grossmann, S. and Schranner, R., *Z. Physik* **B30**, 325, 1978.
Haag, G. *Z. Phys. B* **29**, 153, 1978.
Haag, G. and Hangii, P. *Z. Phys. B* **34**, 411, 1979.
Haag, G., Weidlich, W. and Aber, P., *Z. Phys. B* **26**, 207, 1977.
Haake, F., *Z. Phys. B* **48**, 31, 1982.
Haken, H., *Synergetics: an Introduction*, Springer-Verlag, Berlin, Heidelberg, New York, 1978.
Hangii, P., *Z. Naturforsch.* **33a**, 1380, 1978.
Hangii, P., Rossel, F. and Trauntmann, D., *Z. Naturforsch.*, 402, 1978.
Hangii, P. and Thomas, H., *Phys. Rep.* **88**, 207, 1982.
Hanggi, P., Grabert, H., Talkner, P. and Thomas, H., *Phys. Rev. A* **29**, 371, 1984.
Horsthemke, W. and Brenig, L., *Z. Physik* **B27**, 341, 1977.
Hasegawa, H., Mizuno, M. and Mabuchi, M., *Prog. Theor. Phys.* **67**, 98, 1982.
Helstrom, C. W., *IRE Transactions of information theory II-5*, 139, 1959.
Hongler, M. O. and Zheng, W. M., *J. Stat. Phys.* **29**, 375, 1982.
Hunt, P. M., Hunt, K. L. C. and Ross, J., *J. Chem. Phys.* **79**, 3765, 1983.
Hunt, K. L. C. and Ross, J., *J. Chem. Phys.* **75**, 976, 1981.
Janssen, H. K., *Z. Phys.* **270**, 67, 1974.
Kaneko, K., *Prog. Theo. Phys.* **66**, 129, 1981.
Keilson, J., *Appl. Probl.* **1**, 247, 1964.
Keilson, J. and Ross, H., in *IMS Table Series*, Eds. E. L. Harper and D. B. Owen, Vol. 3, 1975.
Kitahara, K., *Adv. Chem. Phys.* **29**, 85, 1976.
Kociszewski, A., *Acta Phys. Polon. A* **60**, 291, 1981.

Kramers, H. A., *Physica* **7**, 284, 1940.
Kubo, R., Matsuo, K. and Kitahara, K., *J. Stat. Phys.* **9**, 51, 1973.
Kubo, R., *J. Math. Phys.* **4**, 174, 1963.
Kurtz, T. G., *J. Appl. Prob.* **8**, 344, 1971.
Langouche, F., Roekaerts, D. and Tirapegui, E., *Physica A* **97A**, 193, 1979.
Lefever, R. and Nicolis, G., *Membranes, Dissipative Structures and Evolution*, Wiley, New York, 1975.
Lemarchand, H., *Physica A* **101A**, 518, 1980.
Leschke, H. and Schmutz, M., *Z. Phys. B* **27**, 85, 1977.
Lindenberg, K., Seshadri, V., *J. Chem. Phys.* **71**, 4075, 1979.
Lindenberg, K., Shuler, K. E., Freeman, J. and Lie, T. J., *J. Stat. Phys.* **12**, 217, 1975.
Lovenda, B. H., *Phys. Lett. A* **71**, 304, 1979.
Malchup, S. and Onsager, L., *Phys. Rev.* **19**, 1512, 1953.
Mandl, P., *Rev. Math. Pures Appl.* **8**, 273, 1963.
Mandl, P., *Analytical Treatment of One Dimensional Markov Processes*, Springer-Verlag, New York, 1968.
Marchesoni, F., *Phys. Letts. A* **101**, 11, 1984.
Marchesoni, F. and Grigolini, P., *Z. Phys. B* **55**, 257, 1984.
Maruyama, G. and Tanaka, H., *Mem. Fac. Sci. Kyusyu Univ. Ser.*, **11** (2), 117, 1957.
Matheson, I., Walls, D. F. and Gardiner, C. W., *J. Stat. Phys.* **12**, 21, 1975.
Matsuo, K., *J. Stat. Phys.* **17**, 169, 1977.
Matsuo, K., Lindenberg, K. and Shuler, K. E., *J. Stat. Phys.* **19**, 65, 1978.
Montroll, E. W. and Shuler, K. E., *Adv. Chem. Phys.* **1**, 361, 1958.
Moreau, M., *Physica A* **90A**, 419, 1978.
Mori, H., Morita, T. and Mashiyama, K. T., *Prog. Theo. Phys.* **63**, 1865; **64**, 500, 1980.
Muhlschlegel, B., *Path Integrals and Their Applications in Quantum Statistical and Solid State Physics*, Eds. G. J. Papadopoulos and J. T. Devreese, Plenum Press, 1978.
Nakajima, S., *Prog. Theor. Phys.* **20**, 948, 1958.
Newell, G. F., *J. Math. Mech.* **11**, 481, 1962.
Nicolis, G. and Turner, J. W., *Physica A* **89**, 326, 1977.
Nitzan, A., Ortoleva, J. and Deutch, J. and Ross, J., *J. Chem. Phys.* **61**, 1056, 1974.
Onsager, L. and Machlup, S., *Phys. Rev.* **91**, 1505, 1953.
Oppenheim, L., Shuler, K. E. and Weiss, G. H., *Physica A* **85A**, 195, 1977.
Owedyk, J. *Acta Phys. Polon. A* **363**, 317, 1983.
Owedyk, J., *Z. Phys. B* **54**, 183, 1984.
Pikios, C. and Luss, D., *Chem. Eng. Sci.* **32**, 191, 1977.
Procaccia, I., Mukamel, S. and Ross, J., *J. Chem. Phys.* **68**, 3244, 1978.
Ravikumar, V., Kulkarni, B. D. and Doraiswamy, L. K., *AIChE J.* **30**, 649, 1984.
Risken, H., Vollmer, D. H. and Morsch, M., *Z. Phys. B* **40**, 343, 1981.
Risken, H., *The Fokker–Planck Equation, Methods of Solution and Applications*, Ed. H. Haken, Springer-Verlag, New York, 1984.
San Miguel, M. and Sancho, J. M., *J. Stat. Phys.* **22**, 605, 1980.
Schlogl, D., *Z. Physik.* **253**, 147, 1972.
Schnakenberg, J., *Z. Physik B* **38**, 341, 1980.
Seshadri, V., West, B. J. and Lindenberg, K., *J. Chem. Phys.* **72**, 1145, 1980.
Seshadri, V., West, B. and Lindenberg, K., *J. Sound and Vibrat.* **68**, 553, 1980a.
Shibata, F., Takahashi, Y. and Hashitsume, N., *J. Stat. Phys.* **17**, 171, 1977.
Siegert, J., *Phys. Rev.* **81**, 617, 1951.
Siegert, J., *Ann. Math. Stat.* **24**, 624, 1953.
Skinner, J. L. and Woyles, P. G., *Physica A* **96**, 569, 1979.
Steyn-Ross, M. L. and Gardiner, C. W., *Phys. Rev. A* **29**, 2834, 1984.

Stone, L. D., Belkin, B. and Snyder, M. A., *J. Math. Anal. Appt.* **30**, 448, 1970.
Stratonovich, R. L., *Topics in the Theory of Random Noise*, Gordon and Breach, New York, 1967.
Suzuki, M., *Adv. Chem. Phys.* **46**, 195, 1981.
Suzuki, M., *J. Stat. Phys.* **20**, 163, 1979.
Suzuki, M., *Physica* (*Utrecht*) **117A**, 103, 1983.
Takoudis, G. G., Schmidt, L. D. and Aris, R., *Surface Science* **105**, 325, 1981.
Tambe, S. S., Kulkarni, B. D. and Doraiswamy, L. K., *Chem. Eng. Sci.* **40**, 1943, 1985a.
Tambe, S. S., Ravikumar, V., Kulkarni, B. D. and Doraiswamy, L. K., *Chem. Eng. Sci.* **40**, 1951, 1985b.
Tambe, S. S., Kulkarni, B. D. and Doraiswamy, L. K., *Chem. Eng. Sci.* **40**, 1959, 1985c.
Tambe, S. S., Kulkarni, B. D. and Doraiswamy, L. K., *Chem. Eng. Sci.* **40**, 2293, 1985d.
Tambe, S. S., Ravikumar, V., Kulkarni, B. D. and Doraiswamy, L. K., *Chem. Eng. Sci.* **40**, 2297, 1985e.
Tambe, S. S., Ravikumar, V., Ponnani, K. M. and Kulkarni, B. D., *Chem. Eng. J.* (communicated), 1985f.
Tambe, S. S. and Kulkarni, B. D., presented at *2nd Indo-Soviet Seminar on Catalysis*, held at Regional Research Laboratory, Hyderabad, India, during 25–28 Nov. 1986.
Titulaer, U. M., *Z. Phys. B* **50**, 71, 1983.
Titulaer, U. M., *Physica A* **31**, 321, 1980, 100, 251, 1978.
Tokuyama, U. and Mori, H., *Prog. Theo. Phys.* **55**, 411, 1976.
van Kampen, N. G., *Physica* **74**, 215, 239, 1974.
van Kampen, N. G., in *Adv. Chem. Phys.*, Ed. I. Prigogine and S. A. Rice, Vol. 34, Wiley, New York, 1976.
van Kampen, N. G., *Stochastic Processes in Physics and Chemistry*, 1981.
van Kampen, N. G., *J. Stat. Phys.* **17**, 71, 1977.
Weaver, D. L., *Am. J. Phys.* **50**, 1038, 1982.
Weidlich, W. and Haag, G., *Z. Phys. B* **39**, 81, 1980.
Weiss, G. H., *Adv. Chem. Phys.* **13**, 1, 1966.
Weiss, V., *Phys. Rev. A* **25**, 2444, 1982.
West, B. J., Lindenberg, K. and Seshadri, V., *J. Chem. Phys.* **72**, 1151, 1980.
West, B. J., Bulsara, A. R., Lindenberg, K., Seshadri, K. and Shuler, K. F., *Physica* **97A**, 211, 1979.
Wiegel, F. W., *Physica* **33**, 734, 1967.
Wiener, H., *Acta Math.* (*Uppsala*) **55**, 117, 1930.
Wilemski, J., *J. Stat. Phys.* **14**, 153, 1976.
Zwanzig, R., *J. Chem. Phys.* **33**, 1338, 1960.

5 Reaction-Diffusion Systems

5.1 Introduction

In the previous chapter we were concerned with the stochastic behaviour of complex reaction systems that involve multistationarity, instability or operation near the transition points. The treatment assumed that the system as a whole is uniform and that the local effects of reaction spread through it by diffusion at a rate faster than reaction. It is possible, however, that in certain situations dispersion by diffusion may no longer be able to bring about uniformity in the reaction mixture. The result is obvious. We then have a situation where the reaction proceeds in local zones at different rates and a spatial dependence comes into existence. A situation of this type has already been discussed in Section 3.3 of Chapter 3 in connection with the modification of the usual birth–death formalism commonly used for modelling of reactions.

To recapitulate the main points, we assume that dynamical processes of the reaction–diffusion system, subject to appropriate choice of the state variables, define a Markov process. Alternatively, a complete phase-space description may be provided to ensure the Markovian character. The latter approach, though rigorous, leads to the necessity of solving a complex set of equations that often prove to be practically difficult. An intermediate choice for the reaction–diffusion system seeks a straightforward extension of the homogeneous theory where the space is discretized into cells inside which the homogeneous reaction takes place while the transport of matter is accounted for by allowing coupling between cells. The reaction or reactive event leads to the creation or destruction of particles of different species within the cell, while the particle population within the cell is enhanced or decreased due to the dispersing action of diffusion. Both the reaction and diffusion processes can then be formulated in terms of the Markovian birth–death formalism. The number of particles in each cell represents the stochastic variables and the evolution of the system is governed by a multivariate master equation.

The general form of the multivariate master equation is given by Equation (2.116). As we shall mostly be dealing with nonlinear reacting systems that exhibit bifurcations and instabilities, the analytical solutions to such master equations are often difficult. It is necessary then to devise approximation schemes. The present chapter discusses some of these schemes by considering appropriate examples which also illustrate the role of diffusion in systems undergoing multiple steady state transitions and transitions of time-periodic or space-dependent solutions around a single uniform steady state solution. As a first step towards this, it is desirable to study the properties of the stochastic diffusion operator in the absence of reaction. We shall therefore begin with the case of pure diffusion in Section 5.2 and subsequently consider the reaction–diffusion problem involving multistationarity and instability.

5.2 Properties of the Stochastic Diffusion Operator

In a system where the space dependence is overcome by discretising into cells, and for diffusion alone to be responsible for the generation or removal of particles from a given cell, the instantaneous state of the system is described by a multivariate joint probability $P(X_1, \ldots X_N, t)$ denoting the probability of finding X_i particles of species X in the ith cell, $i = 1, \ldots N$. In formulating the multivariate master equation describing the evolution of this joint probability, we would need information regarding the boundary conditions. Thus, a set of cells may have reservoirs at the boundaries that sustain a constant concentration of particles. Alternatively, one may have a situation where the cells form a closed circle implying a periodic boundary condition or a situation of no flux at the boundary. The forms of the diffusion operator for the three situations differ.

To illustrate this point let us consider the first case with fixed boundary conditions. Denoting $\{X\} = [X_1, \ldots X_N]$ and $(X_i, X_j, X') =$

$(X_i, X_j$, all other X except $i, j)$ we obtain

$$\frac{\partial P(\{X\},t)}{\partial t} = \mathscr{D}\Big\{\sum_{i=2}^{N} [(X_i + 1)P(X_{i-1} - 1, X_{i+1}, X', t) - X_i P(X,t)$$

$$+ (X_{i-1} + 1)P(X_{i-1} + 1, X_i - 1, X', t) - X_{i-1}P(X,t)]$$

$$+ (X_1 + 1)P(X_1 + 1, X', t) - X_1 P(X,t) +$$

$$(X_N + 1)P(X_N + 1, X', t)$$

$$- X_N P(X,t) + Q_0[P(X_1 - 1, X', t) - P(X', t)]$$

$$+ Q_{N+1}[P(X_N - 1, X', t) - P(X', t)]\Big\} \tag{5.1}$$

where Q_0 and Q_{N+1} denote the particle numbers at the two boundaries, $\mathscr{D}$ the rate of exchange between adjacent cells defined in terms of Fick's diffusivity by the relation $\mathscr{D} = D/l^2$, where l represents the length of each cell. The equation can be written more compactly by resorting to the generating function defined as

$$\psi(s_1, \dots s_N, t) = \sum_{x_1=0}^{\infty} \dots \sum_{x_N=0}^{\infty} s_1^{X_1} \dots s_N^{X_N} P(X_1, \dots X_N, t) \tag{5.2}$$

Equation (5.1) then becomes

$$\frac{\partial \psi}{\partial t} = \mathscr{D}\mathscr{L}\psi \tag{5.3}$$

where the diffusion operator can be identified as

$$\mathscr{L} = \sum_{j=1}^{N} (s_{j+1} + s_{j-1} - 2s_j)\frac{\partial}{\partial s_j} + Q_0(s_1 - 1) + Q_{N+1}(s_N - 1) \tag{5.4}$$

with $s_0 = s_{N+1} = 1$.

Equation (5.3) represents a general form where the operator changes according to the boundary conditions. Thus for periodic and zero flux boundary conditions one would obtain

$$\mathscr{L} = s^T A \frac{\partial}{\partial s} \tag{5.5}$$

$$\mathscr{L} = s^T B \frac{\partial}{\partial s} \tag{5.6}$$

where A and B represent the $N \times N$ matrix defined as

$$A = \begin{bmatrix} -2 & 1 & 0 & & & 0 & 1 \\ 1 & -2 & 1 & 0 & & & 0 \\ & & & & & & \\ 0 & & & 0 & 1 & -2 & 1 \\ 1 & 0 & & & 0 & 1 & -2 \end{bmatrix} \tag{5.7}$$

$$B = \begin{bmatrix} -1 & 1 & 0 & & & & 0 \\ 1 & -2 & 1 & 0 & & & 0 \\ & & & & & & \\ 0 & & & 0 & 1 & -2 & 1 \\ 0 & & & & 0 & 1 & -1 \end{bmatrix}$$

Solution to Equation (5.3) with operator $\mathscr{L}$ defined as in Equations (5.4)–(5.6) for the three cases of constant, periodic and zero flux boundary conditions have been obtained by van den Broeck, Horsthemke and Malek-Mansour (1977). For the case of constant concentration of particles at the boundaries the result indicates that the initial probability would evolve to a multi-Poissonian distribution after a time of the order of $[|\mathscr{D}\lambda_1|]^{-1}$ where $\lambda_1 = 2[\cos(\pi/N + 1) - 1]$. For sufficiently large values of N, the relaxation time is of the order of Nl^2/D. In the other two cases for a fixed number of particles M, the initial probability evolves to a multinomial distribution with the mean number of particle given as $\langle X \rangle = M/N$. In the event of M and N tending to infinity at a constant ratio of M/N, the multinomial distribution takes the Poissonian form per cell.

The properties of the diffusion process alone established, we now turn to the problem of the combined effect of reaction and diffusion. The general master equation can be written in more compact form in terms of the generating function ψ as

$$\frac{\partial \psi}{\partial t} = \mathscr{D}\mathscr{L}\psi + R\psi \tag{5.8}$$

where R refers to the reaction operator that brings about a change of ψ due to chemical reaction. Once the model reaction scheme is settled this equation can be solved, at least in principle, to obtain explicit information regarding the probability function, which can subsequently be used to compute quantities such as the local variance or the correlation function between the cells. The main difficulty associated with Equation (5.8) is that the equation on expansion yields an infinite hierarchy of moment equations. The usual procedure suggests the truncation of this hierarchy at the level of second-order moments, which together with the averages forms a closed set of equations that can be used to calculate quantities like correlation function between the cells.

An especially suitable case arises when the characteristic time scales for the evolution due to chemical reaction (τ_C) and diffusion (τ_D) differ significantly. Thus for diffusion to be the leading process we can write $\tau_C\mathscr{D} \sim \tau_C/\tau_D \gg 1$. We can introduce a dimensionless perturbation parameter $\varepsilon = 1/\tau_C\mathscr{D} \ll 1$ into Equation (5.8) which can be rescaled as ($\tau = t/\tau_C$).

$$\frac{\partial\psi}{\partial\tau} = \mathscr{L}\psi + \varepsilon\tau_C R\psi \tag{5.9}$$

Let us now develop a perturbation expansion of Equation (5.9). Restricting the development to the case of fixed boundary condition, we set

$$\psi = \sum_{n=0}^{\infty} \varepsilon^n \psi^{(n)} \tag{5.10}$$

Inserting Equation (5.10) in (5.9) evaluated at the steady state, we get

$$\mathscr{L}\psi^{(0)} = 0 \tag{5.11}$$

$$\mathscr{L}\psi^{(n)} = -\tau_C R\psi^{(n-1)} \tag{5.12}$$

with

$$\psi^{(0)}\Big|_{s=1} = 1, \qquad \psi^{(n)}\Big|_{s=1} = 0$$

Equation (5.11) that corresponds to the zeroth order approximation can be recognised as the case of pure diffusion which has been already discussed and leads to the multi-Poissonian form

$$\psi^{(0)} = \exp\left[\sum_{j=1}^{N} \langle X_j\rangle^{(0)}(s_j - 1)\right] \tag{5.13}$$

Referring to the general relation between the moments and the generating function $[\langle X_j\rangle = (\partial\psi/\partial s_j)_{s=1}]$, we can calculate the mean as

$$\langle X_j\rangle^0 = Q_0 + (Q_{N+1} - Q_0)j/N + 1, \qquad j = 1,\dots N \qquad (5.14)$$

The effect of chemical reaction is expressed through Equation (5.12) which in principle can again be solved. The incorporation of nonlinear reaction causes a departure from the Poisson-like regime and the perturbative nature of this effect can be analysed as indicated above. For general systems where the characteristic times for the two processes are of comparable magnitude, no such perturbation parameter can be designed and recourse to other methods becomes necessary.

The mathematical complexities associated with these kinds of problems are enormous. Most investigations have therefore used truncation procedures. Thus Lemarchand and Nicolis (1976) considered the case of Schlogl's reaction scheme

$$A + 2X \underset{2}{\overset{1}{\rightleftharpoons}} 3X; \qquad X \underset{4}{\overset{3}{\rightleftharpoons}} B$$

with rate constants and concentrations of A and B chosen such that the system is operated at the nonequilibrium point that is similar to the second-order phase transition point. The system is supposed to consist of cells and the multivariate master equation for the probability $P(X_1, r_1, \dots x_k, r_k)$ of finding X_i particles in cell r_i can be written as

$$\begin{aligned}\frac{\partial P(\{X,r\};t)}{\partial t} &= \sum_i \{\lambda(X_i - 1)P(\dots; X_i - 1, r_i; \dots t) - \lambda(X_i)P(\{X,r\};t)\\ &\quad + \mu(X_i - 1)P(\dots; X_i + 1, r_i; \dots t) - \mu(X_i)P(\{X,r\};t)\}\\ &\quad + D/2\sum_i \{(X_i + 1)P(\dots; X_i + 1, r_i; X_{i-1} - 1, r_{i+1}; \dots t)\\ &\quad - X_i P(\{X,r\};t)\\ &\quad + (X_i + 1)P(\dots; X_{i-1} - 1, r_{i-1}; \dots r_i; \dots t) \qquad (5.15)\end{aligned}$$

where $\lambda(X)$ and $\mu(X)$ correspond to the rates for the processes of generation $(X \to X + 1)$ and loss $(X \to X - 1)$ and can be identified from the previous discussion:

$$\begin{aligned}\lambda(X) &= 3A^{-1}X(X - 1) + A(1 + \delta')\\ \mu(X) &= A^{-2}X(X - 1)(X - 2) + (3 + \delta)X\end{aligned} \qquad (5.16)$$

D in this equation refers to the stochastic diffusion coefficient that is related to the conventional Fick's diffusion coefficient. As stated earlier, Equation (5.15) should be supplemented with appropriate boundary conditions. Lemarchand and Nicolis (1976) used the periodic boundary condition which implies that the cells 1 and $N + 1$ are identical, thus forming a closed ring of cells. Since analytical solution to Equation (5.15) is not possible in view of the nonlinear nature of $\lambda(X)$ and $\mu(X)$, it is necessary to approximate the description with the hierarchy equations for the moments of the distribution. As stated earlier, by truncating the series at the second moment we obtain a closed set of equations. The first of these equations corresponds to the deterministic description, while the second equation gives information regarding the correlation function. The results obtained indicate that the spatial correlation function $\langle \delta X(r_i)\, \delta X(r_k)\rangle$ is practically zero, if the system is operated well below the bifurcation point. However, as the transition point is approached, long range correlations develop. These correlations, like the macroscopic variable, also develop spatial patterns. The development of spatial correlation is an important observation, particularly since the elementary interactions within the system are of short range. This is therefore a consequence of the nonequilibrium nature and nonlinearity of the system.

An analytical expression for the correlation length of the system using the mean field theory has been obtained by Malek-Mansour and Houdard (1979):

$$\langle \delta X(r_i)\, \delta X(r_j)\rangle = \langle X\rangle\, \delta^{k_r}_{r_i,r_j} + \frac{8\alpha(\alpha^{|r_i - r_j|} + \alpha^{N - |r_i - r_j|})}{D(\alpha^2 - 1)(\alpha^N - 1)} \tag{5.17}$$

where

$$|r_i - r_j| = 0, 1, \ldots N - 1, \qquad \delta = \delta' \geqslant 0$$

and

$$\alpha = 1 + \delta/D + [(1 + \delta/D)^2 - 1]^{1/2} \tag{5.18}$$

This equation also indicates the divergence of the correlation length as $|\delta|^{-1/2}$ in the neighbourhood of the critical point as the number of cells tends to a large value. Malek-Mansour and Houdard (1979) have numerically solved the case taking 21 cells arranged in a periodic chain with $\delta = \delta'$ and $A = 20$. The numerical results indicate that the approximation corresponding to Equation (5.11) is reasonable for systems relatively far from the critical point. As the critical point is approached the agreement with the mean field theory becomes unsatisfactory [Equation (5.17)], suggesting its inadequacy. Malek-Mansour and Houdard (1979) have proposed another approximation beyond the mean field limit. Equation (5.15), summed over all possible X_i, can be written at steady state as

$$P(X_i+1, r_i) = \frac{P(X_i, r_i)\{\lambda(X_i) + \frac{1}{2}D[E(X_{i+1}|X_i) + E(X_{i-1}|X_i)]\}}{\{\mu(X_i+1) + D(X_i+1)\}} \tag{5.19}$$

where $E(.)$ represents the conditional expectation

$$E(X_i|X_j) = \sum_q \frac{qP(X_i = q, r_i; X_i, X_j)}{P(X_j, r_j)} \tag{5.20}$$

As a first approximation it is logical to assume that the conditional expectation of X_i in cell r_i given the value X_j in cell r_j should linearly depend on X_j. We can thus write

$$E(X_i|X_j) = \langle X(r_j)\rangle + \frac{\langle \delta X(r_i)\,\delta X(r_j)\rangle}{\langle[\delta X(r_j)]^2\rangle}[X(r_j) - \langle X(r_j)\rangle] \tag{5.21}$$

Use of Equation (5.21) in the stationary master equation leads to the following equation for spatial correlation:

$$\langle \delta X(r_i)\,\delta X(r_j)\rangle = \langle X\rangle\,\delta^{k_r}_{r_i,r_j} + \left[\frac{\alpha\beta}{D(\alpha^2-1)}\right]\alpha^{-|r_j-r_i|} \tag{5.22}$$

where

$$\alpha = 1 + \gamma/D + [(1+\gamma/D)^2 - 1]^{1/2}$$

$$\beta = 2\langle X\rangle\left[\frac{\langle\lambda(X)\rangle}{\langle X\rangle} - \gamma\right] \tag{5.23}$$

$$\gamma = \frac{\langle X[\mu(X) - \lambda(X)]\rangle}{\langle(\delta X)^2\rangle}$$

The approximate relation (5.22) correctly predicts the numerical results, suggesting its validity even in the neighbourhood of the critical point. While the approximations given by Equation (5.17) or (5.22) provide an easy way of obtaining solutions and are convenient to use, they do not occur naturally and often recourse to numerical methods becomes necessary.

In this connection the numerical schemes proposed by Gillespie (1976), Turner (1977), Hanusse (1977) and Hanusse and Blanche (1981) are extremely helpful. Thus Hanusse (1977) has carried out the Monte Carlo simulation of a variety of chemical schemes incorporating the effect of diffusion. The results of the analysis indicate the effect of diffusion on the onset of instabilities. For a system showing multiple steady state transition, the multiple steady state and hysteresis effects are completely lost in presence of diffusion, if the values of reaction rates are larger than the diffusion rates. For the reverse situation, i.e. when the diffusion coefficients are larger than the chemical rates, two stationary distributions around the macroscopic steady states may appear. The transition between the stationary states involves a nucleation process that is preferentially initiated near the boundaries between the states. Mou (1978) also considered the problem of nucleation in reaction–diffusion systems and investigated the influence of stochastic fluctuations on first-order type transitions between two stable steady states using the birth–death type of master equation. Earlier Nicolis *et al.* (1976) used a Langevin approach to characterise the nucleation process in reaction–diffusion systems.

We refer the readers to the general multivariate Fokker–Planck equation developed in Appendix 5.A where approximate relations for the first two moments is also presented. These relations are, however, valid for the homogeneous systems but can be extended to include the effect of diffusion. Following Gardiner (1976) and Grossmann (1976) the diffusion matrix can now be written as

$$\alpha_{2,ij}(r, r') = [\alpha_{2,ij} + D\delta_{ij}\nabla r \cdot \nabla r x_i(r)] \tag{5.24}$$

where $\alpha_{2,ij}$ refer to local chemical fluctuating forces as defined by Equation (8) in Appendix 5.A and the second term quantifies the contribution associated with the random walk of the particles. The correlation function $\sigma_{ij}(r, t; r', t) = \langle \delta x_i(r, t)\, \delta x_j(r', t)\rangle$ [Equation (10) of Appendix 5.A] likewise gets modified to

$$\frac{\partial \sigma_{ij}(r, t; r', t)}{\partial t} = 2\alpha_{2,ij}(r, r') + \eta_{im}(r)\sigma_{mj}(r, t; r', t) + \eta_{jm}(r')\sigma_{im}(r, t; r', t) \tag{5.25}$$

where the diffusion is now contained in the relaxation matrix

$$\eta_{ij}(r) = \eta_{ij} + D_i\delta_{ij}\nabla_r^2$$

The modification of the results for the homogeneous systems to incorporate the effects of diffusion as envisaged in Equations (5.24)–(5.26) often yields good results for a large volume system that is operating at a point not too near the instability.

Homogeneous reacting systems exhibiting the multiple steady state type transitions corresponding to the Schlogl reaction schemes have been analysed for the role of diffusion this way by Gardiner *et al.* (1976) who established the laws corresponding to the divergence of the correlation length and enhancement of fluctuations for the first and second order transition cases. The case of temporal and spatial oscillations in reaction–diffusion system such as a Brusselator has also been investigated to obtain static and dynamic correlations [see for example, Chaturvedi *et al.*, 1977; Mashiyama *et al.*, 1975; Richter *et al.*, 1979].

The case of a system exhibiting limit cycle oscillations under macroscopically homogeneous conditions has been investigated numerically in presence of diffusion by Nicolis *et al.* (1976). Again, for given values of the reaction and diffusion coefficients, the results reveal the existence of a critical number of neighbouring cells above which the instability may disappear.

Studies on the role of diffusion in the behaviour of systems continue, and different approaches such as the mean field approach (Prigogine *et al.*, 1975; Mou, 1978; Ebeling and Schimansky-Geier, 1980), computer simulations and approximations of the master equation as noted earlier are being extensively used. There are still many pendant problems regarding these systems—particularly those concerned with the divergence of fluctuations near the critical point, role of dimensionality of the system, behaviour of temporal and spatial correlations, etc. for which no clear-cut answers exist.

5.3 Systematic Analysis of Reaction–Diffusion Systems (Nicolis and Malek-Mansour, 1980)

In the previous section the general analysis of reaction–diffusion systems was presented. In this section we consider specific cases of the Schlogl models. It may be noted that these models have been extensively used in the earlier chapters to exemplify the basic theory involved in absence of diffusion.

The multivariate master equation derived on the assumption of diffusion modelled as a random walk between adjacent cells and reaction in each cell modelled as a birth–death process can be written following the development in the previous section. Our interest in this section is to solve this equation for the following Schlogl's models to ascertain the role of diffusion:

$$A + 2X \rightleftharpoons 3X \qquad A + X \rightleftharpoons 2X$$
$$\underset{\text{I}}{X \rightleftharpoons B} \qquad \underset{\text{II}}{B + X \rightleftharpoons C}$$

The macroscopic equations for the first model can be written as

$$\frac{\partial x_r}{\partial \tau} = -x_r^3 - \delta x_r + (\delta' - \delta) + D_1 \nabla^2 x_r \tag{5.26}$$

where

$$(1 + x_r) = X_r/\Delta V, \qquad k_1 A/k_2 = 3\Delta V$$
$$k_3/k_2 = (3 + \delta), \qquad k_4 B/k_2 = (1 + \delta')\Delta V \tag{5.27}$$
$$\tau = k_2 t, \qquad \mathscr{D}_1 = \mathscr{D}_x/k_2$$

and x_r represents the concentration of species X in the cell of volume ΔV centred at the spatial coordinate r. For systems possessing natural or periodic boundary conditions, the last term in Equation (5.26) does not alter the homogeneous solutions and no new solutions are generated. This equation possesses a bifurcation point at $\delta = \delta' = 0$ and it is known that as δ, δ' move to negative values along the line $\delta' = \delta'$, a bifurcation from the trivial solution $x_r = 0$ to the uniform solutions $x_r = \pm\sqrt{-\delta}$ occurs. Also, in the multiplicity region for $\delta \neq \delta'$ one encounters the phenomenon of hysteresis. It is of interest to examine the behaviour of this system in presence of fluctuations and for this purpose we can convert the master equation into a generating function representation. The transformed equation is obtained as

$$\sum_r (1 - S_r) S_r^2 \left(\frac{1}{\Delta V^2} \frac{\partial^3 \psi}{\partial s_r^3} - \frac{3}{\Delta V} \frac{\partial^2 \psi}{\partial S_r^2} \right)$$
$$+ \sum_r (1 - S_r) \left[(3 + \delta) \frac{\partial \psi}{\partial S_r} - (1 + \delta')\Delta V \psi \right]$$
$$+ \frac{D}{2d} \sum_{r\lambda} (S_{r+\lambda} - S_r) \frac{\partial \psi}{\partial S_r} = 0 \tag{5.28}$$

where d refers to the spatial dimensionality and

$$\psi = \sum_{x_r} \prod_r S_r^{X_r} p(\{X_r\}; t) \tag{5.29}$$

It is convenient to recast Equation (5.28) in an alternative form by redefining the function ψ in Equation (5.29) as

$$\psi = \prod_r \exp\{\Delta V(1 + x_r)(S_r - 1)\}\psi' \tag{5.30}$$

For the particular case of symmetric bifurcation ($\delta = \delta'$, where $x_r = 0$), we can then obtain from Equation (5.28)

$$\sum_r (1 - S_r)S_r^2\left(-2\Delta V\psi' - 3\frac{\partial\psi'}{\partial S_r} + \frac{1}{\Delta V^2}\frac{\partial^3\psi'}{\partial S_r^3}\right)$$
$$+ \sum_r (1 - S_r)\left[2\Delta V\psi' + (3 + \delta)\frac{\partial\psi'}{\partial S_r}\right] - \frac{D}{2d}\sum_{r\lambda}(1 - S_r)\left(\frac{\partial\psi'}{\partial S_{r+\lambda}} - \frac{\partial\psi'}{\partial S_r}\right) \tag{5.31}$$

with

$$\psi'\{S_r = 1\} = 1, \qquad \left(\frac{\partial\psi'}{\partial S_r}\right)_{S_r=1} = O(1)$$

A systematic analysis of Equation (5.31) requires proper identification of a perturbation parameter. Despite the lack of such a clearly discernible parameter, one could venture to define it in anticipation of the existence of a nonequilibrium transition where, due to the long range correlations between the spatial cells, one could augment ΔV to macroscopic dimensions. Near the bifurcation point we therefore get

$$\varepsilon = \frac{1}{\Delta V} \ll 1 \qquad \text{with} \quad \delta = \varepsilon^l \delta_i + \cdots \tag{5.32}$$

Further, in the vicinity of $S_r = 1$, one would expect to get all the macroscopically relevant information which will be chosen as the solution of the phenomenological Equation (5.26). It is appropriate then to express S_r around unity as

$$S_r = 1 + \varepsilon^m q_r, \qquad 0 < m < 1 \tag{5.33}$$

With these transformations Equation (5.31) becomes

$$\sum_r -q_r(1 + 2\varepsilon^m q_r + \varepsilon^{2m} q_r^2)\varepsilon^{-2m+2} \frac{\partial^3 \psi'}{\partial q_r^3}$$
$$+ \sum_r -q_r(2\varepsilon^m q_r + \varepsilon^{2m} q_r^2)\left(-2\varepsilon^{-1+m}\psi' - 3\frac{\partial \psi'}{\partial q_r}\right) + \sum_r -q_r \delta \frac{\partial \psi'}{\partial q_r}$$
$$+ \frac{D}{2d}\sum_{r\lambda}\left(\frac{\partial \psi'}{\partial q_{r+\lambda}} - \frac{\partial \psi'}{\partial q_r}\right) = 0 \tag{5.33}$$

where the diffusion can be scaled as

$$D = \varepsilon^n D_1 + \cdots \tag{5.34}$$

Subject to choice of exponents l, m and n, Equation (5.33) yields different results. Thus for (i) $n > \max[l, 2(1-m), 2m-1]$, $l = 2(1-m) = 2m-1$, we obtain the results that correspond to the neglect of spatial fluctuations to zeroth order of the perturbation analysis; (ii) for $n < \min\{l, 2(1-m), 2m-1\}$, these have diffusion as a dominating process and chemical reaction can be treated as a perturbation (this situation has been discussed in the earlier sections); and finally (iii) for $2(1-m) > 2m-1$ or $m < 3/4$ and $2m-1 = l = n$, we obtain

$$\sum_r q_r\left[\delta\frac{\partial \psi'}{\partial q_r} - \frac{D}{2d}\sum_\lambda\left(\frac{\partial \psi'}{\partial q_{r+\lambda}} - \frac{\partial \psi'}{\partial q_r}\right)\right] = \varepsilon^{2m-1}\sum_r 4q_r^2\psi' \tag{5.35}$$

which when transformed into a cumulant generating function $[\psi' = \exp(1/\varepsilon)\omega]$ yields the Gaussian approximation. We thus see that Gaussian representation is a natural consequence when m is small enough or for the case where the system is far from the bifurcation point.

Near to the bifurcation point the results derived above become invalid and the higher order terms of the perturbation expansion diverge. To overcome this problem one could develop a singular perturbation that retains to lowest order in the perturbation parameter ε both the diffusion and reaction forms. Thus for a situation where $m = 3/4$ and $l = n = 1/2$, if we call $\psi'(0)$ as the lowest order approximation, then we obtain from Equation (5.33) the following approximations:

$$\frac{\partial^3 \psi'(0)}{\partial q_r^3} + \delta_1\frac{\partial \psi'(0)}{\partial q_r} - \frac{D}{2d}\sum_\lambda\left(\frac{\partial \psi'(0)}{\partial q_{r+\lambda}} - \frac{\partial \psi'(0)}{\partial q_r}\right) = 4q_r\psi'(0) \tag{5.36}$$

Taking the Mellin–Fourier transform,

$$\psi'(\{q_r\}) = \int_{-\infty}^{\infty} \cdots \int_{-\infty}^{\infty} \{d\theta_r\} \exp\left[\left(\sum_r q_r \theta_r\right)\right] R(\{\theta_r\}) \qquad (5.37)$$

Once the solution is known the successive moments of the probability distribution, such as concentration correlation function, can be easily derived:

$$G(r, r') = N \int_{-\infty}^{\infty} \cdots \int_{-\infty}^{\infty} \{dZ_r\} Z_r Z_{r'} \exp\left\{ -\frac{1}{4} \sum_r \varepsilon^{-1} \right\}$$
$$\times \left\{ \left[\delta \frac{Z_r^2}{2} + \frac{Z_r^4}{4} + \frac{D}{8d} \sum_\lambda (Z_{r+\lambda} - Z_r)^2 \right] \right\}, \qquad Z_r = \theta_r \varepsilon^{1/4}$$
$$(5.38)$$

This result has an important message: in presence of diffusion, the solution of the multivariate master equation is not Gaussian even before the point of bifurcation.

The treatment thus far ignored the higher order corrections that may be important for situations beyond the bifurcation point. Following a procedure as above, even these higher order corrections can be similarly worked out. The analysis indicates that the second moment of the probability distribution or the correlation function as in Equation (5.38) is given correctly to dominant order. The case with higher order moments may, however, be different.

The analysis for a symmetric case ($\delta = \delta'$) as performed above can be redone for a nonsymmetric case ($\delta \neq \delta'$) where x_r in Equation (5.30) is no more equal to zero. Instead of Equation (5.31) we now obtain

$$\sum_r (1 - S_r) S_r^2 \left[\frac{1}{\Delta V^2} \frac{\partial^3 \psi'}{\partial S_r^3} + \frac{3}{\Delta V} (x_r) \frac{\partial^2 \psi'}{\partial S_r^2} \right.$$
$$+ (3(1 + x_r)^2 - 6(1 + x_r)) \frac{\partial \psi'}{\partial S_r}$$
$$\left. + \Delta V((1 + x_r)^3 - 3(1 + x_r)^2)\psi' \right]$$
$$+ \sum_r (1 - S_r) \left\{ (3 + \delta) \frac{\partial \psi'}{\partial S_r} + \Delta V[(1 + x_r)(3 + \delta) - (1 + \delta')]\psi' \right\}$$
$$- \frac{D}{2d} \sum_{r\lambda} (1 - S_r) \left(\frac{\partial \psi'}{\partial S_{r+\lambda}} - \frac{\partial \psi'}{\partial S_r} \right) = 0 \qquad (5.39)$$

Setting $x_r = \varepsilon^g h$, we then have, instead of Equation (5.33),

$$\sum_r -q_r[1 + O(\varepsilon^m)]\left[\varepsilon^{2(1-m)}\frac{\partial^3\psi'}{\partial q_r^3} + 3\varepsilon^{1-m+g}h\frac{\partial^2\psi'}{\partial q_r^2} + (3\varepsilon^{2g}h^2 + \delta_1\varepsilon^l)\frac{\partial\psi'}{\partial q_r}\right]$$
$$+ \sum_r -q_r[2\varepsilon^m q_r + O(\varepsilon^{2m})]\left[3\left(-1 + \varepsilon^{2g}h^2\frac{\partial\psi'}{\partial q_r}\right) - (2\varepsilon^{-1+m} + 3\varepsilon^g h)\psi'\right]$$
$$+ \frac{D}{2d}\sum_{r\lambda}\varepsilon^n q_r\left(\frac{\partial\psi'}{\partial q_{r+\lambda}} - \frac{\partial\psi'}{\partial q_r}\right) = 0 \tag{5.40}$$

The equations corresponding to Equations (5.36) and (5.37) can be obtained following a similar procedure:

$$\frac{\partial^3\psi_{(0)}}{\partial q_r^3} + 3h\frac{\partial^2\psi'_{(0)}}{\partial q_r^2} + (\delta_1 + 3h^2)\frac{\partial\psi'_{(0)}}{\partial q_r} - \frac{D}{2d}\sum_\lambda\left(\frac{\partial\psi_{(0)}}{\partial q_{r\lambda}} - \frac{\partial\psi'_{(0)}}{\partial q_r}\right) = 4q_r\psi'_{(0)} \tag{5.41}$$

and

$$R_{(0)} = N\exp\left\{-\frac{1}{4}\sum_r\left[(\delta_1 + 3h^2)\frac{\theta_r^2}{2} + h\theta_r^3 + \frac{\theta_r^4}{4} + \frac{D}{8d}\sum_\lambda(\theta_{r+\lambda} - \theta_r)\right]^4\right\} \tag{5.42}$$

The differences between equations for the symmetric and nonsymmetric cases are clearly seen. Equation (5.41) contains an additional second derivative term and the coefficient of the first derivative term is modified. Likewise in Equation (5.42) an additional cubic term appears.

Much in the same way as for Schlogl's first model, we can proceed to analyse the second model. The macroscopic equation for this case can be written as

$$\frac{dx}{\partial\tau} = -x^2 + \delta x + C \tag{5.43}$$

where

$$x = X/\Delta V, \qquad \tau = k_2 t$$

The solution to Equation (5.43) for $C \neq 0$ can be written as

$$x = [\delta + (\delta^2 + 4C)^{1/2}]/2 \tag{5.44}$$

while for $C = 0$ but $\delta \geqslant 0$ we have two solutions $x = 0$ (unstable) and $x = \delta$ (stable). The point $\delta = 0$ thus represents a bifurcation point, and our interest as before lies in evaluating the role of fluctuations in its neighbourhood. The master equation for this situation can be written in terms of the generating function representation:

$$\sum (1 - S_r)\left\{S_r \frac{1}{\Delta V} \frac{\partial^2 \psi}{\partial S_r^2} + [(\beta + \delta)(1 - S_r) - \delta] \frac{\partial \psi}{\partial S_r} C \Delta V \psi\right\} + \frac{D}{2d} \sum_r (S_r - 1) \sum_\lambda \left(\frac{\partial \psi}{\partial S_{r+\lambda}} - \frac{\partial \psi}{\partial S_r}\right) = 0 \tag{5.45}$$

Using the scaling relations as before, viz.

$$\varepsilon = \frac{1}{\Delta V} \ll 1, \qquad S_r = 1 + \varepsilon^m q_r, \qquad 0 < m < 1 \tag{5.46}$$

and the auxiliary function

$$\psi = \psi' \prod_r \exp[\Delta V x (S_r - 1)] \tag{5.47}$$

the master equation [Equation (5.45)] can be recast as

$$(1 + \varepsilon^m q_r)\varepsilon^{1-m} \frac{\partial^2 \psi'}{\partial q_r^2} + [2x - \delta + \varepsilon^m (2x - \beta - \delta) q_r] \frac{\partial \psi'}{\partial q_r} - \varepsilon^{2m-1}(\beta x - C) q_r \psi' - \frac{D}{2d} \sum_\lambda \left(\frac{\partial \psi'}{\partial q_{r+\lambda}} - \frac{\partial \psi'}{\partial q_r}\right) = 0 \tag{5.48}$$

Following the method as elaborated in the earlier case for the situation when $C \neq 0$, we obtain a multi-Gaussian distribution. Also, for the case of $C = 0$, where from a macroscopic viewpoint we have two solutions $x = 0$ and δ, it can be easily shown that provided $l < m$ the solution is described by a Gaussian approximation. For the situations when $l > m$ or $l = m$, the system cannot admit a detailed balance stationary solution. The present case has an absorbing boundary at $x = 0$ and remains as the only permissible solution.

5.4 Analysis of Reaction–Diffusion Systems Using the Langevin Formulation

In the previous section the role of internal noise in modifying the behaviour of the reaction–diffusion system was studied through a formulation of master equation which regards the reaction and diffusion process to be a random phenomenon and constructs the Markov process in the appropriate phase-space. This approach which is quite sound and appealing, however, has the important limitation of mathematical complexity. This rather unsatisfactory nature of the approach has led several investigators to adopt alternative methods for examining the influence of noise on reaction–diffusion systems. Acquaintance with the phenomenological macroscopic equation easily suggested a way to take account of such fluctuations by adding to it the appropriately corrected Langevin forces. The numerical calculations based on Monte Carlo methods (Hanusse and Blanche, 1981) have thus provided considerable insight into the behaviour of these systems. The more recent works of Ritcher, Procaccia and Ross (1980) and Walgraef, Dewel and Borckmans (1982) provide the state-of-the-art in this area. At the analytical level, methods such as cumulant expansions, the Fokker–Planck and Langevin equations (Malek-Mansour *et al.*, 1981), etc. have also been used.

In the Langevin approach the set of reaction–diffusion rate equations are suitably appended by a stochastic source term to account for fluctuations in the system. Here again, it is convenient to discretise the set by dividing the system into cells of homogeneous composition. The set of concentrations at each spatial point then becomes the set of stochastic variables. The set of equations can be solved to obtain the complete probability distribution. In practice, as we noted earlier, the quantities of interest are the static $[\langle \delta X_i(t)\, \delta X_j(t) \rangle]'$ and dynamic $[\langle \delta X_i(t)\, \delta X_j(t+\tau) \rangle]$ correlations and these can be calculated even without solving the equation completely using the standard techniques.

As an illustration one could begin with the nonlinear Fokker–Planck equation (which, as we saw earlier in Chapter 4, is also the asymptotic representation of the full master equation) that can be obtained from the stochastic differential equation as

$$\frac{\partial p(x,t)}{\partial t} = -\frac{\partial}{\partial x_i}[f_i(\mathbf{x})p(\mathbf{x},t)] + \frac{1}{V}\frac{\partial}{\partial x_i}\frac{\partial}{\partial x_k}[\mathscr{D}_{ik}(\mathbf{x})p(\mathbf{x},t)] \tag{5.49}$$

An equation for the variance $\sigma_{ik}(t) = \langle \delta x_i(t)\, \delta x_k(t) \rangle$ can then be easily obtained as

$$\frac{d\sigma_{ik}}{dt} = \frac{2}{V}\mathscr{D}_{ik}(\mathbf{x}) + \frac{\partial f_i}{\partial x_i}\sigma_{ik} + \frac{\partial f_k}{\partial x_k}\sigma_{il} \tag{5.50}$$

For the system in stable stationary state $f(\mathbf{x}_0) = 0$ and we obtain the static correlations

$$\Lambda\sigma + \sigma\Lambda^T = -\frac{2}{V}\mathscr{D}(x_0) \tag{5.51}$$

where $\Lambda = d\mathbf{f}|\partial\mathbf{x}$ represents the relaxation matrix. Equations (5.50) and (5.51), valid for homogeneous systems, can easily be generalised to include diffusion following the works of Gardiner (1976) and Grossmann (1976). For this case the diffusion matrix can be written as

$$\mathscr{D}_{ij}(r, r') = [\mathscr{D}_{ij} + D\delta_{ij}\nabla r\,.\,\nabla r x_i(r)]\,\delta(r, r') \tag{5.52}$$

where $\mathscr{D}_{ij}$ refer to local chemical fluctuating forces and the second term quantifies the contribution associated with the random walk of the particles. The correlation function likewise gets modified to

$$\sigma_{ij}(r, t; r', t) = \langle \delta x_i(r, t)\, \delta x_j(r', t) \rangle$$

$$\frac{\partial \sigma_{ij}(r, t; r', t)}{\partial t} = 2\mathscr{D}_{ij}(r, r') + \eta_{im}(r)\sigma_{ij}(r, t; r', t)$$
$$+ \eta_{jm}(r')\sigma_{im}(r, t; r', t) \tag{5.53}$$

where the diffusion is now contained in the relaxation matrix

$$\eta_{ij}(r) = \eta_{ij} + D\delta_{ij}\nabla^2 r \tag{5.54}$$

The modification of the results for the homogeneous systems to incorporate the effects of diffusion as envisaged in Equations (5.52)–(5.54) often yields good results for a large volume system that is operating at a point not too near the instability.

Homogeneous reacting systems exhibiting multiple steady state type transitions corresponding to Schlogl's reaction schemes have been analysed for the role of diffusion this way by Gardiner *et al.* (1976) who established the laws corresponding to the divergence of the correlation length and enhancement of fluctuations for the first- and second-order transitions. The case of temporal and spatial oscillations in reaction–

diffusion systems such as a Brusselator has been also investigated to obtain static and dynamic correlations (see, for example, Chaturvedi *et al.*, 1977; Mashiyama, Ito and Ohta, 1975; Ritcher, Procaccia and Ross, 1980).

5.5 Compartmental Model for Reaction–Diffusion Systems and Its Stochastic Analogue

5.5.1 Analysis of Macroscopic System and Its Stochastic Analogue Using Master Equation

A simple and practically more useful approach can be based on the compartmental model of Ebeling and Malchow (1979), where the continuous reaction–diffusion system is discretised by requiring that the system consist of n boxes where chemical reaction proceeds with diffusive exchange occurring between the cells. Thus a one-dimensional system described by

$$\frac{\partial x(r,t)}{\partial t} = F[x(r,t)] + D\nabla^2 x(r,t) = F(x,t) \tag{5.55}$$

may be approximated by the equation

$$\frac{dx_i(r,t)}{dt} = f[x_i(t)] + \sum_{j=1}^{n} D_{ij}[x_j(t) - x_i(t)] \tag{5.56}$$

where $x_i(t)$ is the concentration in cell i at time t and D_{ij} the diffusion coefficient between the cells i and j. For convenience one can treat D_{ij} as different from zero only for adjacent cells and for cells of length ΔL we have $D_{ij} = D/(\Delta L)^2$. D_{ij} equals zero for all other cells in the configuration.

In view of the fact that all reaction–diffusion systems are gradient systems, one could think of them as derived from a potential V. Thus Equation (5.56) can be written as

$$\frac{dx}{dt} = -\frac{\partial}{\partial x} V(x_1, \ldots x_n), \qquad n = \text{number of cells} \tag{5.57}$$

where V can be identified as

$$V = \sum_{i=1}^{n} \phi(x_i) + \frac{1}{2} \sum_{i,j=1}^{n} D_{ij}(x_j - x_i)^2$$

and (5.58)

$$\phi(x) = -\int_{x_0}^{x} f(x)\,dx$$

It is known that the stationary solutions of Equation (5.57) are the local extrema of V and the solution is structurally stable if it is a nondegenerate quadratic point of V, that is provided $|\partial^2 V/\partial x_i \partial x_j|_{x=x_s} \neq 0$. The stationary state is structurally unstable if this takes a value equal to zero.

To illustrate the method let us assume the case of two compartments for the Schlogl reaction model and write the macroscopic equations

$$\frac{dx_i}{dt} = -x_i^3 + Ax_i^2 - Bx_i + C + D(x_j - x_i), \qquad i, j = 1, 2, \quad i \neq j \tag{5.59}$$

The stationary solutions to these equations are the intersection points of

$$x_i = x_j - \frac{1}{D} f(x_j), \qquad i, j = 1, 2; \quad i \neq j$$

The potential V can be written for $B = 2/9$, $C = 0$ as

$$\begin{aligned} V(x_1, x_2) &= \phi(x_1) + \phi(x_2) + \tfrac{1}{2}D[(x_1 - x_2)]^2 \\ &= \tfrac{1}{4}(x_1^4 + x_2^4) + \tfrac{1}{2}(B - \tfrac{1}{3})(x_1^2 + x_2^2) \\ &\quad + \tfrac{1}{3}(B - 3C - \tfrac{2}{9})(x_1 + x_2) + \tfrac{1}{2}D(x_1 - x_2)^2 \end{aligned} \tag{5.61}$$

The following derivatives are obtained:

$$\frac{\partial V}{\partial x_i} = -\frac{\partial x_i}{\partial t} = x_i^3 + (B + D - \tfrac{1}{3})x_i + \tfrac{1}{3}(B - 3C - \tfrac{2}{9}) - Dx_j$$

$$\frac{\partial^2 V}{\partial x_i \partial x_j} = \begin{vmatrix} 3x_1^2 + B + D - \frac{1}{3} & -D \\ -D & 3x_2^2 + B + D - \frac{1}{3} \end{vmatrix} \tag{5.62}$$

The characteristic bifurcation may then be obtained using the condition that Equations (5.61) and (5.62) are equal to zero. The typical bifurcation maps for the situations (i) $f(x_m) = -f(x_n)$ and (ii) $f(x_m) \neq -f(x_n)$

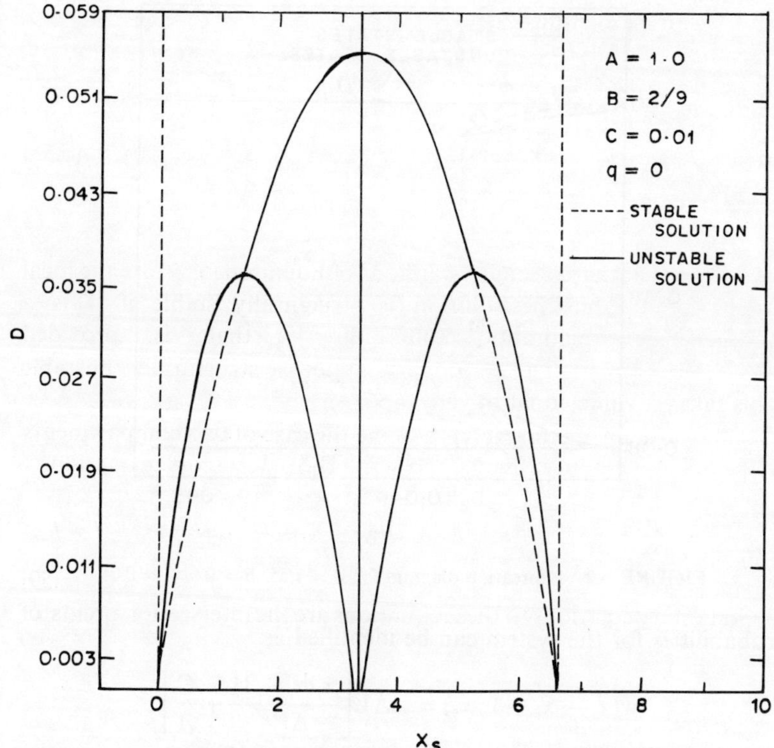

FIGURE 5.1 Steady state solutions (X_s) for a compartment model showing the effect of exchange coefficient (D).

where x_m and x_n are the solutions of $f'(x) = 0$ are shown in Figures 5.1 and 5.2. The figures clearly reveal the existence of critical values of D, below which the two-cell model admits existence of five and nine solutions. The stability of the various solutions, as can be obtained using the linear stability analysis, is also indicated in the figures. Of special interest in Figure 5.2 is the case of lower values of D where the model reveals the existence of a stable inhomogeneous branch of the solution.

The construction of the stochastic analogue for this compartmentalised situation is now relatively simple. The transition

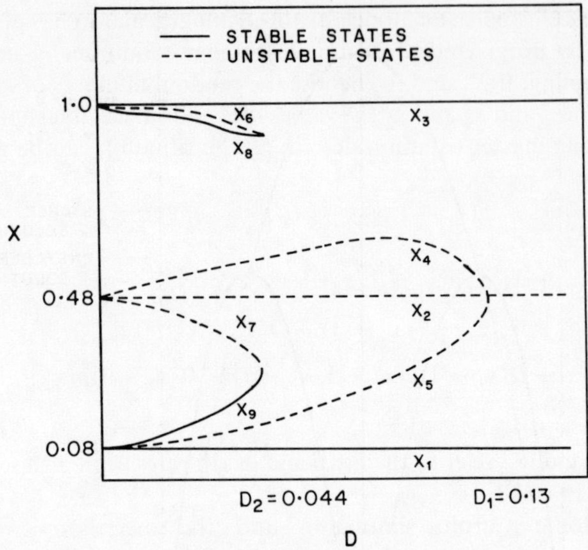

FIGURE 5.2 Bifurcation diagram for $A = 1.55$, $B = 0.6$, $C = 0.04$.

probabilities for the system can be identified as

$$w(x_i \rightarrow x_i + 1; x_j) = A\Delta V\left[\frac{x_i(x_i - 1)}{\Delta V^2} + \frac{C}{A}\right]$$

$$w(x_i \rightarrow x_i - 1; x_j) = x_i\left[\frac{(x_i - 1)(x_i - 2)}{\Delta V^2} + B\right] \tag{5.63}$$

$$w(x_i \rightarrow x_i - 1; x_j \rightarrow x_j + 1) = Dx_j$$

where i and j refer to two cells and A, B, C and D are the coefficients in the macroscopic Equation (5.59). Frankowicz and Gudowska-Nowak (1982) have applied the standard Monte Carlo procedure to calculate the time evolution of the model system. Summing over all times the system spent in a certain state, it is then possible to obtain the probability distribution function. The analysis reveals some interesting points. The stable inhomogeneous states of the macroscopic analysis appear as transient structures during the stochastic evolution, and these facilitate the transition between the homogeneous steady states. Also the transition time from one state to another depends on the extent of diffusion and shows a minimum.

The two-cell stochastic model for this Schlogl reaction has also been examined by Borgis and Moreau (1984) under conditions of small diffusion coupling. If X_1 and X_2 denote the random numbers of species x in cells 1 and 2 and x_1 and x_2 the values the random numbers take, then the following master equation describing the situation can be written:

$$\begin{aligned}\frac{dp(x_1, x_2)}{dt} &= w_{x_1-1}p(x_1-1, x_2) - (w_{x_1} + \bar{w}_{x_1})p(x_1, x_2) \\ &\quad + \bar{w}_{x_1+1}p(x_1+1, x_2) + w_{x_2-1}p(x_1, x_2-1) - (w_{x_2} + \bar{w}_{x_2})p(x_1, x_2 \\ &\quad + \bar{w}_{x_2+1}p(x_1, x_2+1) - D(x_1+x_2)p(x_1, x_2) \\ &\quad + D(x_1+1)p(x_1+1, x_2-1) + D(x_2+1)p(x_1-1, x_2+1)\end{aligned} \tag{5.64}$$

where w_{x_1} and $\bar{w}_{x_1}$ refer to the birth and death rates from a state with x molecules in cell 1.

The Monte Carolo simulation and the macroscopic studies mentioned above suggest that in this system inhomogeneous states occur for values of D less than a certain critical value, whereas for large diffusion rates the two-cell system behaves as a single-cell system showing only the homogeneous solution. In the probabilistic sense, reference to Figure 5.2 suggests that we would have four peaks of the probability distribution corresponding to the four stable states for lower values of D. Two of these states are the homogeneous stable states while the other two are inhomogeneous. On a short time scale the four regions can be considered as independent with the distribution relaxing independently. On the long time scale a slow exchange between them occurs so as to finally realise the stationary distribution. The typical result just described suggests the hypothesis that in each cell the exchange between the regions (i.e. four stable states) occurs on a time scale that is much slower than the relaxation in each of these regions and amounts to a quasi-stationarity hypothesis. One could use this information to simplify the evolution equation. Borgis and Moreau (1984) utilised this to provide a simpler solution to the problem. Their analysis shows that all the earlier results obtained using the Monte Carlo simulation technique can be recovered easily.

In a variation of the study Tambe *et al.* (1985f) modified the Schlogl reaction scheme by considering the case of an autocatalytic reaction followed by nonelementary reaction. The typical case analysed considered the last step to follow a L–H kinetic scheme resulting in the following macroscopic description of the time–cell model:

$$\frac{dx_1}{dt} = -x_1^3 + Ax_1^2 - \frac{Bx_1}{1+qx_1} + C + D(x_2 - x_1) \tag{5.65}$$

$$\frac{dx_2}{dt} = -x_2^3 + Ax_2^2 - \frac{Bx_2}{1+qx_2} + C + D(x_1 - x_2) \tag{5.66}$$

where

$$A = \frac{k_1[A]}{k_{-1}}, \qquad B = k_2/k_{-1}, \qquad C = \frac{k_{-2}B}{k_{-1}}$$

Typical results obtained for the macroscopic system are presented in Figure 5.3 and described below.

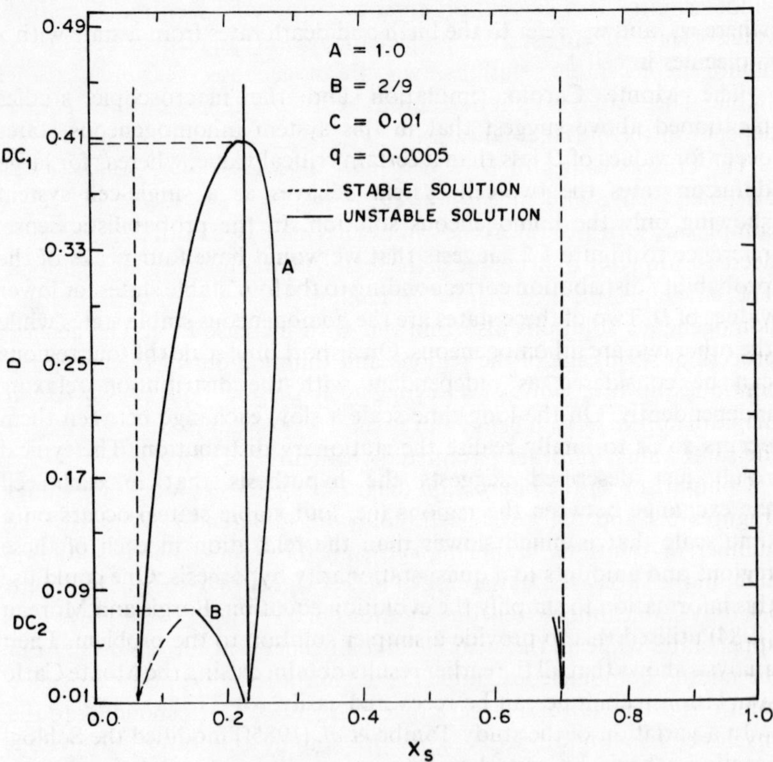

FIGURE 5.3 Steady state solutions (X_s) for a compartment model showing the effect of exchange coefficient (D).

A scrutiny of the curves in Figure 5.3 reveals several interesting features. Beyond a certain value of D ($>DC_1$), the diffusive exchange phenomenon has no influence on the solutions, and the heterogeneous system, beyond this value of D, continues to possess the original homogeneous solutions (2 stable, 1 unstable). For values of D less than this critical value, a region of five solutions, two of which are inhomogeneous and three homogeneous, begins. Both the inhomogeneous solutions are, however, always unstable (saddle points) in the entire range of D for which they appear. As the value of D is lowered further, a second critical value of D ($=DC_2$) is reached at which the region of nine solutions begins. The additional four inhomogeneous solutions originate from a multiple singular point of saddle-stable node. Two of these solutions are of the stable node while the other two are saddle points. As a general observation, it may be seen that a critical value of D ($=DC_2$) exists below which the system can possess, besides the homogeneous stable solutions, two inhomogeneous solutions that are stable. The actual solution reached by the system would of course depend on the initial conditions.

The critical values of D, which mark the onset of regions for the existence of five and nine solutions, depend on the values of q in a complex way. In general, as the value of q is increased both DC_1 and DC_2 decrease implying a shrinkage in the region of inhomogeneous solutions. The envelopes marked A and B joining the lower and the middle homogeneous solutions also thin out with increase in q. The general process of shrinkage of the regions and thinning out of the envelopes continues up to a value of q beyond which a new type of bifurcation diagram appears. Such behaviour is shown, for example, for $q = 0.5$ for the same values of other parameters in Figure 5.4.

Comparison of the bifurcation diagrams for these ranges of q values (Figures 5.3 and 5.4) reveals the following differences. Primarily the number of solutions branching off from the homogeneous solutions differ in the two cases. In the case of higher values of q, such as in Figure 5.4, only stable inhomogeneous solutions branch off from the lower and upper states while all the unstable inhomogeneous solutions do so from the middle homogeneous state. The region for the occurrence of nine solutions extends over a larger range of D values. Corresponding to the envelope A in Figure 5.3, an envelope joining the lower and middle homogeneous solutions exists. However, while this envelope in Figure 5.3 represents the unstable inhomogeneous solutions in the entire range

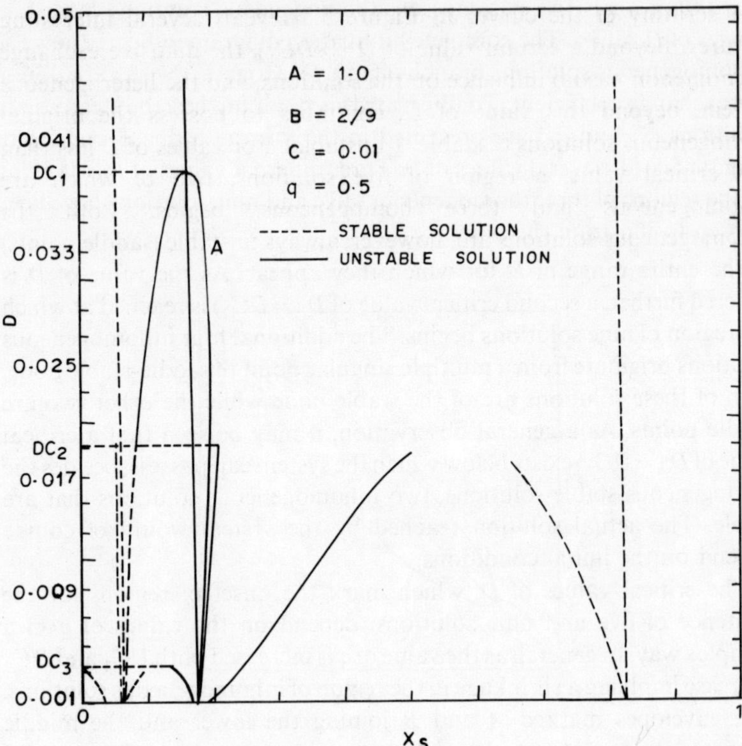

FIGURE 5.4 Steady state solutions (X_s) for a compartment model showing the effect of exchange coefficient (D).

of D values for which they exist, in Figure 5.4 part of this envelope emanating from the lower homogeneous solution represents a stable inhomogeneous solution in the region where nine solutions prevail. As we enter the region of nine solutions and decrease the value of D further, we come across a third critical value of D ($=DC_3$) where new solutions branching off from the lower state emerge. Both the inhomogeneous solutions are stable and one of them lies below the lower homogeneous solution. The existence of inhomogeneous solutions that are stable is known in several cases; however, the present case provides perhaps the only example so far of its existence outside the region of homogeneous solutions (in fact below the lower homogeneous solution).

The deterministic analysis of the reaction function for various values of q thus reveals the existence of inhomogeneous solutions in certain ranges of D values. The complete bifurcation diagram indicates how the inhomogeneous solutions branch off and fuse together with variation of the exchange coefficient. To illustrate this point further, the steady state solutions (x_1 and x_2) are presented in the form of a phase–plane plot (Figure 5.5) for several values of D. Three different values of D belonging to the regions of three (homogeneous), five and nine solutions have been chosen to illustrate the point.

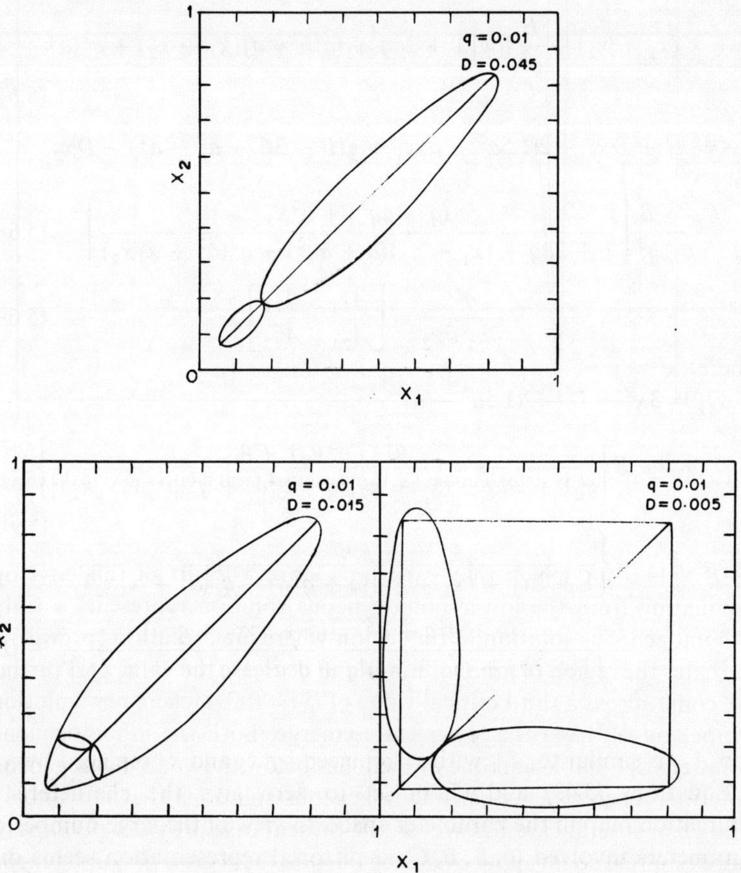

FIGURE 5.5 Phase plane plot for different values of D. The points of intersections denote the solution.

The parametric representation of the bifurcation map for the two-compartment model involves the use of the potential V. For the assumed reaction function $f(x)$ the potential V may be obtained from Equation (5.58) which can be appropriately differentiated to yield the first and second derivatives. For mapping the bifurcation contours it is necessary to set these derivatives to zero and replace x by $(x + a)$, where a refers to the middle homogeneous solution that is unstable. Using this procedure we obtain the following relations between the parameters:

$$V = \tfrac{1}{4}(x_1^4 + x_2^4) + (1.5a^2 - a)(x_1^2 + x_2^2) + \tfrac{1}{3}[B - 3C + a^3 - a^2] \times (x_1 + x_2) - \frac{B}{q^2}\log[1 + 2aq + (q^2a + q)(x_1 + x_2) + q^2(a^2 + x_1x \tag{5.67}$$

$$\frac{\partial V}{\partial x_1} = x_1^3 + Dx_1 - 2(1.5a^2 - a)x_1 + \tfrac{1}{3}(B - 3C + a^3 - a^2) - Dx_2 - \frac{B}{q^2}\left[\frac{(q + aq^2) + q^2x_2}{1 + 2aq + (x_1 + x_2)(q + aq^2) + q^2(a^2 + x_1x_2)}\right] \tag{5.68}$$

$$\frac{\partial V}{\partial x_1\,\partial x_2} = \begin{bmatrix} A_{11} & A_{12} \\ A_{21} & A_{22} \end{bmatrix} \tag{5.69}$$

where

$$A_{11} = 3x_1^2 + D - 2(1.5a^2 - a) + B\left\{\frac{q^2x_2 + q + q^2a}{q[1 + 2aq + (x_1 + x_2)(q + q^2a) + q^2(a^2 + x_1x_2)]}\right\}^2 \tag{5.70}$$

$$A_{21} = A_{12} = \left\{\frac{q^2}{1 + 2aq + (x_1 + x_2)(q + q^2a) + q^2(a^2 + x_1x_2)} - \frac{(q + q^2a + q^2x_1)(q + q^2x_2 + q^2a)}{[1 + 2aq + (x_1 + x_2)(q + q^2a) + q^2(a^2 + x_1x_2)]^2} - D\right\} \tag{5.71}$$

and A_{22} is similar to A_{11} with x_1 replaced by x_2 and x_2 replaced by x_1.

Equations (5.69) and (5.70) set to zero give the characteristic bifurcation map in the parameter space. In view of the large number of parameters involved (q, D, B, C), a pictorial representation seems difficult; however, for the simple case of $q = 0$, $D = 0$, we can obtain the map shown in Figure 5.6. Clearly, the bifurcation corresponds to the

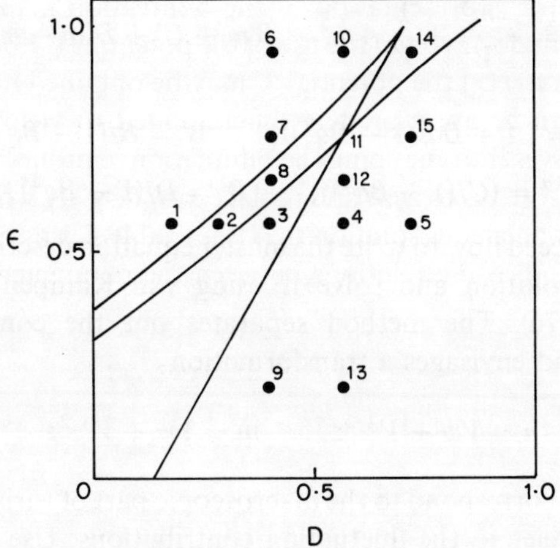

FIGURE 5.6 Characteristic bifurcation map in parameter space *B–C*.

cusp or fold type of catastrophe and crossing of the lines indicates annihilation or generation of new state. As the value of q increases the region in the B–C space shrinks until for sufficiently large value of q it disappears altogether.

Stochastic Analysis of the Two-Compartment Model

In order to analyse the effect of fluctuations on the final solutions obtained, we reformulate the deterministic equations thus:

$$\dot{n} = \frac{-n^3}{V^2} + \frac{A}{V}n^2 - \frac{B_n}{1 + \dfrac{q}{V}n} + CV + D(m - n) \tag{5.72}$$

$$\dot{m} = \frac{-m^3}{V^2} + \frac{A}{V}m^2 - \frac{B_m}{1 + \dfrac{q}{V}m} + CV + D(n - m) \tag{5.73}$$

where $n = x_1 V$ and $m = x_2 V$ and V refers to the appropriate volumes of the system. For a large volume system ($q/V \ll 1$), Equations (5.72) and (5.73) can be written as

$$\dot{n} = -\frac{n^3}{V^2} + \frac{1}{V}A'n^2 - B'n + C' + D'(m - n) \tag{5.74}$$

$$\dot{m} = -\frac{m^3}{V^2} + \frac{1}{V}A'm^2 - B'm + C' + D'(n - m) \tag{5.75}$$

where

$$A' = (A + Bq/(1 - Bq^2)), \qquad B' = B/(1 - Bq^2),$$

$$C' = (C/(1 - Bq^2)), \qquad D' = D/(1 - Bq^2).$$

One could proceed now to write the master equation describing the joint probability evolution and solve it using van Kampen's method of expansion (1976). The method separates out the contributions of fluctuations and envisages a transformation

$$n = V\phi + V^{1/2}z, \qquad m = V\psi + V^{1/2}z' \tag{5.76}$$

where ϕ and ψ correspond to the macroscopic part of variables n and m, and z and z' refer to the fluctuating contributions. Use of Equations (5.76) in the master equation gives rise to the following equations describing the evolution of the macroscopic and fluctuating variables:

$$\frac{d\phi}{dt} = -\phi^3 + A'\phi^2 - B'\phi + C' + D'(\psi - \phi) \tag{5.77}$$

$$\frac{d\psi}{dt} = -\psi^3 + A'\psi^2 - B'\psi + C' + D'(\phi - \psi) \tag{5.78}$$

$$\frac{d\langle z\rangle}{dt} = -[3\phi^2 - 2A'\phi + B' + D']\langle z\rangle - D'\langle z'\rangle \tag{5.79}$$

$$\frac{d\langle z'\rangle}{dt} = -[3\psi^2 - 2A'\psi + B' + D']\langle z\rangle - D'\langle z\rangle \tag{5.80}$$

The formulation of the problem as above affords a more transparent view of the effect of q on the system behaviour. The effect now gets incorporated in the altered definition of the characteristic parameters and the original Schlogl model is recovered. Also, the macroscopic equations as obtained here are exactly identical to the original deterministic equations, so that the analysis in the earlier section is also valid for the macroscopic equations. The influence of fluctuations can be calculated by analysing Equations (5.79) and (5.80) which being linear can be exactly solved around the steady macroscopic states. The equations indicate exponential decay of fluctuations so that the macroscopic state also corresponds to the states of the deterministic equations.

In order to calculate the approach rate of the macroscopic steady states, we can calculate the eigenvalues of Equations (5.77) and (5.78) as

$$\lambda_{1,2} = \tfrac{1}{2}[3(\phi)_{st}^2 + \psi_{st}^2 - 2A'^2(\phi_{st} + \psi_{st}) + 2(B' + D') \pm \{2A'(\phi_{st} - \psi_{st}) - 3(\phi_{st}^2 - \psi_{st}^2) + 4D'^2\}^{1/2}] \qquad (5.81)$$

where ϕ_{st}, ψ_{st} refer to the steady state values of the macroscopic states. The relaxation time for the steady state is then simply obtained as $\tau = 1/\lambda$. Referring to Figures 5.3 and 5.4 it is observed that for a given value of D (and other fixed set of parameter values) one could have three, five or nine possible macroscopic states. Substituting the appropriate steady state values in Equation (5.81) we can obtain the relaxation time for that state. In view of the finite eigenvalues, it is expected that the approach rate to the steady state would be normal and the phenomenon of critical slowing down (see Tambe *et al.*, 1984b) may not be observed for this system. The fluctuation equations [Equations (5.79) and (5.80)] around the macroscopic states also bear identical eigenvalues implying a normal rate of decay of fluctuations.

The realisation that for large volume systems the influence of a nonelementary reaction step, such as the one considered here, can be incorporated in the characteristic coefficients of the rate function for the Schlogl model, affords an easy treatment of this more complex case. In fact the entire analysis of Schlogl's model now becomes applicable to this case with altered definitions of the coefficients. While it is not our intention to rederive these findings here, a special case concerning coexistence of solutions deserves mention.

In the presence of diffusive gradients the basic Schlogl model is known to possess three homogeneous solutions, two of which can coexist. The criterion for the coexistence is derived on the basis of identical values of the potential ϕ_1 for the two states. In the present case also one can calculate the potentials ϕ_1 using (5.58) for the different states and equate them to obtain the following simple criterion for their coexistence:

$$C' = (A' - 2)\left[\frac{B'}{1 + q(A' - 2)} - 2(A' - 2)\right] \qquad (5.82)$$

It is seen that the influence of the parameter q is explicitly indicated in this relation.

5.5.2 Stochastic Analogue Using Langevin Equation

In the previous section the stochastic model was formulated using the master equation formation. In the present section we shall use the Langevin approach where appropriate random forces are merely added to the phenomenological equations. For simplicity we shall again consider the system analysed in the earlier section. The phenomenological equation for a one-compartment reaction–diffusion system is given by

$$\frac{dx}{dt} = f(x) + D\nabla x \tag{5.83}$$

or, in terms of the potentials as in the previous section, by

$$\frac{dx}{dt} = -\frac{\partial V}{\partial x} + D\nabla x = -\frac{\delta W}{\delta x} \tag{5.84}$$

Now we add to this phenomenological equation an external noise and obtain the following Langevin equation:

$$\frac{dx}{dt} = f(x) + D\nabla x + \xi(r, t) \tag{5.85}$$

Here $\xi(r, t)$ is a stochastic force with the properties

$$\langle \xi(r, t) \rangle = 0$$

$$\langle \xi(r, t)\xi(r', t') \rangle = 2\varepsilon\delta(t - t')\delta(r - r') \tag{5.86}$$

where ε refers to the intensity of noise.

One can easily write the Fokker–Planck equation for this, the stationary solution of which can be readily obtained for the natural boundaries as

$$p(x, \infty) = \exp\left\{\frac{1}{\varepsilon}[W_0 - W(x)]\right\} \tag{5.87}$$

where W_0 refers to the normalisation constant.

Let us now apply this methodology to Schlogl's reaction scheme analysed in the earlier section, and for the sake of convenience restrict the development to the two-compartment case. The model Equation

(5.85) can be discretised as

$$\frac{dx_i}{dt} = f(x_i) + D(x_j - x_i) + \xi_i(x, t), \qquad i, j = 1, 2 \tag{5.88}$$

with

$$\langle \xi_i(t) \rangle = 0$$

$$\langle \xi_i(t)\xi_i'(t) \rangle = 2\varepsilon\delta(t - t')\delta_{ij} \tag{5.89}$$

The potential W, as defined in Equation (5.85), can be obtained from the previous section

$$W(x_1, x_2) = V(x_1) + V(x_2) + \tfrac{1}{2}D(x_1 - x_2)^2 \tag{5.90}$$

The stationary probability distribution can then be written using Equation (5.87). Often, in these problems, we are interested in the probability distribution of the mean value of the concentration x which can be written as $x = (x_1 + x_2)/2$. Substituting for x_2 in terms of x and x_1, the probability distribution of x can be obtained as

$$\begin{aligned} p(x, \infty) &= \int_0^{2x} dx_1\, p(x_1, 2x - x_1, \infty) \\ &= \int_0^{2x} dx_1 \exp\left\{\frac{1}{\varepsilon}[W_0 - w(x_1, 2x - x_1)]\right\} \end{aligned} \tag{5.91}$$

Substituting the functional form for Schlogl's model, we finally have

$$\begin{aligned} p(x, \infty) = \int_0^{2x} dx_1 \exp\Bigg\{\frac{1}{\varepsilon}\Bigg[& W_0 - \tfrac{1}{4}(x_1^4 + (2x_1 - x_1)^4) \\ & + \frac{A}{3}(x_1^3 + (2x - x_1)^3) - \frac{B}{2}(x_1^2 - (2x - x_1)^2) \\ & + 2Cx - 2D(x_1 - x)\Bigg]\Bigg\} \end{aligned} \tag{5.92}$$

The solution to this equation has been presented by Malchow *et al.* (1980) and Figure 5.7 shows the $p_{st}(x)$–x curves for different extents of noise parameters t and D.

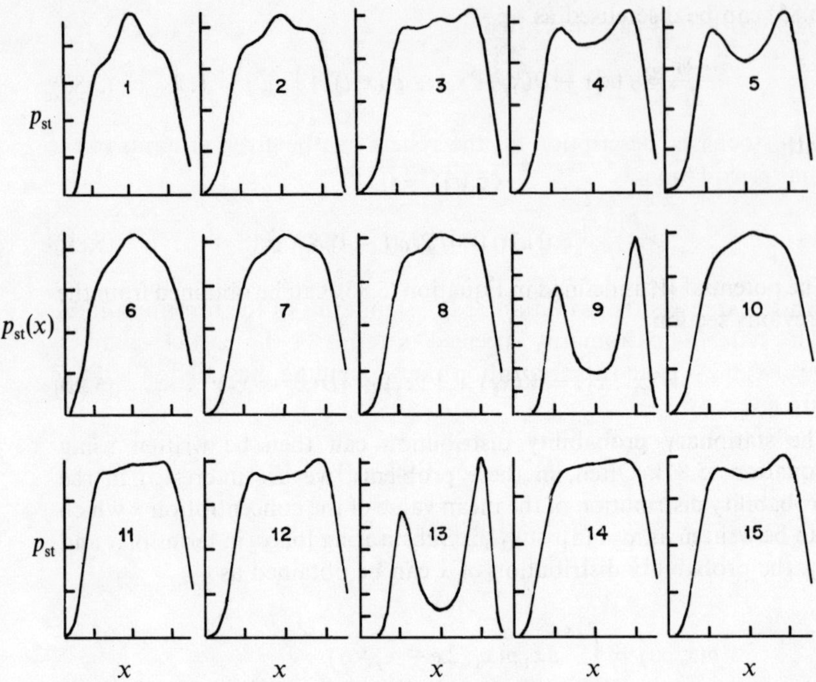

FIGURE 5.7 Probability density functions corresponding to points marked in Figure 5.6.

The results for the two-compartment model can be easily generalised for n compartments, although the calculations become more involved.

Appendix 5.A

General Nonlinear Fokker–Planck Equation and its Moments

Consider a general multistep reaction scheme

$$\mu_{ri}A_i + \nu_{rj}X_j \rightleftharpoons \mu'_{ri}A_i + \nu'_{rj}X_j, \qquad r = 1, \ldots m$$

where μ, ν and the corresponding primed parameters represent the stoichiometric coefficients for the rth step and the subscripts i, j identify the component in the reaction scheme. The rates for the forward and backward steps can be written as

$$R_r = k_r \prod_{i,j} A_i^{\mu_{ri}} X_j^{\nu_{ri}}, \qquad R'_r = k'_r \prod_{i,j} A_i^{\mu'_{ri}} X_j^{\nu'_{ri}} \tag{1}$$

The stochastic description for the system can be given in terms of the master equation as

$$\frac{\partial P(i,t)}{\partial t} = \sum_j W_{ij}(X)P(j,t) - W_{ji}(X)P(i,t) \tag{2}$$

which expresses the evolution of $P(i, t)$ in terms of the transition rates. The state i is customarily specified in terms of the particle numbers $i = (X_i, \ldots X_n)$ and the transition rates assuming mass action kinetics are given by

$$W_{XX'} = \sum_{r=1}^{m} W_{XX'}^{(r)} + W_{X'X}^{(r)} \tag{3}$$

where the forward and backward transition rates are given by

$$W_{XX'}^{(r)} = R_r \sum_{j=1}^{n} \delta(X_j, X'_j + \nu'_{rj} - \nu_{rj}) \tag{4}$$

$$W_{X'X}^{(r)} = R'_r \sum_{j=1}^{n} \delta(X_j, X'_j - \nu'_{rj} + \nu_{rj}) \tag{5}$$

Using the scaled quantities $\tilde{W}_{XX'} = VW_{XX'}$ with $x_j = X_j/V$ in the master equation and expanding it in powers of V^{-1} leads to the nonlinear Fokker–Planck equation

$$\frac{\partial P(X,t)}{\partial t} = -\frac{\partial}{\partial X_i}[\alpha_{1,j}(X)P(X,t)] + \frac{1}{V}\frac{\partial}{\partial x_i}\frac{\partial}{\partial x_j}[\alpha_{2,ij}(X)P(X,t)] \tag{6}$$

where $\alpha_{1,i}$ and $\alpha_{2,ij}$ refer to the drift and diffusion terms and can be identified from the macroscopic description as

$$\alpha_{1,i}(X) = \sum_r (\nu'_{ri} - \nu_{ri}) \frac{R_r - R'_r}{V} \tag{7}$$

$$\alpha_{2,ij}(X) = \frac{1}{2}\sum_r (\nu'_{ri} - \nu_{ri})(\nu'_{rj} - \nu_{rj}) \frac{R_r + R'_r}{V} \tag{8}$$

and define the first and second moment of transition rates. Equation (7) represents the multivariate analog of the simple one-dimensional

Fokker–Planck equation and is subject to the same criticism of inadequate approximation of the multivariate master equation. While the corrective measures to take care of this inadequacy have been dealt with in Section 4.3.1, in the present form the zeroeth order in $1/V$ generates the macroscopic equation

$$\frac{dx_j}{dt} = \alpha_{1,j}(X) \tag{9}$$

To first order, linearizing around this we can obtain the equation for the variance $\sigma_{ij}(t) = \langle \delta x_i(t)\, \delta x_j(t) \rangle$

$$\frac{\partial \sigma_{ij}(t)}{\partial t} = \frac{2}{V} \alpha_{2,ij}(X) + \frac{\partial \alpha_{1,j}}{\partial x_m} \sigma_{mj}(t) + \sigma_{im} \frac{\partial \alpha_{1,j}}{\partial x_m} \tag{10}$$

For a system in stationary state, $\alpha_1(X_0) = 0$ and the static correlations are given by

$$\eta\sigma + \sigma\eta^T = -\frac{2}{V} \alpha_2(x_0) \tag{11}$$

where $\eta = d\alpha_1/dx$ is the relaxation matrix. Equation (11) is a set of $(n)(n+1)/2$ linear equations for a n component system. The approximation yields good results for large volume systems operated not close to the instability point.

REFERENCES

Borgis, D. and Moreau, M., *J. Stat. Phys.* **37**, 631 (1984).

Chaturvedi, S., Gardiner, C. W., Matheson, J. S. and Walls, D. F., *J. Stat. Phys.* **17**, 469 (1977).

Ebeling, W. and Malchow, H., *Ann. Physik* **36**, 121 (1979).

Ebeling, W. and Schimansky-Geier, L., *Proceedings 6th International Conference on Thermodynamics*, Merseburg, Germany, 1980.

Frankowicz, M. and Gudowska-Nowak, E., *Physica* **116A**, 331 (1982).

Gardiner, C. W., *J. Stat. Phys.* **15**, 451 (1976).

Gardiner, C. W., McNeil, K. J., Walls, D. F. and Matheson, J. S., *J. Stat. Phys.* **14**, 307 (1976).

Gillespie, D. T., *J. Comput. Phys.* **22**, 403 (1976).

Grossmann, B., *J. Chem. Phys.* **65**, 2007 (1976).

Hanusse, P., *J. Chem. Phys.* **67**, 1282 (1977).

Hanusse, P. and Blanche, A., *J. Chem. Phys.* **74**, 6148 (1981).

Lemarchand, H. and Nicolis, G., *Physica* **82A**, 521 (1976).

Malchow, H., Ebeling, W., Feistel, R. and Schimansky-Geier, L., *Ann. Physik* **40**, 151 (1983).

Malek-Mansour, M. and Houdard, J., *Phys. Lett.* **70A**, 366 (1979).
Malek-Mansour, M., van den Broeck, C., Nicolis, G. and Turner, J. W., *Ann. Phys.* (*N.Y.*) **131**, 282 (1981).
Mashiyama, H., Ito, A. and Ohta, T., *Prog. Theor. Phys.* **54**, 1050 (1975).
Mou, C. Y., *J. Chem. Phys.* **68**, 1385 (1978).
Nicolis, G., Malek-Mansour, M., Kitahara, K. and van Nypelseer, A., *J. Stat. Phys.* **14**, 417 (1976).
Nicolis, G. and Malek-Mansour, M., *J. Stat. Phys.* **22**, 495 (1980).
Prigogine, I., Lefever, R., Turner, J. S. and Turner, J. W., *Phys. Lett.* **51A**, 317 (1975).
Ritcher, P. M., Procaccia, I. and Ross, J., *Adv. Chem. Phys.* **XLIII**, 217 (1980).
Tambe, S. S., Ravikumar, V., Kulkarni, B. D. and Doraiswamy, L. K., *Chem. Eng. Sci.* **40**, 2303 (1986f).
Turner, J. S., *J. Phys. Chem.* **81**, 2379 (1977).
van den Broeck, C., Horsthemke, W. and Malek-Mansour, M., *Physica* **89A**, 359 (1977).
van Kampen, N. G., *Adv. Chem. Phys.* **34**, 245 (1976).
Walgraef, D., Dewel, G. and Borckmans, P., *Adv. Chem. Phys.* **49**, 311 (1982).

PART III

Role of External Noise

6 Theoretical Development

6.1 Introduction

THE STOCHASTIC approach followed in Part II assumed the internal fluctuations to be Markovian and described them in terms of a discrete birth and death process via the master equation, or as a continuous diffusion process via the Fokker–Planck equation. The stochastic approach in the present case leads to another type of equation—the stochastic equation or the equivalent Langevin equation—which has also been discussed to some extent in the preceding chapters. To recapitulate, we note that a given Langevin equation can always be converted into an equivalent Fokker–Planck equation which can also therefore be used to describe the external fluctuations. In a general Langevin equation

$$\frac{dx_i}{dt} = F(x_i) + G(x_i)\xi_i \tag{6.1}$$

where x_i represents the macroscopic variable, $F(x_i)$ usually a nonlinear function and $G(x_i)$ the coupling of the variable x_i to the external noise ξ_i; the function $G(x_i)$ decides whether the coupling is additive or multiplicative depending on whether it is constant or otherwise. As a general observation, one could easily appreciate that the two types of coupling would have different properties. Thus an additive noise would not affect the stability of the deterministic equation associated with Equation (6.1), while the multiplicative noise may well change the stability properties of the system. In fact, one could expect a shift in the existing patterns of a deterministic system, or even a completely new type of transition originating due to the presence of the external noise. In what follows we shall exemplify this feature of the role of external fluctuations by considering specific examples in Section 6.2.

A common assumption regarding the statistical properties of the noise concerns its Gaussianity with no correlation existing between different times (white noise). In real situations, however, white noise is more of an idealisation and finite correlation time effects may become important (leading to coloured noise). The influences of coloured noise as well as of deviations from Gaussianity are considered later in this chapter.

6.2 Analysis Based on Various Phenomenological Models

The present section is concerned with the analysis of several model reaction schemes incorporating the effect of external fluctuations. To begin with we consider the case of a simple Schlogl's model:

$$A + X \rightleftharpoons 2X$$

$$B + X \longrightarrow C$$

which, for constant concentrations of species A and B and continuous removal of species C, leads to a first-order differential equation with quadratic rate term. To analyse a general class of such systems we begin with the phenomenological equation

$$\frac{dx}{dt} = -x^2 + \beta x \tag{6.2}$$

where β represents a positive or negative coefficient. The deterministic Equation (6.2) always possesses a stationary solution that corresponds to extinction of species x ($x = 0$). In addition, $x = \beta$ also represents another solution which exists only if $\beta > 0$. The stability of these solutions, as obtained by linearising Equation (6.2) around the states, indicates that $x = 0$ is a stable solution if $\beta < 0$, while $x = \beta$ is a stable solution if $\beta > 0$. The coefficient β in the equation comprises factors such as the concentrations of species A and B, their input–output rates, etc. and may as such involve certain fluctuations around its mean value. To represent this effect of environment we should associate a stochastic differential equation with Equation (6.2) taking β as a random variable. Let us suppose β to be a white Gaussian noise with a mean value of $\bar{\beta}$ and variance equal to σ^2. The stochastic differential equation can then be written as

$$\frac{dx}{dt} = [-x^2 + \bar{\beta}x] + \sigma x \frac{dW}{dt} = F(x) + G(x)\frac{dW}{dt} \tag{6.3}$$

where W is a Wiener process and its derivative a white Gaussian noise with zero mean and δ-correlated variance.

As noted earlier it is necessary to interpret this stochastic equation in the Stratonovich sense, to keep to the correct physical picture. The mathematical operations are, however, developed for the Ito form. It is necessary therefore to convert this equation into the Ito form:

$$\begin{aligned}\frac{dx}{dt} &= [F(x) + \tfrac{1}{2}G(x)G'(x)] + G(x)\frac{dW}{dt} = F_1(x) + G(x)\frac{dW}{dt} \\ &= [-x^2 + \bar{\beta}x + \tfrac{1}{2}\sigma^2 x] + \sigma x \frac{dW}{dt}\end{aligned} \tag{6.4}$$

The equation possesses a stationary solution ($x = x_0$) when

$$F_1(x_0) = G(x_0) = 0 \tag{6.5}$$

The obvious solution satisfying this is when $x_0 = 0$. As regards the stability of this solution, we shall inquire into it a little later after ascertaining all other solutions. To examine other solutions to this problem we convert Equation (6.4) into a Fokker–Planck equation. Note that Equation (6.4) is the Ito representation of the stochastic equation. The corresponding Fokker–Planck equation would take the form

$$\frac{\partial p}{\partial t} = -\frac{\partial}{\partial x} F_1(x)p + \frac{1}{2}\frac{\partial^2}{\partial x^2} G^2(x)p \tag{6.6}$$

or, in explicit terms,

$$\frac{\partial p}{\partial t} = \frac{\partial}{\partial x}[x^2 - (\bar{\beta} + \tfrac{1}{2}\sigma^2)x]p + \tfrac{1}{2}\sigma^2 \frac{\partial^2}{\partial x^2} x^2 p \tag{6.7}$$

Equation (6.7) possesses the stationary solution

$$p_{\mathrm{st}}(x) = \frac{N}{G^2(x)} \exp \int^x \frac{2F_1(x)}{G^2(x)}\, dx \tag{6.8}$$

$$p_{\mathrm{st}}(x) = \left[\frac{(2/\sigma^2)^{(2\bar{\beta}/\sigma^2)}}{\Gamma(2\bar{\beta}/\sigma^2)}\right] x^{(2\bar{\beta}/\sigma^2)-1} \exp(-2x/\sigma^2), \qquad \bar{\beta} > 0 \tag{6.9}$$

where the first term represents the normalisation factor and exists provided $\bar{\beta} > 0$. The stationary solution $p_{st}(x)$ can therefore be considered as a probability density. As regards the stability of this solution one could construct the Liapunov function, which for the one-dimensional case can be generally written as

$$\Phi[p(x,t)] = \int dx\, p(x,t) \ln \frac{p(x,t)}{p_{st}(x)} + \phi \tag{6.10}$$

Using the known stability theorems one can then ascertain the stability of the solution [note: ϕ is defined by $p_{st}(x) = N \exp(-\phi(x))$].

A close look at the stationary probability density reveals the following points. At $x = 0$, the stationary probability density $p_{st}(x)$ always diverges for $0 < \bar{\beta} < \sigma^2/2$. Also $p_{st}(x)$ becomes zero for $\bar{\beta} > \sigma^2/2$.

Let us now return to the question of the stability of the solution $x_0 = 0$. For this purpose we construct a function

$$u = \int_0^\delta \exp\left\{\int_y^\delta \frac{2F_1(z)}{G^2(z)}\, dz\right\} dy \tag{6.11}$$

where the integral is similar to that in Equation (6.8). The function thus measures the effect of small deviations and the situation is stable if u remains finite. Substituting the actual expressions for the functions $F_1(z)$ and $G^2(z)$ and carrying out the integrations, one finds that u remains finite provided $\bar{\beta} < 0$. This implies that the solution $x_0 = 0$ is stable if $\bar{\beta} < 0$ and unstable if $\bar{\beta} > 0$.

Let us now summarise the results of the anslysis which have been presented in greater detail by Horsthemke and Malek-Mansour (1976). We note that

(i) if $\bar{\beta} < 0$, $x = 0$ is a stable solution,
(ii) $\bar{\beta} = 0$ is a transition point since $x = 0$ is stable for $\bar{\beta} < 0$ and unstable for $\bar{\beta} > 0$,
(iii) $p_{st}(0)$ diverges for $0 < \bar{\beta} < \sigma^2/2$, and
(iv) $p_{st}(0) = 0$ for $\bar{\beta} > \sigma^2/2$.

The analysis reveals the existence of two transition points corresponding to $\bar{\beta} = 0$ and $\bar{\beta} = \sigma^2/2$. At the former point the stable state changes to an unstable state typical of a soft transition, while at the latter point the divergence of $p_{st}(0)$ disappears with its value dropping to zero. The case corresponds to hard transition and originates due to the external noise.

In order to illustrate this more clearly we open up the Fokker–Planck equation (Equation 6.7) at steady state to obtain

$$[x^2 - (\bar{\beta} + \tfrac{1}{2}\sigma^2)x]p + \sigma^2 xp + \tfrac{1}{2}\sigma^2 x^2 \frac{\partial p}{\partial x} = 0 \tag{6.12}$$

where for the extremal value of $x = x_m$ the last term can be ignored. The resulting equation is quadratic in nature

$$x_m^2 - (\bar{\beta} + \tfrac{1}{2}\sigma^2)x_m + \sigma^2 x_m = 0 \tag{6.13}$$

and possesses the solution

$$x_{m1} = 0, \qquad x_{m2} = \bar{\beta} - \tfrac{1}{2}\sigma^2, \qquad \text{if} \quad \bar{\beta} > \sigma^2/2 \tag{6.14}$$

We note from Equation (6.14) that x_{m1} represents the maximum value when $0 < \beta < \sigma^2/2$, while x_{m2} takes over when $\bar{\beta} > \sigma^2/2$. The increasing presence of external fluctuations thus diminishes the most probable value with respect to the deterministic value.

The mean and variance also can be calculated to obtain

$$\langle x \rangle = \int x p_{st}(x)\, dx = \bar{\beta}$$

$$\langle x^2 \rangle = \int x^2 p_{st}(x)\, dx = \bar{\beta}(\bar{\beta} + \sigma^2/2) \tag{6.15}$$

$$\langle (y - \langle y \rangle)^2 \rangle = \bar{\beta}\sigma^2/2$$

The calculations of mean, variance, etc. indicate that the $\langle x \rangle$–$\bar{\beta}$ curve is identical to the deterministic x–$\bar{\beta}$ curve and does not reflect the change in character of the stationary distribution at $\bar{\beta} = \sigma^2/2$. The x_m–$\bar{\beta}$ curve exhibits, however, the transition at $\bar{\beta} = \sigma^2/2$ and not the deterministic one.

The simple model system analysed above in some detail brings out certain features of external influence. The same methodology can be applied to other model systems, and for brevity we present below the main results. Model systems with the phenomenological description

$$\frac{dx}{dt} = -x^3 + \beta x^2 - x \tag{6.16}$$

have been analysed by Horsthemke and Malek-Mansour (1976):

$$A + 2X \rightleftharpoons 3X$$
$$B + 2X \longrightarrow C$$
$$X \longrightarrow D$$

which can be considered as an extended Schlogl's model. The stationary solutions of Equation (6.16) can be obtained as

$$x_1 = 0, \qquad x_2 = \beta/2 + [(\beta^2/4) - 1]^{1/2},$$
$$x_3 = \beta/2 - [(\beta^2/4) - 1]^{1/2} \tag{6.17}$$

Clearly the solutions x_2, x_3 exist only if $\beta > 2$. As regards the stability of these solutions, the analysis reveals $x_1 = 0$ as the stable state for all values of β and x_2 and x_3 as stable and unstable states, respectively, when they exist. The typical x–β curve for the deterministic case is shown in Figure 6.1. As before the parameter β contains the influence of environment, and to provide a stochastic description corresponding to Equation (6.16) we assume β to be a random variable corresponding to a Gaussian white noise with mean value of $E(\beta) = \bar{\beta}$ and $E(\beta^2) = \sigma^2$. The stochastic differential equation can now be written as

$$\frac{dx}{dt} = -x^3 + \bar{\beta}x^2 - x + \sigma x^2 \frac{dW}{dt} \tag{6.18}$$

which should be interpreted in the Stratonovich sense. The

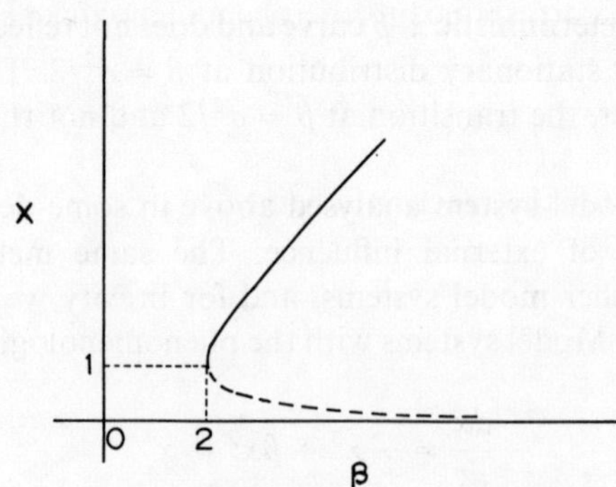

FIGURE 6.1 Deterministic solutions to Equation (6.16)
------ unstable, ——— stable.

corresponding Ito form can be written as

$$\frac{dx}{dt} = [(\sigma^2 - 1)x^3 + \bar{\beta}x^2 - x] + \sigma x^2 \frac{dW}{dt} \tag{6.19}$$

which possesses the stationary solution $x = 0$. As regards its stability, employing Equation (6.11) we find the solution to be stable for all values of $\bar{\beta}$. The result is the same as in the deterministic case. To ascertain the other stationary solutions we convert Equation (6.19) into a Fokker–Planck equation

$$\frac{\partial p}{\partial t} = -\frac{\partial}{\partial x}[(\sigma^2 - 1)x^3 + \bar{\beta}x^2 - x]p + \tfrac{1}{2}\sigma^2 \frac{\partial^2}{\partial x^2} x^4 p \tag{6.20}$$

which possesses the stationary solution

$$p_{st}(x) = N x^{-2-2\sigma^2} \exp[-(2\bar{\beta}/\sigma^2 x) + (1/\sigma^2 x^2)] \tag{6.21}$$

To estimate the normalisation constant we evaluate $\int_0^\infty p_{st}(y)\,dy$, which due to the divergence at zero always tends to ∞. The stationary probability density therefore does not exist. The earlier deterministic result of transition at $\bar{\beta} = 2$ thus disappears in the stochastic description. The reaction step $X \longrightarrow D$ creates a drift towards the boundary zero and is responsible for this result of the stochastic description. The present case illustrates the role of external fluctuations in systems possessing multistationarity.

Another model system for which the role of external fluctuations has been investigated corresponds to the phenomenological equation

$$\frac{dx}{dt} = C - x + \beta x(1 - x) = f(x) \tag{6.22}$$

which is realised for the reaction scheme

$$X \underset{2}{\overset{1}{\rightleftharpoons}} Y$$

$$A + X + Y \xrightarrow{3} 2Y + A$$

$$B + X + Y \xrightarrow{4} 2X + B$$

where the total number of X and Y particles is conserved ($X + Y = N$).

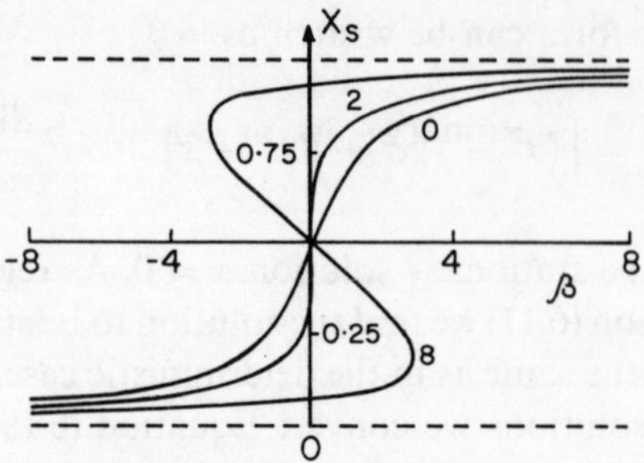

FIGURE 6.2 Stationary solutions for the deterministic and stochastic equations.

The coefficients C and β in Equation (6.22) can be identified as

$$C_1 = k_2/(k_1 + k_2), \qquad \beta = \frac{(k_4B - k_3A)}{N(k_1 + k_2)} \tag{6.23}$$

Arnold, Horthemke and Lefever (1978) used the specific value of $C = 1/2$ and $\beta = [(x_s - 1/2)/x_s(1 - x_s)]$, for which the stationary solution of Equation (6.23) is shown in Figure 6.2. The stability of the solution can be obtained by noting the sign of $f'(x)$ which can be obtained from Equation (6.22) as

$$[f'(x)]_{x_s} = -\frac{x_s^2 - x_s + 1/2}{x_s(1 - x_s)} \tag{6.24}$$

The corresponding stochastic description, subject to β as a fluctuating quantity with the mean value of $\bar{\beta}$ and variance of σ^2, can be written as

$$\frac{dx}{dt} = \{\tfrac{1}{2} - x + \bar{\beta}x(1 - x)\} + \sigma x(1 - x)\frac{dW}{dt} \tag{6.25}$$

or, in the Ito form, as

$$\frac{dx}{dt} = \left\{\tfrac{1}{2} - x + x(1 - x)\left[\bar{\beta} + \frac{\sigma^2(1 - 2x)}{2}\right]\right\} + \sigma x(1 - x)\frac{dW}{dt} \tag{6.26}$$

The corresponding Fokker–Planck equation can be written as

$$\frac{\partial p}{\partial t} = -\frac{\partial}{\partial x}\left\{\tfrac{1}{2} - x + x(1 - x)\left[\beta + \frac{\sigma^2(1 - 2x)}{2}\right]\right\}p + \tfrac{1}{2}\sigma^2\frac{\partial^2}{\partial x^2}x^2(1 - x)^2 p \tag{6.27}$$

which possesses the normalisable stationary solution that can be identified as the stationary probability density.

The extrema of this stationary solution can also be obtained referring to Equation (6.27):

$$\tfrac{1}{2} - x_m + x_m(1 - x_m)\left[\bar{\beta} + \frac{\sigma^2(1 - 2x_m)}{2}\right] + \sigma^2 x_m(1 + 2x_m^2 - 3x_m) = 0 \quad (6.28)$$

which for $\bar{\beta} = 0$ leads to a cubic equation

$$3\sigma^2 x_m^3 - \tfrac{3}{2}\sigma^2 x_m(x_m - 1) - x_m + \tfrac{1}{2} = 0 \quad (6.29)$$

one of the three roots of which is equal to 1/2. The stationary solutions of Equation (6.27) for different values of σ are also shown in Figure 6.2.

Notice that for $\bar{\beta} = 0$ in the deterministic equation we obtain $x_s = 1/2$. The phenomenological equation of the type

$$\frac{dx}{dt} = C + (1 - k_m x)x - \beta x/(1 + x) \quad (6.30)$$

that frequently arises in the Michaelis–Menten type of enzymatic reactions has been analysed by Horsthemke and Lefever (1977). The dimensionless coefficients C, k_m, β, etc. can be properly identified for the case under consideration. Allowing β to fluctuate around a mean value of $\bar{\beta}$, the stochastic description corresponding to (6.30) can be written as

$$\frac{dx}{dt} = [C + (1 - k_m x)x - \bar{\beta}x/(1 + x)] + \sigma \frac{x}{1 + x}\frac{dW}{dt} \quad (6.31)$$

or, assuming (6.31) as the Ito representation, in the equivalent Fokker–Planck equation as

$$\frac{\partial p}{\partial t} = -\frac{\partial}{\partial x}[C + (1 - k_m x)x - \bar{\beta}x/(1 + x)]p(x) + \frac{\sigma^2}{2}\frac{\partial^2}{\partial x^2}(x/(1 + x))^2 p(x) \quad (6.32)$$

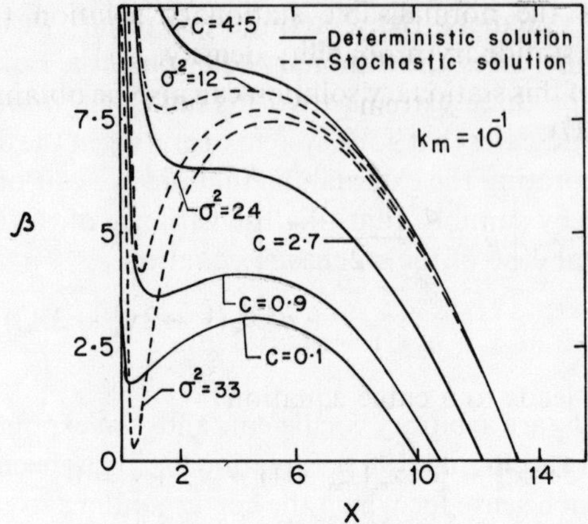

FIGURE 6.3 Stationary solutions of the deterministic and stochastic equations for different parameter values.

The stationary solution to Equation (6.32) can be obtained as

$$p_{st}(x) = N \exp \frac{2}{\sigma^2}\left\{ -\frac{C}{x} + (C + 2 - k_m - \bar{\beta})x + \tfrac{1}{2}(1 - 2k_m)x^2 - \tfrac{1}{3}k_m x^3 + (2C + 1 - \bar{\beta} - \sigma^2)\ln x + \sigma^2 \ln(1 + x)\right\} \tag{6.33}$$

The maxima of the stationary distribution are given by

$$-\left\{C + (1 - k_m x_m)x_m - \frac{\bar{\beta}x_m}{1 + x_m}\right\} + \frac{\sigma^2 x_m^3}{(1 + x_m)^5} = 0 \tag{6.34}$$

In order to compare the results of the deterministic equation with those from its stochastic analogue, the steady state solutions are plotted in Figure 6.3 for several values of β with C as parameter. The deterministic equation indicates that a critical value of C exists below which there are three states for a given set of other parameter values. The results in general are corroborated by the stochastic model which additionally shows certain other features. Thus, for values of C beyond the critical value, the deterministic analysis shows only one state, but the stochastic model [Equation (6.34)] indicates the existence of three states depending on the value of the variance σ^2. For low values of σ

Equation (6.34) predicts the deterministic result with only one maximum. As σ increases, the stationary solution of the Fokker–Planck equation develops three extrema (two maxima, one minimum). This example clearly indicates that, for a system operating at the deterministic point, by incorporating the external fluctuations we can bring about a phase transition by suitably adjusting the variance of the fluctuations.

A more general case of the stochastic equation

$$\frac{dx}{dt} = (a_1 x - a_2 x^\gamma) + (b_1 x^2 + b_2 x^{\gamma+1})^{1/2} \frac{dW}{dt} \tag{6.35}$$

where a_1, a_2, b_1, b_2 are arbitrary coefficients and γ an exponent has been treated by Banai and Brenig (1983). Equation (6.35) has been interpreted in the Stratonovich sense for which the corresponding Ito form can be written as

$$dx = \left\{\left(a_1 + \frac{b_1}{2}\right)x - \left(a_2 - b_2\frac{(\gamma+1)}{4}\right)x^\gamma\right\} dt$$
$$+ (b_1 x^2 + b_2 x^{\gamma+1})^{1/2}\, dW \tag{6.36}$$

The associated Fokker–Planck equation takes the form

$$\frac{\partial p(x,t)}{\partial t} = -\frac{\partial}{\partial x}\left\{\left[\left(a_1 + \frac{b_1}{2}\right)x - \left(a_2 - b_2\frac{(\gamma+1)}{4}\right)x^\gamma\right] p(x,t)\right\}$$
$$+ \frac{1}{2}\frac{\partial^2}{\partial x^2}[(b_1 x^2 + b_2 x^{\gamma+1})^{1/2} p(x,t)] \tag{6.37}$$

Equation (6.36) possesses the following stationary solutions: for $\gamma = 3[F_1(x) = G(x) = 0]$

$$x_1 = 0, \qquad x_2 = (-b_1/b_2)^{1/2} \tag{6.38}$$

The solution x_2 exists provided

$$\frac{b_1}{b_2} = 2a_1/(b_2 - 2b_1) < 0 \tag{6.39}$$

As regards the stability of these solutions we resort to the condition (6.39) which requires consideration of two cases: (1) $b_1, b_2 > 0$, and (2) $b_1 > 0, b_2 < 0$. In the former case δ can take any positive value, while in the latter the square root in Equation (6.36) restricts the range of δ to $[0 - (-b_1/b_2)^{1/2}]$. The stationary point $x =$

$(-b_1 b_2)^{1/2}$ is thus an intrinsic boundary of the system. Evaluating the integral in Equation (6.11) for either of the cases indicates that it is finite for the stationary state $x_1 = 0$ provided

$$(1 + \delta^2 b_2/b_1) > 0, \qquad a_1/b_1 < 0 \tag{6.40}$$

The first condition is always met while the latter condition requires that $a_1 < 0$. The stationary point $x_1 = 0$ is therefore stable for any value of a_2 when $a_1 < 0$ and unstable when $a_1 > 0$.

As regards the second stationary point, for the case when $b_1 > 0$, $b_2 < 0$, we can evaluate the integral in Equation (6.11). The integral is finite (implying stability) provided

$$a_1 > b_1/2, \qquad a_2 > 0$$

or

$$a_1 < b_1/2, \qquad a_2 < 0 \tag{6.41}$$

The results thus far indicate that in the range $-\infty < a_1 \leqslant 0$, both the stationary points x_1 and x_2 are stable provided $a_2 < 0$. A typical plot showing the bistability and hysteresis effects is shown in Figure 6.4. The influence of external noise is clearly seen (the deterministic equation does not show this feature).

In the second case, when $b_1 < 0$ and $b_2 > 0$, the range of x for the process to be real gets restricted to $(-b_1/b_2)^{1/2} \leqslant x \leqslant \infty$. In this range the integral of Equation (6.11) can be integrated to obtain the condition

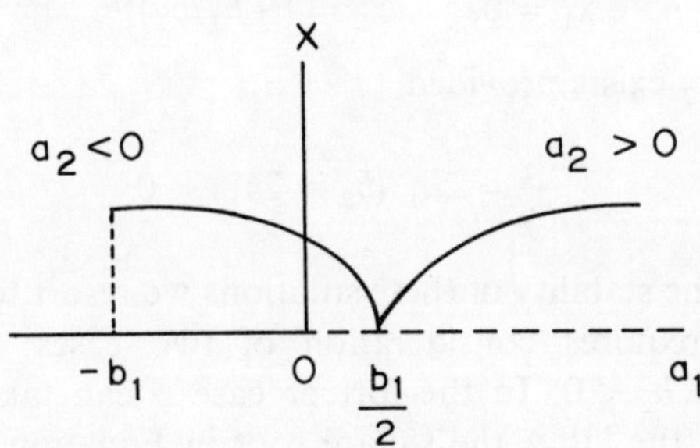

FIGURE 6.4 Stationary solution showing bistability and hysteresis effect.

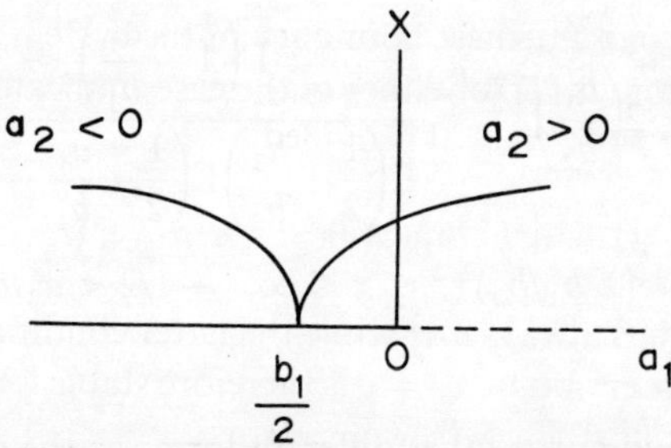

FIGURE 6.5 Stationary solutions satisfying Equation (6.42).

for it to be finite:

$$a_1 < -|b_1|/2, \quad a_2 < 0$$

or

$$a_1 > -|b_1|/2, \quad a_2 > 0 \tag{6.42}$$

A typical plot for this range is shown in Figure 6.5.

We now turn to the Fokker–Planck equation to identify other stationary solution(s). These are

$$p_{\mathrm{st}}(x) = N x^{(2a_1/b_1 - 1)} \left| 1 + \frac{b_2 x^2}{b_1} \right|^{(a_2/b_2) - (a_1/b_1) - (1/2)} \tag{6.43}$$

where the normalisation constant takes the following values in different regions:

$$N = 2\left(\frac{b_1}{b_2}\right)^{-(a_1/b_1)} \frac{\Gamma\left(\frac{1}{2} - \frac{a_2}{b_2} + \frac{a_1}{b_1}\right)}{\Gamma\left(\frac{a_1}{b_1}\right)\Gamma\left(\frac{a_1}{b_1} - \frac{a_2}{b_2}\right)}$$

$$b_1 > 0, \quad b_2 > 0, \quad 0 \leqslant x \leqslant \infty \tag{6.44a}$$

$$N = 2\left(-\frac{b_1}{b_2}\right)^{-(a_1/b_1)} \frac{\Gamma\left(\frac{1}{2} - \frac{a_2}{b_2}\right)}{\Gamma\left(\frac{a_1}{b_1}\right)\Gamma\left(\frac{1}{2} - \frac{a_1}{b_1} - \frac{a_2}{b_2}\right)}$$

$$b_1 > 0, \quad b_2 < 0, \quad 0 \leqslant x \leqslant (-b_1/b_2)^{1/2}, \quad a_1 > 0, \quad a_2/b_2 > 1/2 - a_1/b_1 \tag{6.44b}$$

$$N = 2\left(\frac{b_1}{b_2}\right)^{-(a_1/b_1)} \frac{\Gamma\left(1 - \frac{a_1}{b_1}\right)}{\Gamma\left(\frac{1}{2} + \frac{a_2}{b_2}\right)\Gamma\left(\frac{1}{2} - \frac{a_1}{b_1} - \frac{a_2}{b_2}\right)}$$

$$b_1 > 0,\ \ b_2 > 0,\ \ (-b_1/b_2)^{1/2} \leqslant x \leqslant \infty,\ \ -1/2 < a_2/b_2 < 1/2 - a_1/b_1 \tag{6.44c}$$

The distribution curve takes different forms as the parameter values are varied, the schematic representation of which has been provided by Banai and Brenig (1983) and reproduced here in Figure 6.6. The probability maximum as calculated using the stationary Fokker–Planck equation also indicates sharp changes in x_m. The entire phenomenon owes its existence to the presence of noise.

Before we close this section on the role of external fluctuations in model reaction schemes, we shall briefly summarise the results of two other schemes which involve multivariable cases as against the ones treated here thus far. Arnold, Horsthemke and Stucki (1979) considered the case of a Lotka–Volterra oscillator under the influence of an external noise and observed that the macroscopic periodicity of the system is completely disrupted. Mikhailov (1979) considered a system consisting

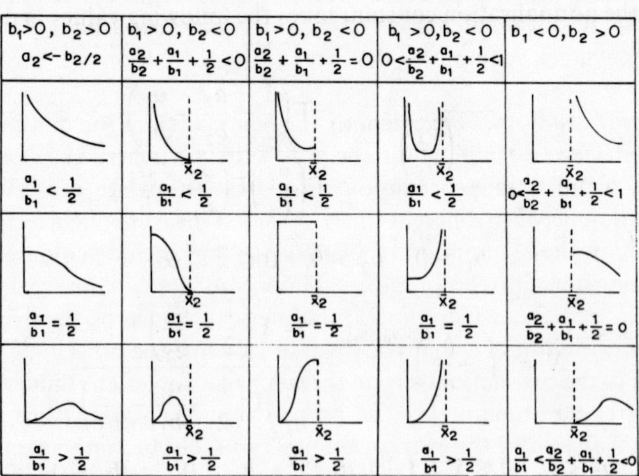

FIGURE 6.6 Several forms of the distribution curve as a function of parameter values.

of two biological species competing for the same food with an additional transport mechanism (diffusion) for one of the species. In the model it was assumed that the food growth fluctuates in space and time and the correlation of this random noise can be represented by a specific form. The analysis indicates that the deterministic result of extinction of one of the species does not hold in presence of fluctuations. A critical value of the noise intensity has been found to exist above which both the species have finite particle numbers under stationary conditions. The general results indicate the importance of external fluctuations.

6.3 Role of External Coloured Noise

The previous section was concerned with the role of external noise on model schemes and indicated that new transition phenomena may be induced solely by its presence. Two tacit assumptions regarding the nature of the fluctuations were made in these developments. The first assumption concerns its Gaussianity—a choice testified on the grounds that we now need to specify only the first two moments of the stochastic force, the higher order cumulants being zero. The second assumption regarding the fluctuations is that of white noise. The problem then becomes mathematically very convenient since the process is Markovian and can be formulated as a diffusion equation—usually called the Fokker–Planck equation—describing the evolution of probability density.

The idealisations, convenient as they may be from the mathematical point of view, do not reflect the true situation where the noise may exhibit a finite time-correlation as against the white noise with no correlation between different times. As a first measure of correction one may retain the assumption of Gaussianity but include coloured noise with finite time-correlation effects. This introduces two time scales in the problem under consideration. One time scale concerns the dynamical time of evolution of the physical variable, while the other time scale is related to the correlation-time of the stochastic force. The finite width of the correlation-time makes the process non-Markovian. The general theory available for the Markovian processes thus becomes inapplicable to this situation. One of the tasks therefore has been to develop suitable schemes that render the evolution of the probability density in a form that is analogous to the Fokker–Planck equation. The different techniques developed towards this end along with their applications are discussed in Section 6.3.1. The correction to the assumption of Gaussianity is dealt with in Section 6.3.2.

6.3.1 Evolution of Probability Density in Presence of External Coloured Noise

A. General Development for Single Variable Systems

In seeking the influence of coloured noise on the behaviour of systems we are mainly interested in two aspects of it. The first addresses the question whether the results obtained using white noise qualitatively correspond to those obtained using coloured noise. The second concerns the quantitative estimation of the effect. The general schemes which have been used towards this end basically follow two strategies. In one, a special class of coloured noise, for which exact solutions are possible, is examined to ascertain its role. As a limiting case the general results also reduce to that of white noise (see, for example, Pawula, 1967, 1977; Kitahara, Horsthemke and Lefever, 1979; Kitahara *et al.*, 1980; Horsthemke and Lefever, 1981).

The approach is, however, restrictive due to the fact that the type of coloured noise analysed does not occur frequently in practice. The dichotomous Markov noise is one such noise most commonly analysed. In the second approach more practical forms of coloured noise, such as the one described by an Ornstein–Uhlenbeck process, are analysed. In this approach exact results are possible only for simple models and frequent recourse to approximation schemes is necessary. The two most common approximations concern the limiting cases: The correlation time of the noise is much longer than the characteristic time of the system and vice versa. The latter being practically more interesting, we shall confine ourselves to this situation.

Our intention in this section is clear. For a given physical situation described in terms of the phenomenological equation, we would like to construct an equation that describes the evolution of the probability density when external coloured noise is present. The coloured noise is described as an Ornstein–Uhlenbeck process and as a practical reality we assume that the correlation time of the noise is much shorter than the characteristic time of the system. Over a small interval of time, this implies that the stochastic force varies considerably while the physical variable remains almost invariant.

To illustrate, we shall begin with a one-dimensional phenomenological equation

$$\frac{dx}{dt} = F(x) + \alpha G(x) \tag{6.45}$$

where α represents an arbitrary parameter that depends on the fluctuating environment and fluctuates according to $\alpha \rightarrow \bar{\alpha} + \xi(t)$ where $\xi(t)$ is the external noise modelled as an Ornstein–Uhlenbeck process with zero mean and finite correlation given by

$$\langle \xi(t)\xi(t') \rangle = \gamma(t, t') = \left(\frac{D}{\tau}\right) exp - \{(t - t')/\tau\}, \qquad t > t' \tag{6.46}$$

τ in this equation refers to the correlation time and D the strength of fluctuations. The Langevin equation corresponding to Equation (6.45) then becomes

$$\frac{dx}{dt} = F(x) + \alpha G(x) + G(x)\xi(t) = F_1(x) + G(x)\xi(t) \tag{6.47}$$

We wish now to derive a Fokker–Planck-like equation for the probability density associated with Equation (6.47). For this purpose we write the corresponding stochastic Liouville equation

$$\frac{\partial \rho(x, t)}{\partial t} = -\frac{\partial}{\partial x}[F(x) + \alpha G(x) + G(x)\xi(t)]\rho(x, t) \tag{6.48}$$

where $\rho(x, t)$ refers to the probability density for each sample realisation of $\xi(t)$ and represents the average over the initial conditions of $\delta[x(t) - x]$. The probability density $p(x, t)$ of the process is the average of $\rho(x, t)$ over the distribution $\xi(t)$

$$p(x, t) = \langle \rho(x, t) \rangle = \langle \delta(x(t) - x) \rangle \tag{6.49}$$

Substituting Equation (6.49) in (6.48) we obtain

$$\frac{\partial p(x, t)}{\partial t} = -\frac{\partial}{\partial x}[F(x) + \alpha G(x)]p(x, t) - \frac{\partial}{\partial x}G(x) \times \langle \xi(t)\, \delta[(x(t) - x)] \rangle \tag{6.50}$$

where the quantity $\langle \ldots \rangle$ can be further expanded to

$$\langle \xi(t)\,\delta[x(t)-x]\rangle = \int_0^t dt'\,\gamma(t,t')\left\langle \frac{\delta[\delta(x(t)-x)]}{\delta\xi(t')}\right\rangle$$

$$= -\frac{\partial}{\partial x}\int_0^t dt'\,\gamma(t,t')\left\langle \delta(x(t)-x)\frac{\delta x(t)}{\delta\xi(t')}\right\rangle \tag{6.51}$$

Substituting this in Equation (6.50) we obtain

$$\frac{\partial p(x,t)}{\partial t} = -\frac{\partial}{\partial x}[F(x)+\alpha G(x)]p(x,t) + \frac{\partial}{\partial x}G(x)\frac{\partial}{\partial x} \times \int_0^t dt'\,\gamma(t,t')\left\langle \delta(x(t)-x)\frac{\delta x(t)}{\partial\xi(t')}\right\rangle \tag{6.52}$$

Equation (6.52) is a general evolution equation for the probability density. The main difficulty associated with this equation is that it is not a closed equation for $p(x,t)$ due to the presence of an average in the last term. The new problem generated is then the estimation of this averaged quantity, which can be done as follows.

A formal integration of Equation (6.47)

$$x(t) = x(0) + \int_0^t ds[F_1 x(t) + G(x(t))\xi(t')] \tag{6.53}$$

Functional differentiation of $x(\xi,t)$ (see Hanggi, 1978), gives

$$\frac{\delta x(t)}{\delta\xi(t')} = G(x(t')) + \int_{t'}^t ds[F_1'(x(s)) + G'(x(s))\xi(s)]\frac{\delta x(s)}{\delta\xi(t')}, \qquad t > t' \tag{6.54}$$

where the primes on F_1 and G denote differentiation with respect to x. Differentiating Equation (6.54) with respect to t leads to a differential equation

$$\frac{\partial}{\partial t}\left(\frac{\partial x(t)}{\partial\xi(t')}\right) = \{F_1'[x(t)] + G'[x(t')]\xi(t)\}\frac{\partial x(t)}{\partial\xi(t')} \tag{6.55}$$

which can be solved subject to the initial condition

$$\left.\frac{\delta x(t)}{\delta\xi(t')}\right|_{t=t'} = G[x(t')] \tag{6.56}$$

that follows from Equation (6.54), with the result

$$\frac{\delta x(t)}{\delta \xi(t')} = G[x(t')] \exp \int_{t'}^{t} ds \{F'_1[x(s)] + G'[x(s)]\xi(s)\} \tag{6.57}$$

The exact estimation of this term is in general impossible, for it requires a knowledge of $x(s)$ which would amount to the solution of the original Langevin equation. It is, however, possible to exactly estimate the average value of the term when the system is linear or at least when it can be reduced to a linear form using suitable approximations. For these cases it is possible to obtain a closed equation for probability density from Equation (6.52), and different approaches such as the cumulant expansion techniques (Lax, 1966; van Kampen, 1976; Kubo, 1963; Mukamel, Oppenheim and Ross, 1977) or the functional methods (Klyatskin and Tatarskii, 1973; Hanggi, 1978; Sancho and Miguel, 1980; Miguel and Sancho, 1981) have been employed for this purpose.

Linear Case

In the event the governing equation is linear as, for example, when $F(x) = ax$ and $G(x) = 1$, Equation (6.57) can be evaluated to obtain

$$\frac{\delta x(t)}{\delta \xi(t')} = \exp[a(t - t')] \tag{6.58}$$

Substituting this in Equation (6.52) we obtain the Fokker–Planck equation

$$\begin{aligned} \frac{\partial p(x,t)}{\partial t} &= -\frac{\partial}{\partial x}[F(x) + \alpha G(x)]p(x,t) + \frac{\partial}{\partial x}G(x)\frac{\partial}{\partial x}D(t)p(x,t) \\ &= -\frac{\partial}{\partial x}[ax + 1]p(x,t) + \frac{\partial^2}{\partial x^2}Dp(x,t) \end{aligned} \tag{6.59}$$

where

$$D(t) = \int_{t'}^{t} dt' \, \gamma(t,t') \exp\{a(t - t')\} \tag{6.60}$$

If the noise analysed is white we have

$$\gamma(t,t') = 2D\,\delta(t - t')$$

and we recover from Equation (6.59) the well-known Stratonovich form of the Fokker–Planck equation for the probability density

$$\frac{\partial p(x,t)}{\partial t} = -\frac{\partial}{\partial x} F_1(x)p(x,t) + D\frac{\partial}{\partial x} G(x)p(x,t) \tag{6.61}$$

Since for a white noise the process is Markovian, the solution of Equation (6.61) is also the transition probability for the process.

Nonlinear Case

The reduction of the general Equation (6.52) to the correct form for the linear case indicates that the equation is of the proper form. In the general nonlinear case we can again obtain a similar Fokker–Planck equation provided the equation can be reduced to a linear form by proper transformations. The necessary and sufficient condition for such a transformation to exist is

$$G(x)\frac{\partial}{\partial x}\frac{F_1(x)}{G(x)} = A = \text{Constant} \tag{6.62}$$

If the above condition is satisfied the transformation

$$g = \int \frac{dx}{G(x)} \tag{6.63}$$

converts the original Langevin equation to the form

$$\frac{dg}{dt} = Ag + B + \xi(t) \tag{6.64}$$

which, being linear, again validates Equation (6.58) for the average.

General Case

In the general case no exact result for the average similar to Equation (6.58) can be found and approximations are necessary. As a first approximation for small values of correlation time, we can expand around the white noise limit to obtain

$$\frac{\delta x(t)}{\delta \xi(t')} = \left.\frac{\delta x(t)}{\delta \xi(t')}\right|_{t\to t'} + \left.\frac{d}{dt'}\frac{\delta x(t)}{\delta \xi(t')}\right|_{t\to t'}(t-t') + \cdots \tag{6.65}$$

where the time derivative as $t \to t'$ can be obtained by differentiating Equation (6.57). Appropriate substitution in Equation (6.52) then gives

$$\frac{\partial p(x,t)}{\partial t} = -\frac{\partial}{\partial x} F_1(x)p(x,t) + D\frac{\partial}{\partial x} G(x)\frac{d}{dx} h(x)p(x,t) + O(\tau^2, e^{-t/\tau}) \tag{6.66}$$

where

$$h(x) = G(x)\left\{1 + \tau G(x)\left[\frac{F_1(x)}{G(x)}\right]'\right\} \tag{6.67}$$

Equation (6.66) is similar to (6.61) except for a factor $h(x)$ which replaces $G(x)$. Note that in deriving Equation (6.66) we made use of the series Equation (6.65) which was truncated for reasons of convenience. This creates some mathematical factors such as non-physical boundaries, and thus the solution of Equation (6.66) may sometimes be misleading. To overcome these difficulties the solution to Equation (6.66) should be obtained in the form

$$p_{st}(x) = p_{st}^0(x) + \tau p_{st}'(x) + O(\tau^2) \tag{6.68}$$

where $p_{st}^0(x)$ represents the white noise distribution and $p_{st}'(x)$ the first correction to it. The final form of the stationary solution to Equation (6.68) can be obtained as

$$p_{st}(x) = p_{st}^0(x)\left\{1 + \tau\left|N - G(x)\left[\frac{F_1(x)}{G(x)}\right]' - \frac{1}{2D}\left[\frac{F_1(x)}{G(x)}\right]^2\right|\right\} \tag{6.69}$$

where the constant N is given by

$$N = -\frac{1}{2D}\int_{x_1}^{x_2} \frac{F_1^2(x)}{G^2(x)} p_{st}^0(x)\,dx \tag{6.70}$$

In the series given by Equation (6.65) we can retain higher orders of τ and write

$$\frac{\partial(x(t))}{\partial \xi(t')} = \sum_{n=0}^{\infty} \frac{(-1)^n}{n!}\left|\frac{d^n}{dt^n}\frac{\delta x(t)}{\delta \xi(t')}\right|_{t\to t'} (t-t')^n \tag{6.71}$$

Using (6.71) instead of (6.65) and proceeding as before we finally obtain an equation for $p(x,t)$ in the form of a Kramers–Moyal expansion

$$\frac{\partial p(x,t)}{\partial t} = \sum_{i=0}^{\infty} \frac{\partial^i}{\partial x^i} K_i(x)p(x,t) \tag{6.72}$$

Truncating this equation after the second term gives

$$\frac{\partial p(x,t)}{\partial t} = -\frac{\partial}{\partial x} F_1(x)p(x,t) + D\frac{\partial}{\partial x} G(x)\frac{\partial}{\partial x} H(x)p(x,t) \quad (6.73)$$

where

$$H(x) = F_1(x)\left|1 + \tau F_1(x)\frac{\partial}{\partial x}\right|^{-1} \frac{G(x)}{F_1(x)} \quad (6.74)$$

The formal procedure of obtaining this approximate relation indicates that the equation should be valid for lower values of D.

The general development of the equations for the probability density presented here has been described in greater detail through a series of papers by Sancho and Miguel (see, for example, Sancho and Miguel, 1980; Sancho *et al.*, 1982; Miguel and Sancho, 1981). The extension of the Fokker–Planck approximation developed here for the single variable case has also been extended to an n-dimensional non-Markovian Langevin equation by Miguel and Sancho (1981). Garrido and Sancho (1982) have developed an ordered cumulant technique to solve the stochastic differential equation.

In a similar approach Lindenberg and West (1983) employed a renormalisation scheme, and for a Langevin equation of the type

$$\frac{dx}{dt} = F_1(x) + \varepsilon\xi(t) \quad (6.75)$$

obtained a second-order equation with a state-and-time-dependent diffusion coefficient in the form

$$\frac{\partial p(x,t)}{\partial t} = -\frac{\partial}{\partial x} F_1(x) + \frac{\partial^2}{\partial x^2} D(x,t)p(x,t) \quad (6.76)$$

$$D(x,t) = \varepsilon^2 \int_0^t ds\, \gamma(s) \sum_{n=0}^{\infty} \frac{\tau^n}{n!} Q_n(x) \quad (6.77)$$

where $Q_n(x)$ obeys the recursion relation

$$Q_n(x) = F_1'(x)Q_{n-1}(x) - F_1(x)Q_{n-1}'(x), \qquad Q_0(x) = 1 \quad (6.78)$$

Several approaches outlined above attempt to describe the temporal evolution of the system that is non-Markovian due to the presence of coloured noise. As an alternative view it is also possible to treat the system as a two-component Markovian process. The state variable and the external fluctuating parameters are the obvious two components, and give rise to a bivariate Fokker–Planck equation. The difficulty, however, lies in the fact that in general the stationary probability distribution for such cases cannot be calculated exactly. Horsthemke and Lefever (1980) have developed a method that permits a systematic calculation of the stationary probability density in the form of a perturbation expansion in terms of a small parameter related to the correlation time of the noise.

All the techniques discussed above have been illustrated by considering appropriate examples. The common examples analysed involved the phenomenological equations showing cubic and quadratic forms of nonlinearity

$$\text{(i)}\quad \frac{dx}{dt} = \alpha x(t) - \beta x^3(t) + x(t)\xi(t) \tag{6.79}$$

$$\text{(ii)}\quad \frac{dx}{dt} = \frac{1}{2} - x + \beta x(1 - x) \tag{6.80}$$

The general analysis, not repated here for reasons of brevity, corroborates the results for the white noise in the appropriate limits, but at the same time indicates new transitions that occur as a consequence of finite correlation time.

B. Exactly Solvable Single and Two Variable Systems

From the discussion presented thus far it appears that the probability distribution of the process $x(t)$ can be obtained only approximately. While this is the case in many practical situations, a class of systems exists for which the probability distribution can be exactly evaluated. Thus let us again consider the case of a single variable stochastic differential equation such as Equation (6.47) where the functions $F_1(x)$ and $G(x)$ are some arbitrary functions of the variable x, and α is the parameter that fluctuates ($\alpha = \bar{\alpha} + \xi(t)$). As before we assume that the noise parameter $\xi(t)$ is defined as an Ornstein–Uhlenbeck process with mean and correlation time as defined in Equation (6.46). We assume that the functions $F_1(x)$ and $G(x)$ are regular and the two-dimensional Markovian process (x, ξ) is described by

$$dx = [F_1(x) + \bar{\alpha}G(x) + \xi G(x)]\, dt = A(x, \xi)\, dt \tag{6.81}$$

$$d\xi = -\varepsilon^{-2}\xi \, dt + \varepsilon^{-1}\sigma \, dW \tag{6.82}$$

For the class of systems that are exactly solvable, i.e. for which the exact probability distribution function can be derived, we require that the functional relation

$$\frac{d}{dx}\left[\frac{F_1(x)}{G(x)}\right] = \frac{C}{G(x)} > 0 \tag{6.83}$$

holds where C represents some arbitrary constant. The equation is simple, and knowing $F_1(x)$ or $G(x)$ one can find the combination of these functions which would possess exactly solvable distribution. Thus for given $F_1(x)$, $G(x)$ can be calculated as

$$G(x) = C_0 F_1(x) \exp\left(-C\int \frac{dx}{F_1(x)}\right) \tag{6.84}$$

or, if $G(x)$ is the known function, then the $F_1(x)$ that would be compatible with this function $G(x)$ can be obtained as

$$F_1(x) = G(x)\left(C_1 + C\int \frac{dx}{G(x)}\right) \tag{6.85}$$

Of course C_0 and C_1 in these equations are some arbitrary constants.

We can write the associated Fokker–Planck equation for the two-variable process (x, ξ) defined by Equation (6.82) as

$$\frac{\partial p(x, \xi, t)}{\partial t} = -\frac{\partial}{\partial x}[A(x, \xi)p(x, \xi, t)] + \varepsilon\frac{\partial}{\partial \xi}[\xi p(x, \xi, t)] + \tfrac{1}{2}\varepsilon^2\sigma^2 \frac{\partial^2 p(x, \xi, t)}{\partial \xi^2} \tag{6.86}$$

The solution of Equation (6.86) can be obtained in the form $p(x, \xi, t)$ which can be used to calculate the probability distribution $p(x, t)$ from the equality $p(x, t) = \int_{-\infty}^{\infty} p(x, \xi, t)\, d\xi$. The explicit form of this function can be easily obtained as (Luczaka, 1984)

$$p(x, t) = [C/G(x)][2\pi\phi(t)]^{-1/2} \exp[-q^2(x)/2\phi(t)] \tag{6.87}$$

where

$$q(x) = \frac{F_1(x)}{G(x)} + \alpha - \left[\frac{F(x_0)}{G(x_0)} + \alpha\right] e^{Ct}$$

and

$$\phi(t) = \frac{\varepsilon C\sigma^2}{2(C-\varepsilon)} + \frac{\varepsilon C\sigma^2}{2(C+\varepsilon)}\exp(2Ct) - \frac{\varepsilon C^2\sigma^2}{C^2-\varepsilon^2}\exp(C-\varepsilon)t \tag{6.88}$$

Equation (6.87) represents the exact probability distribution. As can be verified the stationary probability exists provided $C > 0$ and is given by

$$p_{st}(x) = \lim_{t\to\infty} p(x,t) = \left[\frac{C}{G(x)}\right]\left[\frac{2\pi\varepsilon C\sigma^2}{2(C-\varepsilon)}\right]^{-1/2} \times \exp -\left[\left\{\frac{F_1(x)}{G(x)} + \alpha\right\}^2 \bigg/ \frac{C\varepsilon\sigma^2}{C-\varepsilon}\right] \tag{6.89}$$

The most probable value x_m corresponds to the extremal point of the stationary distribution and is given by

$$\phi(t)\frac{d}{dX}G(x) + Cq(x) = 0 \tag{6.90}$$

As analysed earlier, when the solution of this equation bifurcates, we have a phase transition in the system. The definitions of $\phi(t)$ and $q(x)$ as noted in Equation (6.88) suggest that the noise transition can be brought about either by (i) the parameters ε, σ, (ii) the model parameter such as α, C, (iii) the initial value of the process x_0, or (iv) the time-dependent bifurcation parameter $\phi(t)$ itself. The white noise limit can be easily obtained when we let $\varepsilon \to \infty$. Note that the stationary probability distribution exists only if $C > 0$. For $C < 0$, the probability distribution vanishes, i.e. $p_{st}(x) = 0$.

Another class of systems which are also exactly solvable arises when we consider the noise to be a two-step Markov process or a dichotomous noise. For brevity we shall consider the following equations describing the physical system:

$$\frac{dx}{dt} = f(x) + g(x)\xi(t) \tag{6.91}$$

where $\xi(t)$ is now a dichotomous two-step Markov process with zero mean and autocorrelation

$$\langle\xi(t)\xi(t')\rangle = \Delta^2 \exp[-\varepsilon(t-t')] \tag{6.92}$$

The dichotomous noise can obviously take only two possible values $\pm\Delta$ with equal probability and jumps with probability $\varepsilon/2(dt)$ for dt. Our intention now is to obtain an exact stationary probability distribution $p_{st}(x)$. We begin with the stochastic Liouville equation as before, and write for the density $\rho(x, t)$ corresponding to Equation (6.91):

$$\frac{\partial \rho(x,t)}{\partial t} = -\frac{\partial}{\partial x}[f(x) + g(x)\xi(t)]\rho(x,t) \tag{6.93}$$

Taking the average over $\xi(t)$ and using the fact that $p(x, t) = \langle \rho(x, t) \rangle$ we arrive at

$$\frac{\partial p(x,t)}{\partial t} = -\frac{\partial}{\partial x} f(x)p(x,t) - \frac{\partial}{\partial x} g(x)p_1(x,t) \tag{6.94}$$

where $p_1(x, t) = \langle \xi(t)\rho(x, t) \rangle$.

The dependence of $\rho(x, t)$ on $\xi(t)$ requires the evaluation of a functional derivative in Equation (6.94). Following the earlier procedure outlined in the case of Gaussian noise, Sancho (1984) developed the following formal differential equation for the probability density $p(x, t)$:

$$\frac{\partial p(x,t)}{\partial t} = -\frac{\partial}{\partial x} f(x)p(x,t) + \Delta^2 \frac{\partial}{\partial x} g(x)B(x,t) \tag{6.95}$$

where $B(x, t)$ is given as

$$B(x,t) = \int_0^t \exp\left\{-\left(\varepsilon + \frac{\partial}{\partial x} f(x)\right)(t - t')\right\} \frac{\partial}{\partial x} g(x)p(x,t') \tag{6.96}$$

It is not always possible to obtain the complete transient solution to Equation (6.95). The stationary solution, if it exists, can be obtained as

$$p_{st}(x) = \frac{Ng(x)}{\Delta^2 g^2(x) - f^2(x)} \exp\left\{\varepsilon \int^x \frac{dx'\, f(x')}{\Delta^2 g^2(x') - f^2(x')}\right\} \tag{6.97}$$

Equation (6.97), derived for a coloured dichotomous process, reduces to the appropriate white noise limit where we would have $\exp\{-\varepsilon(t - t')\} \simeq (2/\varepsilon)\, \delta(t - t')$. The exact transient solution to Equation (6.95) exists for cases where the following condition is satisfied (Sancho, 1984):

$$g(x)f'(x) - g'(x)f(x) = g^2(x)\frac{d}{dx}\left(\frac{f(x)}{g(x)}\right) = Cg(x) \tag{6.98}$$

where C again represents some arbitrary constant. The formal Equation (6.95) for cases satisfying this condition can be rewritten in the standard form as a hyperbolic second-order partial differential equation:

$$\left[\frac{\partial}{\partial t^2} + (f^2(x) - \Delta^2 g(x))\frac{\partial^2}{\partial x^2} + 2f(x)\frac{\partial^2}{\partial t\,\partial x} + (\varepsilon + C + 2f'(x))\frac{\partial}{\partial t}\right.$$
$$+ [(\varepsilon + C)f(x) - 3\Delta^2 g(x)g'(x) + 3f(x)f'(x)]\frac{\partial}{\partial x}$$
$$\left. + [(f(x)f'(x))' - \Delta^2(g(x)g'(x))' + (\varepsilon + C)f'(x)]\right] p(x,t) \tag{6.99}$$

Equation (6.99) represents a non-Fokker–Planck master differential equation for the probability density of the process that yields the exact solution to the stochastic Equation (6.91) provided the condition (6.98) is satisfied. Sancho (1984) has demonstrated this by considering the pure diffusion process where $dx/dt = \xi(t)$ and for the linear case where $dx/dt = -\alpha x + \xi(t)$. The exact results for these cases reveal that the shape of the probability distribution in presence of coloured noise can be drastically different from that known for the white noise limit and that certain features exclusive to the presence of colour (ε) can be noted.

We were so far concerned with exactly solvable cases that could be modelled as single variable systems, to examine the effects of noise. Quite similar effects may be expected even for more complex cases such as two variable systems. The two-dimensional systems, however, differ from the single-dimensional systems in that they can possess new types of stationary states such as the limit cycle behaviour. It is of interest then to analyse these systems with a view to analytically evaluating the role of fluctuations. Towards this end we begin with a simple two-dimensional system possessing the macroscopic equations

$$\frac{dx}{dt} = f(x,y,t), \qquad \frac{dy}{dt} = g(x,y,t) \tag{6.100}$$

the solution of which describes a closed trajectory in the phase plane. Several examples of the functions $f(x, y, t)$ and $g(x, y, t)$ that lead to exactly solvable elliptical limit cycles within the context of chemical reactions have been presented by Escher (1979). For such cases it is possible to define a single function $H(x, y, t)$ that converts the macroscopic Equation (6.100) into the form

$$\frac{dx}{dt} = \frac{\partial H(x,y)}{\partial y} + f(H,\alpha)\frac{\partial H(x,y)}{\partial x}$$
$$\frac{dy}{dt} = -\frac{\partial H(x,y)}{\partial x} + f(H,\alpha)\frac{\partial H(x,y)}{\partial y} \tag{6.101}$$

where α represents some externally controlled parameter. It is possible that the parameter α is constant about a mean value $(\bar{\alpha})$ with variable σ. We are obviously modelling the fluctuations in the parameter α as a Gaussian white noise and its incorporation into Equation (6.101) leads to the following stochastic differential equations:

$$\frac{dx}{dt} = \frac{\partial H}{\partial y} + a^2 f(H,\bar{\alpha})\frac{\partial H}{\partial x} + a\sigma_1\frac{\partial H}{\partial x}g(H)\xi(t) \tag{6.102}$$

$$\frac{dy}{dt} = -\frac{\partial H}{\partial x} + b^2 f(H,\bar{\alpha})\frac{\partial H}{\partial y} + b\sigma_1\left(\frac{\partial H}{\partial y}\right)g(H)\xi(t) \tag{6.103}$$

In these equations $g(H)$ represents the term that is associated with the fluctuating parameter α in the function $f(H,\alpha)$.

We shall specifically consider the case when $a = b = 1$. A mathematical lemma asserts that the stochastic equations remain exactly solvable provided the following condition is met (Ebeling and Engel-Herbert, 1980):

$$\left(\frac{\partial H}{\partial x}\right)^2 + \left(\frac{\partial H}{\partial y}\right)^2 = h(H) \tag{6.104}$$

With the use of Equation (6.104) the two-dimensional systems [Equations (6.102) and (6.103)] can be written in single variable form as

$$\frac{dH}{dt} = \frac{\partial H}{\partial x}\frac{dx}{dt} + \frac{\partial H}{\partial y}\frac{dy}{dt} = h(H)[f(H,\bar{\alpha}) + \sigma_1 g(H)\xi(t)] \tag{6.105}$$

One can now proceed to write the corresponding Fokker–Planck equation and the stationary probability density for the observance of the natural boundary condition:

$$p_{st}(H) = \frac{N}{h(H)g(H)}\exp\left(\frac{2}{\sigma_1^2}\int\frac{f(H,\bar{\alpha})}{h(H)g^2(H)}\,dH\right) \tag{6.106}$$

N in this equation refers to the normalisation constant.

The class of systems defined by Equations (6.102) and (6.103) with $a \neq b$ has been considered by Hongler and Ryter (1978) who, using the method of mixed canonical-dissipative dynamics, obtained the following equation for the probability density:

$$p_{st}(H) = \frac{N}{g(H)} \exp\left(\frac{2}{\sigma^2} \int \frac{f(H, \bar{\alpha})}{g^2(H)} dH\right) \quad (6.107)$$

The above analysis gives a clear insight into the role of external noise in the behaviour of an oscillating system. To elucidate this further, Ebeling and Engel-Herbert (1980) considered the simple case of a limit cycle described as a circle ($H^2 = x^2 + y^2$) where x and y follow the macroscopic equations:

$$\begin{aligned} \frac{dx}{dt} &= y + \frac{x}{\sqrt{x^2 + y^2}}[\alpha - (x^2 + y^2)] \\ \frac{dy}{dt} &= -x + \frac{y}{\sqrt{x^2 + y^2}}[\alpha - (x^2 + y^2)] \end{aligned} \quad (6.108)$$

The probability density as per Equation (6.106) is readily obtained as

$$p_{st}(H) = N(2H)^{-1/2} \exp\left\{\frac{2}{\sigma_1^2}[\bar{\alpha}(2H)^{1/2} - \tfrac{1}{3}(2H)^{3/2}]\right\} \quad (6.109)$$

which is characterised by the following equation for maximum:

$$(2H)^{1/2}(\bar{\alpha} - 2H) - \sigma_1^2/2 = 0 \quad (6.110)$$

This equation possesses two solutions for $\sigma_1 > 0$ indicating that an extra solution is generated due to the presence of noise. Stability analysis indicates that, in addition to the deterministic stable limit cycle, incorporation of σ has induced the generation of unstable limit cycle which grows with σ. At a critical value of σ the stable and unstable limit cycles coalesce and no stochastic solutions are permitted beyond this value of σ. In recent years this approach has been used to analyse chemical systems of interest. The results for these systems will be discussed in Chapter 7.

C. General Analysis for Noise Entering in a Nonlinear Fashion

The previous sections (A and B) dealt with stochastic equations where the noise enters in a linear way in the governing formulation. Section A gave the general analysis of this problem and Section B presented some exact results for the situation. In the present section we shall be concerned with cases where noise enters in a nonlinear way. Such situations occur, for example, when we are considering the variations in the input temperatures. The rate constants being exponentially dependent upon the temperature, any variations or fluctuations in the temperature would induce exponential variations or fluctuations in the rate constant. It is also possible, in certain physical situations, that the parameters which enter the phenomenological description are quadratic. Any fluctuations in such parameters would therefore entail consideration of quadratic noise. The general methodology outlined in Section A is applicable to such cases but the final formal results differ. In the present section we shall consider these types of noises and obtain the probability distribution expression for the process.

To be specific we consider the model system described by

$$\frac{dx}{dt} = f(x) + e^{\alpha} g(x) \tag{6.111}$$

where the parameter α fluctuates about the mean value $\bar{\alpha}$ as $\alpha = \bar{\alpha} + \xi(t)$ and $\xi(t)$ represents the noise. We shall consider the noise to be Ornstein–Uhlenbeck as well as two-step Markov or dichotomous, and obtain the probability distribution. The two types of noises have the properties of zero mean and autocorrelation as defined previously in Equations (6.46) and (6.92). The stochastic equation corresponding to Equation (6.111) can be written as

$$\frac{dx}{dt} = f(x) + e^{\bar{\alpha}} g(x)\, e^{\xi(t)} \tag{6.112}$$

From $\xi(t)$ let us now define a non-Gaussian process $\eta(t)$ with zero mean as

$$e^{\xi(t)} = \langle e^{\xi(t)} \rangle + \eta(t) \tag{6.113}$$

Substitution of this into Equation (6.112) leads to

$$\frac{dx}{dt} = f(x) + e^{\bar{\alpha}} g(x) \eta(t) + \langle e^{\xi(t)} \rangle\, e^{\bar{\alpha}} g(x) \tag{6.114}$$

The average $\langle e^{\xi(t)} \rangle$ in Equation (6.114) can be evaluated since $\xi(t)$ has been prescribed to have Gaussian statistics, and obtained as equal to

$e^{D/2\tau}$. The equivalent description of Equation (6.114) in terms of the approximate Fokker–Planck equation has been obtained by Sagues (1984) subject to the condition that both D and τ are small quantities although D/τ tends to a finite value. The equation can be written as

$$\frac{\partial p(x,t)}{\partial t} = -\frac{\partial}{\partial x}[f(x) + e^{D/2\tau} e^{\bar{\alpha}} g(x)]p(x,t) + D' \frac{\partial}{\partial x} e^{\bar{\alpha}} g(x) \frac{\partial}{\partial x} e^{\bar{\alpha}} g(x) p(x,t) \tag{6.115}$$

where D' refers to the effective intensity of noise and is defined by

$$D' = \lim_{t \to \infty} \int_0^t dt' \langle \eta(t)\eta(t')\rangle = \tau\, e^{D/\tau} \sum_{n=1}^{\infty} \frac{1}{nn!}(D/\tau)^n \tag{6.116}$$

The stationary probability distribution is given by

$$p_{st}(x) = N[e^{\bar{\alpha}} g(x)]^{-1} \exp\left(\int^x dx' \frac{f(x') + e^{D/2\tau} e^{\bar{\alpha}} g(x')}{D' e^{2\bar{\alpha}} g^2(x')} \right) \tag{6.117}$$

In the event $\xi(t)$ in Equation (6.112) is considered as a dichotomous process we can obtain the exact probability distribution as

$$p_{st}(x) = N\, e^{\bar{\alpha}} g(x)[f^2(x) + 2f(x)\, e^{\bar{\alpha}} g(x) \cosh\Delta + e^{2\bar{\alpha}} g^2(x)]^{-1} \times \exp\left(-\int^x dx' \frac{\varepsilon[f(x') + e^{\bar{\alpha}} g(x') \cosh\Delta]}{f^2(x') + 2f(x')\, e^{\bar{\alpha}} g(x') \cosh\Delta + e^{2\bar{\alpha}} g^2(x')} \right) \tag{6.118}$$

We shall now consider the situation where the noise enters the stochastic description nonlinearly. To be specific we consider the noise to be quadratic and an Ornstein–Uhlenbeck process. The system under consideration is modelled macroscopically by the equation

$$\frac{dx}{dt} = f(x) + \alpha^2 g(x) \tag{6.119}$$

which, for a fluctuating parameter $\alpha = \bar{\alpha} + \xi(t)$, can be written as

$$\frac{dx}{dt} = f(x) + \bar{\alpha}^2 g(x) + 2\bar{\alpha} g(x)\xi(t) + g(x)\xi^2(t) \tag{6.120}$$

The problem seems trivial, in the sense that replacing α^2 by α' in

Equation (6.119) would reduce the case to a simple linear noise. There are, however, important considerations that defy the use of this simple mathematical transformation. The first is that α is usually a precisely defined parameter and our interest lies mainly in investigating the effects of fluctuations on this parameter. Second, α being an external parameter can be precisely controlled. The derived process of $\alpha' = \alpha^2$ depends on the statistics of α and cannot *a priori* be assumed to be Gaussian. In fact, for the assumed process of α as here, α' will certainly not be a Gaussian process and its mean value would depend upon the noise parameters of α (D and τ). It is therefore necessary to consider the quadratic process separately.

Our intention now is to derive a non-Markovian Fokker–Planck equation representing the probability density as has been done in the previous cases. Towards this end we first calculate the mean value of α^2 ($= \langle\alpha^2\rangle$) as

$$\langle\alpha^2\rangle = (\bar{\alpha})^2 + D/\tau \tag{6.121}$$

where the properties of the noise associated with the α process are used. Also the process α^2 can be written as

$$\alpha^2 = \langle\alpha^2\rangle + \eta(t) \tag{6.122}$$

where $\eta(t)$ is a random process that depends on $\xi(t)$. In Equation (6.122) we note [see Equation (6.121)] that the mean value $\langle\alpha^2\rangle$ depends on the noise parameter. This cannot therefore be assumed to take a fixed value. The equations can be used to describe $\eta(t)$ as

$$\eta(t) = \xi^2(t) + 2\bar{\alpha}\xi(t) - D/\tau \tag{6.123}$$

and the correlation function of $\eta(t)$ can be obtained as

$$\langle\eta(t)\eta(t')\rangle = 4\bar{\alpha}^2(D/\tau)\exp\left(-\frac{t-t'}{\tau}\right) + 2\left(\frac{D}{\tau}\right)^2\exp\left(-\frac{2(t-t')}{\tau}\right) \tag{6.124}$$

Notwithstanding this, it is still possible to approximate $\eta(t)$ as a Gaussian process, and especially in the limit of $\tau \to 0$, η can be approximated as a white noise with the effective value of D denoted as D' that is related to the noise parameters of $\xi(t)$ by

$$D' = \int_0^\infty dt'\langle\eta(t)\eta(t')\rangle = 4\bar{\alpha}^2 D + D^2/\tau \tag{6.125}$$

The problem is now trivial. We use Equation (6.122) in (6.119) to obtain the stochastic equation

$$\frac{dx}{dt} = f(x)t(\bar{\alpha}^2 + D/\tau)g(x) + g(x)\eta(t) \tag{6.126}$$

and the associated Fokker–Planck equation (in the Stratonovich sense) as

$$\frac{\partial p(x,t)}{\partial t} = -\frac{\partial}{\partial x}[f(x) + (\bar{\alpha}^2 + D/\tau)g(x)]p(x,t)$$

$$+ D'\frac{\partial}{\partial x}g(x)\frac{\partial}{\partial x}g(x)p(x,t) \tag{6.127}$$

We note some important difference in this equation in comparison with that for the linear case [Equation (6.61)]. There is now a contribution to the drift term and even the diffusion term is modified. A more rigorous derivation of Equation (6.127), which begins with the formulation of the first stochastic Liouville equation followed by evaluation of the functional derivatives, indicates that Equation (6.127) can be obtained provided the conditions D and $\tau \ll 1$ and $D/\tau \rightarrow$ finite are met. In addition, transient terms of the order of $\exp(-t/\tau)$ are ignored (Miguel and Sancho, 1981). The formal stationary solution to this is

$$p_{\text{st}}(x) = \frac{N}{g(x)}\exp\int dx\{f(x) + \bar{\alpha}^2 g(x) + (D/\tau)g(x)/[D(4\bar{\alpha}^2 + D/\tau)g^2(x)]\} \tag{6.128}$$

The maximum occurs at

$$f(x) + \bar{\alpha}^2 g(x) + D/\tau[g(x)] - D\left[4\bar{\alpha}^2 + \frac{D}{\tau}\right]g(x)g'(x) \tag{6.129}$$

We notice that the first two terms correspond to the deterministic case which is modified due to the presence of noise as indicated by the latter terms. Of special importance is the sign of the third term which is opposite to that appearing in the case of linear noise. The corrections to the approximate Equation (6.127) in the event the restriction on D to be smaller is removed can be worked out in a manner similar to that described earlier [Equations (6.69) and (6.74)], and the formal expressions are given by Miguel and Sancho (1981). These authors illustrate the use of these formal expressions by considering examples and show that the behaviour is entirely different from what one would have expected by considering $\alpha^2 = \alpha'$ as a linear noise.

The probability density equation for the case of a system driven by a quadratic noise that is assumed to follow an Ornstein–Uhlenbeck process was obtained by the method of expansion to the lowest nontrivial order in τ. The characteristics of the noise assumed were such that both D and τ were small in value with D/τ tending to be a finite value. This equation corresponds to a Markovian assumption where the random term $\eta(t)$ is considered to be of an effective intensity $D_e = D[4\bar{\alpha}^2 + D/\tau]$. The equations for the joint probability density and probability density coincide in this limit. However, if we want to go beyond the Markovian approximation, the two equations are different. Here we have to follow the non-Markovian dynamics of the governing equations. The detailed derivations for the joint probability density, correlation functions, calculations of relaxation times, etc. are provided by Sagues, Miguel and Sancho (1984) for the cases of an Ornstein–Uhlenbeck dichotomous noise.

Finally a comment concerning the adequacy of the several expressions presented here for calculating the probability density is in order. As we have noted, for systems driven by coloured noise, the underlying dynamics are governed by a non-Markovian process. The exact solutions of these equations can be obtained only in special cases, and in general approximation schemes that invoke the smallness or largeness of some noise parameters are often used. Some of these have already been presented in this section. All such schemes have obvious shortcomings—the chief among them being that the wings of the exact stationary probability distribution are not recovered correctly. All those properties that depend upon the exact stationary probability—such as the mean passage time, activation rates, etc.—cannot therefore be predicted accurately using such approximations. A critical study of some of these approximations has been reported by Hanggi, Marchesoni and Grigolini (1984). The approximations, however, provide a first-order analysis that is often helpful in practical cases.

6.3.2 Evolution of Probability Density for Systems under the Influence of White and Coloured Poisson Noises (Deviations from Gaussian Noise)

The previous sections were concerned with the evolution of probability density for processes under the influence of Gaussian noise. The present section deals with cases where the noise is non-Gaussian. To be specific we shall consider the case of Poisson noises. Gaussian and Poissonian types of noises together cover most of the situations normally encountered. This simple observation arises from the fact that noise is usually induced by a large number of bodies and therefore, according to the microscopic theory, Gaussian. In case the number of bodies is large but finite, we would obviously have the noise in the form of a succession of pulses at random times; it would therefore be Poissonian.

We shall make the assumption that the noise that is considered now is Poissonian; the general one-dimensional stochastic equation may then be written as follows:

$$\frac{dx}{dt} = f(x) + g(x)\xi(t) \tag{6.130}$$

Note that the Poisson noise that is represented as a succession of δ-pulses at random times is a white noise, while for peaked functions we have a shot noise or more generally coloured Poisson noise. This noise can be defined as

$$\xi(t) = \sum_i D_p q(t - t_i) - D_p \sum_i \langle q(t - t_i)\rangle \tag{6.131}$$

where $q(t)$ is some function of t, D_p the intensity of the Poissonian noise and t_i the random times that are selected according to the Poisson law

$$p_n(t) = e^{-\lambda t}\frac{(\lambda t)^n}{n!} \tag{6.132}$$

We note that in case the function $q(t)$ is a delta function we have the white Poisson noise given by

$$\xi_p(t) = \sum_i D_p \delta(t - t_i) - D_p \lambda \tag{6.133}$$

where λ, on comparison with Equation (6.131), clearly represents the mean density of instant t_i in the t space. The cumulants for the white and shot noises can be readily evaluated as

$$\left.\begin{aligned} \langle\langle \xi_p \rangle\rangle &= 0 \\ \langle\langle \xi_p(t_1)\ldots\xi_p(t_n)\rangle\rangle &= \lambda D_p^n \delta(t_1 - t_2)\ldots\delta(t_{n-1} - t_n) \end{aligned}\right\}\text{white noise} \tag{6.134}$$

and

$$\left.\begin{aligned} \langle\langle \xi(t) \rangle\rangle &= 0 \\ \langle\langle \xi(t_1)\ldots\xi(t_n)\rangle\rangle &= \lambda D_p^n \int_{-\infty}^{\infty} q(t_1 - Z)\ldots q(t_n - Z)\, dZ \end{aligned}\right\}\text{shot noise} \tag{6.135}$$

The cumulants given by Equation (6.134) are nonvanishing only at equal times. Also for the special case of $\lambda \to \infty$, $D_p \to 0$ but $\lambda D^2 = 2D$ = finite, we find that only the second cumulant is nonvanishing. We recognise this to be the Gaussian noise. The above transformations thus enable the Gaussian noise to be recovered as one of the limiting forms.

Let us now revert to Equation (6.130). Our intention is to derive an equation for $p(x, t)$. We begin with a simple case where the functions $f(x) = 0$ and $g(x) = 1$ and recognise this to be the case of pure noise. The governing equations may be written for this case as

$$\frac{dx}{dt} = \xi(t) \tag{6.136}$$

where $\xi(t)$ is the coloured Poisson noise defined by

$$\frac{d\xi}{dt} = \frac{1}{\tau}[\xi(t) + \xi_p(t)] \tag{6.137}$$

Equation (6.137) can be interpreted as an output of a linear system whose response to the δ-perturbation is $q(t)$ and has an input $\xi_p(t)$. Note that the process $\xi(t)$ appearing in Equation (6.136) is nonstationary and its cumulants can be written as

$$\begin{gathered} \langle\langle \xi(t) \rangle\rangle - e^{-t/\tau}\xi(0) \\ \langle\langle \xi(t_1) \ldots \xi(t_n) \rangle\rangle = \lambda D_p^n \int_0^\infty q(t_1 - Z) \ldots q(t_n - Z)\, dZ \end{gathered} \tag{6.138}$$

The stochastic Liouville equation associated with Equation (6.136) can be written as

$$\frac{\partial \rho(x, t)}{\partial t} = -\frac{\partial}{\partial x}\xi(t)\rho(x, t) \tag{6.139}$$

which on averaging over all realisations of $\xi(t)$ and initial conditions and taking account of the fact that

$$p(x, t) = \langle \rho(x, t) \rangle = \langle \delta(x(t) - x) \rangle$$

yields the following equation:

$$\frac{\partial p(x, t)}{\partial t} = -\frac{\partial}{\partial x}\langle \xi(t)\rho(x, t) \rangle \tag{6.140}$$

The evaluation of the rhs requires the use of the functional derivatives, and as indicated by Barcons and Garrido (1980) we first arrive at

$$\frac{\partial p(x,t)}{\partial t} = \lambda p[x - D_p(1 - e^{-t/\tau}), t] - \lambda p[x, t] + \lambda D_p(1 - e^{-t/\tau})\frac{\partial}{\partial q}p[x,t] - \frac{\partial}{\partial q}e^{-t/\tau}p(0)p[x,t] \tag{6.141}$$

The equation properly reduces to the Gaussian limiting form. Thus Equation (6.141) presents a generalisation to non-Gaussian cases. Note, however, that such exact equations can be derived only for cases with $f(x) = 0$ and $g(x) = 1$. In the general case we can only obtain an approximate equation, especially when the noise $\xi(t)$ is near-Gaussian and its time-correlation effects are small. In such cases one can again proceed with the associated stochastic-Liouville equation that is averaged over all realisations of $\xi(t)$ and initial conditions to give

$$\frac{\partial p(x,t)}{\partial t} = -\frac{\partial}{\partial x}f(x)p[x,t] - \frac{\partial}{\partial x}g(x)\langle\xi(t)\rho(x,t)\rangle \tag{6.142}$$

The correlation function in the last term is evaluated as usual to finally obtain

$$\frac{\partial p(x,t)}{\partial t} = -\frac{\partial}{\partial x}f(x)p[x,t] + D\frac{\partial}{\partial x}g(x)\frac{\partial}{\partial x}[g(x) - \tau G(x)]p[x,t] - \frac{DD_p}{3}\frac{\partial}{\partial x}g(x)\frac{\partial}{\partial x}[g(x) - \tau G(x)]\frac{\partial}{\partial x}g(x)p[x,t] + O(DD_p^2, \tau^2) \tag{6.143}$$

This approximate relation provides further generalisation to the general case in that by letting $D_p \to 0$ we recover the case of coloured Gaussian noise. Also, for $\tau \to 0$ we recover the equation describing the effect of white Poisson noise. For small intensity (D_p) and correlation time (τ) of the noise, the general solution to Equation (6.143) can be written under stationary conditions as

$$p_{st}(x) = N\exp\left\{\int^x \frac{f(x) - Dg(x)g'(x)}{Dg^2(x)}dx\right\} \times \left[1 + \int^x \left[\frac{f(x) - Dg(x)g'(x)}{Dg^2(x)}\left(\frac{\tau G(x)}{g(x)} + \frac{D_p f(x)}{3Dg(x)}\right)\right]dx + \int^x \left[\tau\frac{G'(x)}{g(x)} + \frac{D_p f'(x)}{3Dg(x)}\right]dx\right] \tag{6.144}$$

where $G(x) = f(x)g'(x) - g(x)f'(x)$. The maxima of the stationary distribution can be obtained as the solution of the equation

$$f(x) - Dg(x)[g'(x) - \tau G'(x)] + \frac{D_p g(x)}{3} f'(x) = 0 \qquad (6.145)$$

Equations (6.144) and (6.145) provide an approximate analytical method of establishing the effects of non-Gaussianity and time-correlation effects of the noise. As can be visualised, situations might arise where the non-Gaussianity may produce effects that are not permissible for the Gaussian case.

REFERENCES

Arnold, L., Horsthemke, W. and Lefever, R., *Z. Physik* **29B**, 367 (1978).
Arnold, L., Horsthemke, W. and Stucki, J. W., *Biom. J.* **21**, 451 (1979).
Banai, N. and Brenig, L., *Physica* **119A**, 512 (1983).
Barcons, F. X. and Garrido, L., *Physica* **117A**, 212 (1983).
Ebeling, W. and Engel-Herbert, H., *Physica* **104A**, 378 (1980).
Escher, C., *Z. Physik* 35B, 351 (1979).
Garrido, L. and Sancho, J. M., *Physica* **115A**, 479 (1982).
Hanggi, P., Marchesoni, F. and Grigolini, P., *Z. Physik* **56B**, 333 (1984).
Hanggi, P., *Z. Physik* **31B**, 407 (1978).
Hongler, M. D. and Ryter, D. M., *Z. Physik 31B*, 333 (1978).
Horsthemke, W. and Malek-Mansour, M., *Z. Physik* **24B**, 307 (1976).
Horsthemke, W. and Lefever, R., *Phys. Lett.* **64A**, 19 (1977).
Horsthemke, W. and Lefever, R., *Biophys. J.* **35**, 415 (1981).
Kitahara, K., Horsthemke, W. and Lefever, R., *Phys. Lett.* **70A**, 377 (1979).
Kitahara, K., Horsthemke, W., Lefever, R. and Inaba, Y., *Prog. Theor. Phys.* **64**, 1233 (1980).
Klyantskin, V. I. and Tatarskii, V. I., *Usp. Fiz. Nauk* **110**, 499 (1973).
Kubo, R. J., *J. Math. Phys.* **4**, 174 (1963).
Lax, M., *Rev. Mod. Phys.* **38**, 359 (1966); **38**, 541 (1966).
Lindenberg, K. and West, B. J., *Physica* **119A**, 485 (1983).
Luczka, J., *Phys. Lett.* **102A**, 401 (1984).
Mikhavilov, A. S., *Phys. Lett.* **73A**, 143 (1979).
Mukamel, S., Oppenheim, I. and Ross, J., *Phys. Rev. A* **17**, 1999 (1977).
Pawula, R. F., *IEE Trans. Inf. Theory* **13**, 33 (1967).
Pawula, R. F., *Int. J. Control* **25**, 283 (1977).
Sagues, F., *Phys. Lett.* **104A**, 1 (1984).
Sagues, F., San Miguel, M. and Sancho, J. M., *Z. Physik* **55B**, 269 (1984).
San Miguel, M. and Sancho, J. M., *Z. Physik* **43B**, 361 (1981).
Sancho, J. M., San Miguel, M., Katz, S. L. and Gunton, J. D., *Phys. Rev. A* **26**, 1589 (1982).
Sancho, J. M., *J. Math. Phys.* **25**, 354 (1984).
Sancho, J. M. and San Miguel, M., *Z. Physik* **36B**, 357 (1980).
van Kampen, N. G., *Phys. Rep.* **24**, 171 (1976).

7 Application to Model Systems

7.1 Introduction

IN THE previous chapter we were concerned with the general formulation and role of external fluctuations in systems operating near the point of transition. The general theory presented therein is applied to a few case studies in the present chapter.

The chapter begins with the analysis of a fluid-bed reactor where fluctuations in the system properties that arise due to the random nature of the bubbles are incorporated. The results are discussed in Section 7.2, where the extension of the basic model to complex reactions and nonisothermal effects are also discussed. Section 7.3 is devoted to a quantitative evaluation of the effect of external noise on reacting systems that possess bistability. The examples analysed in this section include several reaction schemes on the catalyst surface, systems possessing several rate forms of the Langmuir–Hinshelwood type, and systems with systemic type of feedback. In addition, simple examples such as the effect of noise in a deactivating system and the role of temperature fluctuations in modifying the kinetic behaviour of reacting systems are also included. Section 7.4 is devoted to the analysis of reacting systems that possess limit cycle behaviour. Considering a simple example of this type, the role of noise is analytically evaluated. In addition, the example also provides a simple means to compare the effects of different types (white or nonwhite and Gaussian or non-Gaussian) of noises on the system behaviour. The section also analyses known mechanisms, such as the Brusselator, and with the help of examples highlights the role of noise in bringing about noise induced transitions. In fact, in a particularly noteworthy example analysed, the development suggests that the oscillations reported are not an intrinsic feature of the reaction but have their origin in the noise.

7.2 Analysis of Fluid-Bed Reactor

Various models depicting the behavioural features of fluidised beds have appeared in the literature and have been reviewed from time to time (Bukur *et al.*, 1977; Mao and Potter, 1983). A comprehensive treatment of the modelling of fluid-beds also appears in a book by Doraiswamy and Sharma (1984). All these models tend to describe the bed behaviour at various levels of sophistication, and the variations in bed behaviour caused by the fluctuations and randomness of the bubbles are not explicitly accounted for. The chaotic nature of the fluidised bed as observed in laboratory scale experiments suggests that these reactors should be modelled as stochastic systems and Bukur *et al.* (1977) have suggested that, in fact, deterministic modelling cannot describe with any precision the behavioural features of such beds.

The randomness of the bed behaviour arises due to the presence of bubbles which, as is known now, strongly affects the interphase mass transfer and solids mixing in the bed. As a first step in modelling these reactors it is necessary then to account for the fluctuations in these quantities.

To keep things simple, we shall assume that physically the reactor could be described as consisting of two well-mixed cells describing, respectively, the bubble and emulsion phases. The cells are connected to each other through an exchange that represents the overall mass transfer coefficient. Referring to the schematic diagram shown in Figure 7.1, the macroscopic balance equations for the bubble and emulsion phases with reaction located in the latter phase can be written as

$$\frac{dC_b}{d\tau} = \bar{K}(C_e - C_b) + \frac{\bar{F}_b}{\bar{V}_b}(C_{b0} - C_b) \tag{7.1}$$

$$\frac{dC_e}{d\tau} = \bar{K}\frac{\bar{V}_b}{\bar{V}_e}(C_b - C_e) + \frac{\bar{F}_e}{\bar{V}_e}(C - C_e) - R \tag{7.2}$$

where $\bar{F}$, $\bar{V}$ and C refer, respectively, to the volumetric flow rate, total volume and concentration, and subscripts b and e identify, respectively, the bubble and emulsion phases. $\bar{K}$ represents the mass transfer coefficient between the two phases and R the net rate of reaction equal to $kf(C_e)$. The overall relations $\bar{F}_b + \bar{F}_e = \bar{F}$ and $\bar{V}_b + \bar{V}_e = \bar{V}$ are

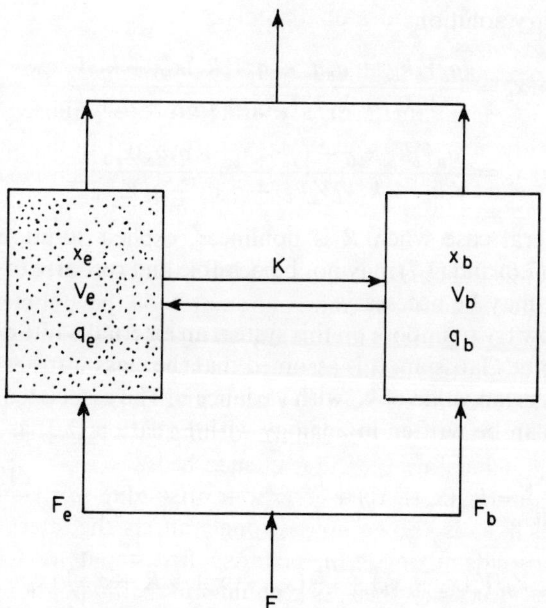

FIGURE 7.1 Schematic model of a fluid-bed reactor.

obviously satisfied. The initial conditions to this set can be defined as $C_0 = C_{b0} = C_{e0}$.

Equations (7.1) and (7.2) can be converted into dimensionless form:

$$\frac{dx_b}{dt} = k_{be}(x_e - x_b) + \frac{q_b}{V_b}(x_{b0} - x_b) \tag{7.3}$$

$$\frac{dx_e}{dt} = k_{be}\frac{V_b}{V_e}(x_b - x_e) + \frac{q_e}{V_e}(x_{e0} - x_e) - R \tag{7.4}$$

by using the transformations

$$\begin{aligned} x_b &= C_b/C_0, \qquad x_e = C_e/C_0, \qquad t = \bar{F}\tau/\bar{V} \\ V_b &= \bar{V}_b/\bar{V}, \qquad V_e = \bar{V}_e/\bar{V}, \qquad k_{be} = \bar{K}\bar{V}/\bar{F} \\ q_b &= \bar{F}_b/\bar{F}, \qquad q_e = \bar{F}_e/\bar{F}, \qquad R = R\bar{V}/\bar{F}C_0 = kf(C_e)\bar{V}/\bar{F}C_0 \end{aligned} \tag{7.5}$$

For linear first-order kinetics $R = k_1 x_e$ and Equations (7.3) and (7.4) give the stationary solutions

$$x_b = \frac{(q_b V_b k_{be} + q_b q_e + q_b V_e k_1)x_{b0} + k_{bc} V_e x_{c0}}{V_b k_{bc} + k_1 V_b V_e k_{bc} + q_b q_e + q_b V_e k_1}$$
$$x_e = \frac{q_b V_b k_{bc} x_{b0} + (q_e V_b k_{bc} + q_b q_e)x_{e0}}{V_b k_{bc} + k_1 V_b V_e k_{bc} + q_b q_e + q_b V_e k_1} \tag{7.6}$$

In the general case when R is nonlinear, explicit solutions such as Equations (7.6) and (7.7) may not be possible and recourse to numerical procedures may be necessary.

Let us now try to impose on this system an external white noise that is assumed to be Gaussian. It is assumed that the mass transfer coefficient possesses a mean value of $\bar{k}_{bc}$ with variance σ. The associated stochastic equations can be written in analogy with Equation (7.3) as

$$\frac{dx_b}{dt} = \bar{k}_{bc}(x_e - x_b) + \frac{q_b}{V_b}(x_{b0} - x_b) + \sigma(x_b - x_e)\frac{dW}{dt} \tag{7.8}$$

$$\frac{dx_e}{dt} = \bar{k}_{bc} V_B/V_e(x_e - x_b) + \frac{q_e}{V_e}(x_{e0} - x_e) - R + \sigma\frac{V_b}{V_e}(x_b - x_e)\frac{dW}{dt} \tag{7.9}$$

The equations can be expressed generally as a vector differential equation of the Langevin form

$$\frac{d\mathbf{X}}{dt} = f(\mathbf{X}, t)\,dt + g(\mathbf{X}, t)\,d\mathbf{w}(t) \tag{7.10}$$

where

$$\mathbf{X} = \begin{bmatrix} x_b \\ x_e \end{bmatrix}$$

$$F(\mathbf{X}, t) = \begin{bmatrix} -(\bar{k}_{bc} + q_b/V_b) & \bar{k}_{bc} \\ \frac{V_b}{V_e}\bar{k}_{bc} & -\bar{k}_{bc}\frac{V_b}{V_e} + \frac{q_e}{V_e} + R_1 \end{bmatrix} \begin{bmatrix} x_b \\ x_e \end{bmatrix} + \begin{bmatrix} q_b x_{b0}/V_b \\ q_e x_{e0}/V_e + \Delta \end{bmatrix} \tag{7.11}$$

$$g(\mathbf{X}, t) = \begin{bmatrix} x_e - x_b \\ \frac{V_b}{V_e}(x_b - x_e) \end{bmatrix}$$

and the reaction rate, for reasons of generality, may be expressed as

$$R = R_1 x_e + \Delta \tag{7.12}$$

Thus, for a cubic rate form $R = -x^3 + \alpha x^2 - \beta x + \Delta$, we have $R_1 = (-x^2 + \alpha x - \beta)$.

As we noted earlier, the Langevin Equation (7.10) could be interpreted in the Ito or Stratonovich sense yielding two different forms of the Fokker–Planck equations. The Ito form can be written as

$$\frac{\partial p(X,t)}{\partial t} = -\sum_{i=1}^{n} \frac{\partial}{\partial x_i} f_i(X,t)p(X,t) + \frac{1}{2}\sum_{i,j=1}^{n} \frac{\partial^2}{\partial x_i\,\partial x_j}[g(\mathbf{X},t)\sigma g(\mathbf{X},t)^T]\,p(\mathbf{X},t) \tag{7.13}$$

Equation (7.13) completely specifies the stochastic process in Equations (7.8) and (7.9) and as such can be solved to obtain the probability density $p(\mathbf{X},t)$. In general, the solution of Equation (7.13) poses considerable technical difficulties and only in the linear case does analytical solution seem possible. As such we are not interested as much in the solution $p(\mathbf{X},t)$ as in the average quantities like mean. These can be obtained straightaway from Equation (7.13) by employing the method of moments:

$$\frac{dE(x_b^k x_e^j)}{dt} = \frac{dE(y)}{dt} = \sum_{i=1}^{n} E\left(f_i \frac{\partial y}{\partial x_i}\right) + \frac{1}{2}\sum_{i,j=1}^{n} E\left\{[g\sigma g^T]_{ij}\frac{\partial^2 y}{\partial x_i\,\partial x_j}\right\} + E\left(\frac{\partial y}{\partial t}\right) \tag{7.14}$$

The stationary solution of this equation gives the stationary moments of the solution of the Ito Equation (7.13). Employing the proper values of the functions f and g, we obtain

$$\begin{aligned}\frac{dm_{kj}}{dt} &= E\{[-(\bar{k}_{bc} + q_b/V_b)x_b + \bar{k}_{bc}x_e + q_b x_{b0}/V_b]kx_b^{k-1}x_e^j\}\\ &+ E\left\{\left[\frac{\bar{k}_{bc}V_b}{V_e}x_b - \left(\frac{V_b\bar{k}_{bc}}{V_e} + \frac{q_e}{V_e} + R_1\right)x_e + \frac{q_e x_{e0}}{V_e}\right]jx_b^k x_e^{j-1}\right\}\\ &+ \tfrac{1}{2}E\{\sigma(x_e - x_b)^2 k(k-1)x_b^{k-2}x_e^j\}\end{aligned}$$

$$- \frac{V_b}{V_e} E\{\sigma(x_e - x_b)^2 kj x_b^{k-1} x_e^{j-1}\}$$

$$+ \frac{1}{2}\left(\frac{V_b}{V_e}\right)^2 E\{\sigma(x_e - x_b)^2 j(j-1) x_b^k x_e^{j-2}\} \tag{7.15}$$

At stationary state the first moments ($k = 1, j = 0, j = 1, k = 0$) are obtained as

$$-(\bar{k}_{bc} + q_b/V_b)m_{10} + \bar{k}_{bc} m_{01} + \frac{q_b x_{b0}}{V_b} = 0 \tag{7.16}$$

$$\frac{\bar{k}_{bc} V_b}{V_e} m_{10} - \left(\frac{\bar{k}_{bc} V_b}{V_e} + \frac{q_e}{V_e} + R_1\right) m_{01} + \frac{q_e x_{e0}}{V_e} = 0 \tag{7.17}$$

Note that it is only possible in linear cases to write the second equation in closed form. If the function R_1 is nonlinear, higher moments would enter the equation coupling them to equations for higher order moments. In general, a systematic procedure for truncation of the hierarchy of these moment equations does not exist. The method therefore becomes inoperable for nonlinear situations. The moments m_{10} and m_{01} of course correspond to the mean values of the concentrations x_b and x_e, respectively. Comparison of Equations (7.16) and (7.17) with the deterministic equations reveals complete identity. The mean concentrations as obtained from the Ito description are therefore identical to the deterministic results. Such an equivalence would of course break down if R_1 is nonlinear.

We can now proceed in a similar way to obtain the second moment from Equation (7.15) as the solution of

$$\begin{bmatrix} m_{20} \\ \\ m_{11} \\ \\ m_{02} \end{bmatrix} \left[\begin{array}{ll} -\left(\bar{k}_{bc} + \dfrac{q_b}{V_b}\right) + \sigma & 2\bar{k}_{bc} - 2\sigma \\ \bar{k}_{bc} V_b/V_e - \sigma \dfrac{V_b}{V_e} & -\left(\bar{k}_{bc} + \dfrac{q_b}{V_b}\right) - \left(\bar{k}_{bc}\dfrac{V_b}{V_e} + \dfrac{q_e}{V_e} + R_1\right) + \dfrac{2V_b\sigma}{V_e} \\ (V_b/V_e)^2 \sigma & 2\bar{k}_{bc}\dfrac{V_b}{V_e} - 2\left(\dfrac{V_b}{V_e}\right)^2 \sigma \end{array} \right.$$

$$\begin{bmatrix} \sigma \\ \bar{k}_{bc} - \dfrac{V_b \sigma}{V_e} \\ -2\left(\dfrac{\bar{k}_{bc} V_b}{V_e} + \dfrac{q_e}{V_e} + R_1\right) + \left(\dfrac{V_b}{V_e}\right)^2 \sigma \end{bmatrix} = - \begin{bmatrix} \dfrac{2q_b x_{b0}}{V_b} m_{10} \\ \dfrac{2q_b x_{b0}}{V_b} m_{01} + \dfrac{q_e x_{e0}}{V_e} m_{10} \\ \dfrac{2q_e x_{e0}}{V_e} m_{01} \end{bmatrix} \tag{7.18}$$

The stationary first and second moments as obtained from the above equations are presented in Figure 7.2 at different values of the strength of fluctuations.

In obtaining these moments we interpreted the Langevin equation [Equation (7.10)] in the Ito sense. If the equation were interpreted according to the Stratonovich convention, the following Fokker–Planck equation would result:

$$\frac{\partial p(\mathbf{X}, t)}{\partial t} = -\sum_{i=1}^{n} \frac{\partial}{\partial x_i}[f_i(\mathbf{X}, t)p(\mathbf{X}, t)] + \frac{1}{2}\sum_{i,k=1}^{n}\sum_{j=1}^{m} \frac{\partial}{\partial x_i}\left[g(\mathbf{X}, t)\sigma \frac{\partial}{\partial x_k} g(\mathbf{X}, t)p(\mathbf{X}, t)\right] \tag{7.19}$$

In obtaining the moments from this equation, a procedure similar to that followed earlier could be adopted. This again results in Equations (7.16) and (7.17) for the first moments and Equation (7.18) for the second moments, with the difference that the coefficients of f_i are now supplemented by the additional factor $\frac{1}{2}\sigma g g'$ arising due to the second term. The solutions of the final equations are also presented in Figure 7.2 for comparison with the Ito moments.

The method of moments to obtain the desired average from the Fokker–Planck equation works well for linear systems. In the case of nonlinear reaction rates, these Fokker–Planck equations will have to be solved numerically. Certain approximation schemes that simplify the calculations may also lead to some useful information. One such approximation may be illustrated by considering the nonlinear rate $r = gx^3(t) - \gamma(x)$. The governing balance equations now read:

$$\frac{dx_b}{dt} = \bar{k}_{bc}(x_e - x_b) + \frac{q_b}{V_b}(x_{b0} - x_b) + \sigma(x_e - x_b)\, dW \tag{7.20}$$

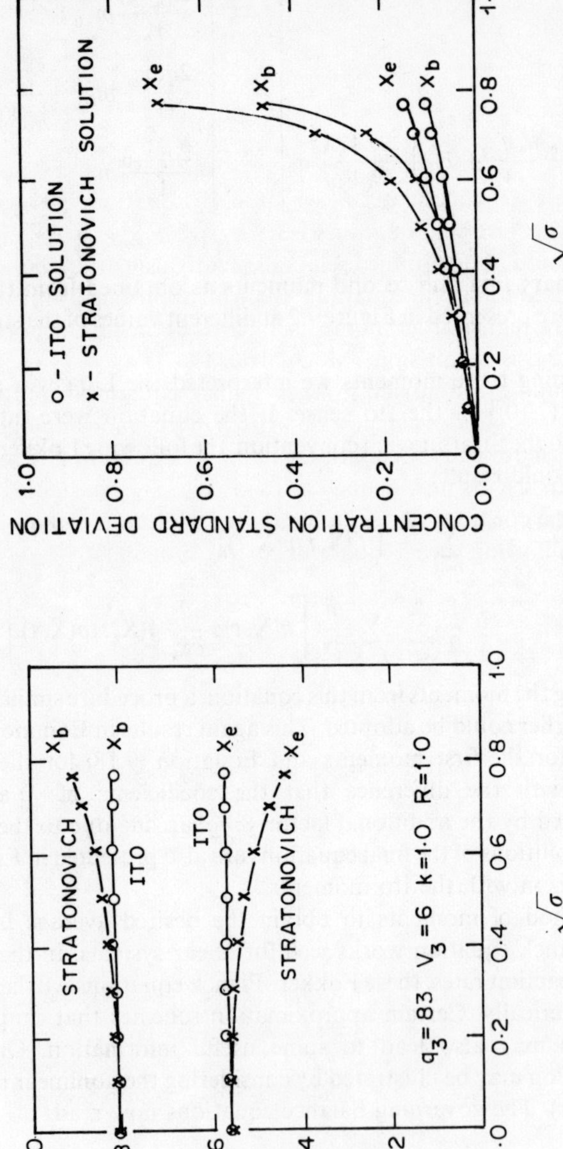

FIGURE 7.2 The Ito and Stratonovich moments

$$\frac{dx_e}{dt} = \frac{V_b}{V_e}\bar{k}_{bc}(x_b - x_e) + \frac{q_e}{V_e}(x_{e0} - x_e) - gx_e^3 + \gamma x_e$$

$$+ \frac{V_b}{V_e}\sigma(x_b - x_e)\,dW \tag{7.21}$$

The simplest self-consistent linearisation of Equation (7.21) consists in replacing the rate in Equation (7.21) by $[-g\langle x_e^2\rangle + \gamma]x_e$ where the average should be determined self-consistently using the solutions of the now linearised Langevin equations. Such an approximation amounts to replacing the higher order moments in terms of the product of the lower moments and is permissible (McQuarrie, 1967).

The Ito and Stratonovich Fokker–Planck equations appropriate to the Langevin equations can now be written in a fashion analogous to the previous case to give an equation similar to Equation (7.15), with the difference that R_1 is now defined as $[-g\langle x_e^2\rangle + \gamma]$. Assigning different values to the coefficient k, j in the moment evolution equation, we can write the following set of equations for the moments:

$$\frac{dm_{10}}{dt} = -\left(\bar{k}_{bc} + \frac{q_b}{V_b}\right)m_{10} + \bar{k}_{bc}m_{01} + \frac{q_b x_{bo}}{V_b} \tag{7.22}$$

$$\frac{dm_{01}}{dt} = \bar{k}_{bc}\frac{V_b}{V_e}m_{10} - \left(\frac{\bar{k}_{bc}V_b}{V_e} + \frac{q_e}{V_e} + gm_{02} - \gamma\right)m_{01} + \frac{q_e x_{e0}}{V_e} \tag{7.23}$$

$$\frac{dm_{20}}{dt} = \left[-2\left(\bar{k}_{bc} + \frac{q_b}{V_b}\right) + \sigma\right]m_{20} + 2(\bar{k}_{bc} - \sigma)m_{11} + \sigma m_{02} + \frac{2q_b x_{bo}}{V_b}m_{10} \tag{7.24}$$

$$\frac{dm_{11}}{dt} = \frac{V_b}{V_e}(\bar{k}_{bc} - \sigma)m_{20} - \left[\bar{k}_{bc} + \frac{q_b}{V_b} + \frac{V_b}{V_e}\bar{k}_{bc} + \frac{q_e}{V_e}\right.$$

$$\left. + gm_{02} - \gamma + \frac{2V_b}{V_e}\sigma\right]m_{11}$$

$$+ \left(\bar{k}_{bc} - \frac{V_b\sigma}{V_e}\right)m_{02} + \frac{q_b x_{bo}}{V_b}m_{01} + \frac{q_e x_{e0}}{V_e}m_{10} \tag{7.25}$$

$$\frac{dm_{02}}{dt} = \left(\frac{V_b}{V_e}\right)^2 \sigma m_{20} + 2\frac{V_b}{V_e}\left(\bar{k}_{bc} - \frac{V_b}{V_e}\sigma m_{11}\right)$$
$$- \left[2\frac{V_b}{V_e}\bar{k}_{bc} + \frac{q_e}{V_e} + gm_{02} - \gamma + \left(\frac{V_b}{V_e}\right)^2 \sigma\right] m_{02} + \frac{2q_e x_{e0}}{V_e} m_{01} \tag{7.26}$$

Equations (7.22)–(7.26) represent a coupled set of moment equations and can be solved to give the various first and second moments. The original nonlinear problem is thus reduced to solution of a coupled moments equation. The procedure illustrated for the Ito form of the Fokker–Planck equation can be extended to the Stratonovich form, and equations similar to the moments equations as above can be obtained. The results of the Ito and Stratonovich forms are presented along with the deterministic results in Figures 7.3 and 7.4.

We now seek to understand the behavioural features of this system in presence of coloured noise. As we might recall from the previous section, coloured noise has a finite correlation time and the system's temporal evolution is a non-Markovian process. This causes considerable mathematical difficulties in that the theory developed for Markovian processes now becomes inapplicable. A simplification is possible when the noise itself is Markovian. In such cases we can treat the system as a multidimensional process, with the state variables and the noise as its components. The process now being Markovian, the methods of solution developed for Markovian processes in general can be straightaway applied, provided the multivariate Fokker–Planck equation describing such processes is amenable to solution. This, however, may not always be the case and suitable methods such as perturbation schemes may have to be developed. Alternatively, numerical procedures may become necessary.

In a second approach one may begin with the non-Markovian description and develop the appropriate differential equation—similar in form to the Fokker–Planck equation for the Markov process—describing the evolution of probability density.

In what follows we shall demonstrate the first approach by considering two different forms, viz. the correlated Gaussian and the Gamma distributed noises, and the second approach by considering an Ornstein–Uhlenbeck process.

To illustrate the use of the first approach we shall begin by assuming

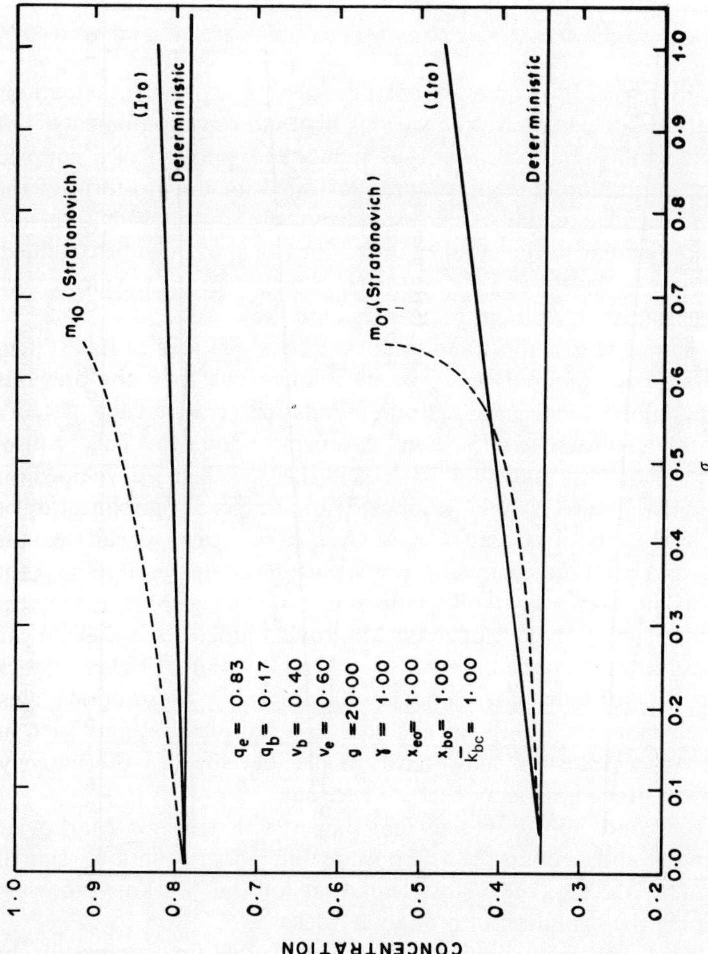

FIGURE 7.3 Ito and Stratonovich first moment.

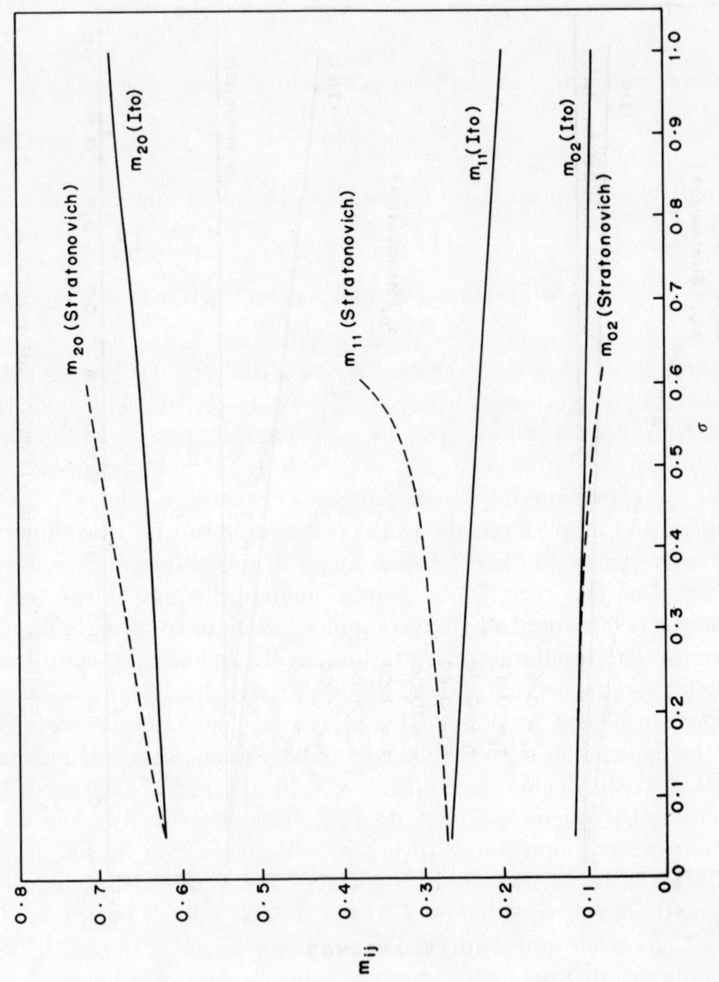

FIGURE 7.4 Ito and Stratonovich second moments.

that the overall interphase mass transfer coefficient of the deterministic model can be replaced as $y = \bar{k}_{bc} + x = k_{bc}$ where x represents a correlated Gaussian noise process defined by

$$\frac{dx}{dt} = -\mu x + \theta\mu\frac{dW}{dt}, \qquad \frac{dy}{dt} = -\mu(y - \bar{k}_{be}) + \theta\mu\frac{dW}{dt} \qquad (7.27)$$

with zero mean and stationary autocorrelation function given by

$$R(\tau) = \tfrac{1}{2}\theta^2\mu\sigma\, e^{-\mu\tau} \qquad (7.28)$$

Alternatively, the process y may be described as a Gamma process defined by

$$\frac{dy}{dt} = [\alpha + 1 - \xi - \mu y] + (2\delta y)^{1/2}\, dW \qquad (7.29)$$

Note that the noise defined by Equations (7.27) and (7.28) is Markovian. The deterministic Equations (7.3) and (7.4) with the overall transport coefficient k_{bc} replaced by y are now supplemented with an additional equation for y, Equation (7.27) or (7.29). The multiplying factor $y(x_b - x_e)$ appearing in the deterministic equations renders them nonlinear, and simple methods such as that of moments for the solution of the corresponding Fokker–Planck equation are no longer applicable. The problem has then to be solved numerically and Ligon and Amundson (1981a) used a hybrid computer technique to solve the set of equations. The resultant concentrations in the bubble and emulsion phases for varying degrees of standard deviations in the interphase mass transport coefficient are presented in Figures 7.5 and 7.6. It is observed from the figures that both the Gaussian and Gamma correlated noise models yield equivalent mean values for identical values of variance and of fluctuation frequencies. The standard deviations for the two models, however, become increasingly divergent with increase in the variance and decrease in the fluctuation frequency. The Gaussian results are consistently higher than for the Gamma model. Also, comparison of Figure 7.2 for white noise with the corresponding Figure 7.5 for coloured noise indicates that the trends observed using the Stratonovich form of the equation are in line with the general results obtained for real noise.

Let us illustrate now the use of the second approach. We assume that the noise is described as an Ornstein–Uhlenbeck process that is defined

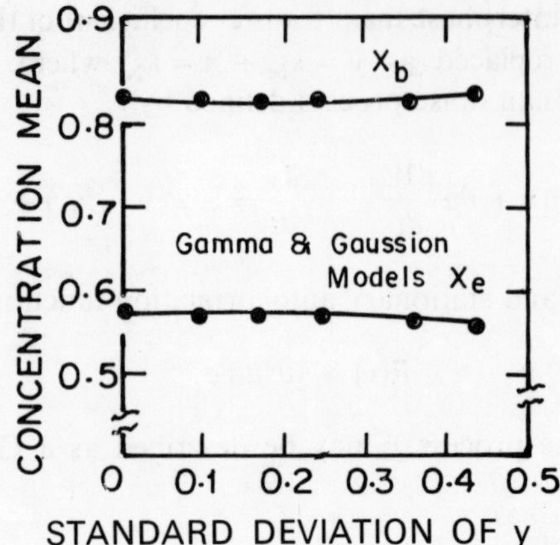

FIGURE 7.5 Bubble and emulsion phase concentrations as functions of standard deviations of the mass transport coefficient.

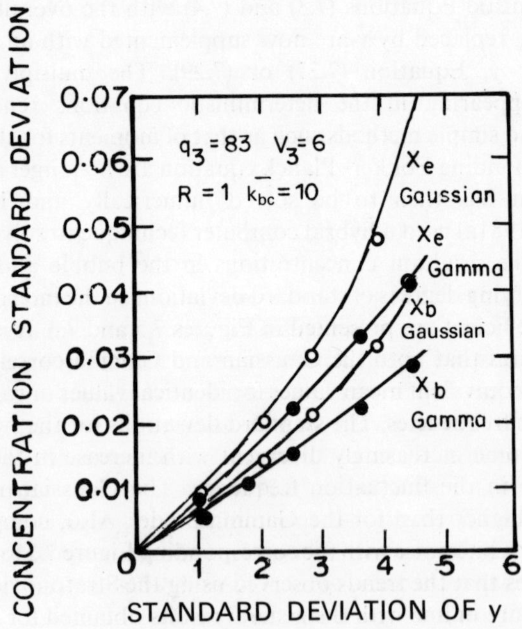

FIGURE 7.6 Bubble and emulsion phase concentration deviations as functions of standard deviation of the mass transport coefficient.

by the differential equation

$$\frac{d\xi}{dt} = -\varepsilon^{-2}\xi + \varepsilon^{-1}D\,dW \tag{7.30}$$

The process has a mean value of zero and correlation as given by Equation (6.46). The general two-dimensional processes defined by the deterministic Equations (7.3) and (7.4) can be written as a Langevin equation

$$\frac{d\mathbf{x}_\mu}{dt} = V_\mu(\mathbf{x}(t)) + g_{\mu\nu}[\mathbf{x}(t)]\xi(t), \qquad \mu = 1, 2 \tag{7.31}$$

where $\xi(t)$ are Gaussian stochastic forces of zero mean and correlation function

$$\langle \xi(t)\xi(t') \rangle = \gamma_{\mu\nu}(t, t') \tag{7.32}$$

We now proceed with the general development presented in Section 6.3.1 and obtain the following Fokker–Planck approximation for the process:

$$\frac{\partial p(\mathbf{x}, t)}{\partial t} = -\frac{\partial}{\partial x_\mu} V_\mu(\mathbf{x})p(\mathbf{x}, t) + D\frac{\partial}{\partial x_\mu} g_{\mu\nu}\frac{\partial}{\partial x_\beta} \times [g_{\beta\nu}(\mathbf{x}) - \tau_\nu M_{\beta\nu}(\mathbf{x})]p(\mathbf{x}, t) \tag{7.33}$$

where

$$M_{\beta\nu} = \frac{V_\beta \delta g_{\beta\nu}}{\delta x_\rho} - (\delta V_\beta/\delta x_\rho)g_{\rho\nu} \tag{7.34}$$

Equation (7.33) is the two-dimensional analogue of the more simple Equation (6.66) of the one-dimensional process. It includes the first-order corrections in the correlation times τ_μ. For the simple case of δ-correlated process, Equation (7.33) reduces to the Stratonovich form of Equation (7.19). The general results of the previous analysis, that the Stratonovich form of the equation gives the correct trends in tune with those obtained for real noise, are therefore already contained in Equation (7.33). The equation in the limiting case thus gives the results of white noise (in the Stratonovich sense). In addition, it contains the effects of finite correlation time τ_μ which have been analysed by Dabke, Kulkarni and Doraiswamy (1986a) and discussed below.

The transient profiles for the concentration in the bubble and emulsion phases are presented in Figures 7.7 and 7.8 for several values of

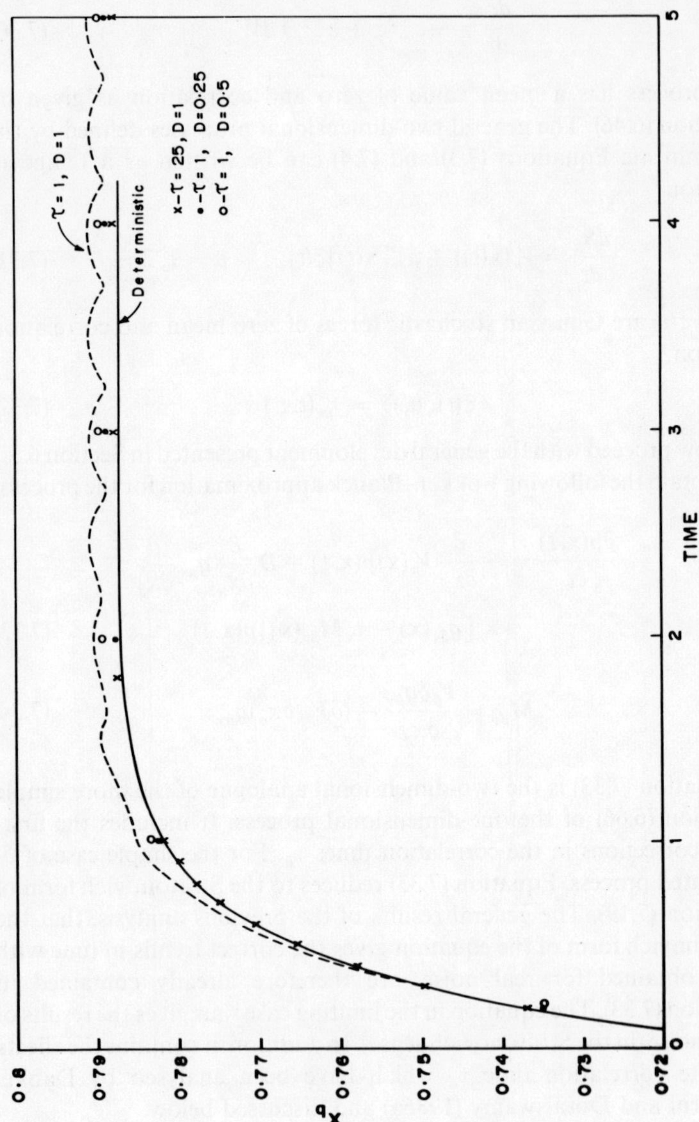

FIGURE 7.7 Concentration profiles in the bubble phase for various values of noise parameters

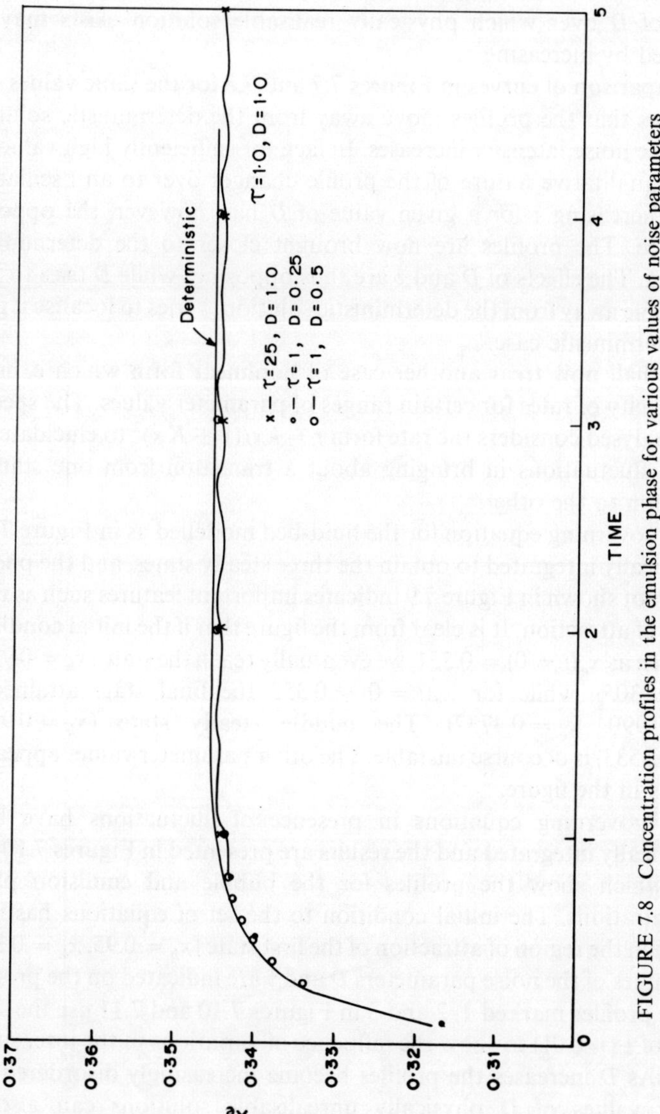

FIGURE 7.8 Concentration profiles in the emulsion phase for various values of noise parameters

D and τ. Clearly it is in general difficult to obtain meaningful solutions for values of $D > 0.7$. This of course depends on the value of τ, and the range of D over which physically realisable solution exists may be extended by increasing τ.

Comparison of curves in Figures 7.7 and 7.8 for the same values of τ indicates that the profiles move away from the deterministic solution when the noise intensity increases. In fact, for sufficiently high values of D, the qualitative nature of the profile changes over to an oscillatory form. Increasing τ for a given value of D has, however, the opposite influence. The profiles are now brought closer to the deterministic solution. The effects of D and τ are thus opposing: while D tries to pull the profile away from the deterministic solution, τ tries to localise it near the deterministic case.

We shall now treat another case of nonlinear form which exhibits multiplicity of rates for certain ranges of parameter values. The specific case analysed considers the rate form $r = kx/(1 + Kx)^2$ to elucidate the role of fluctuations in bringing about a transition from one state of operation to the other.

The governing equation for the fluid-bed modelled as in Figure 7.1 is numerically integrated to obtain the three steady states, and the phase–plane plot shown in Figure 7.9 indicates important features such as their region of attraction. It is clear from the figure that if the initial condition is chosen as $x_e(t = 0) = 0.351$, we eventually reach the state ($x_b = 0.9833$, $x_e = 0.0305$), while for $x_e(t = 0) = 0.352$ the final state attained is ($x_b = 0.990$, $x_e = 0.4732$). The middle steady state ($x_b = 0.9888$, $x_e = 0.3533$) is of course unstable. The other parameter values appear as legends in the figure.

The governing equations in presence of fluctuations have been numerically integrated and the results are presented in Figures 7.10 and 7.11, which show the profiles for the bubble and emulsion phase concentrations. The initial condition to the set of equations has been chosen in the region of attraction of the first state ($x_b = 0.95$, $x_e = 0.351$). The values of the noise parameters D and τ are indicated on the profiles.

The profiles marked 1, 2 and 3 in Figures 7.10 and 7.11 use the same value of τ ($=0.01$) to show the influence of variations in the intensity of noise. As D increases the profiles become increasingly disordered (for higher values of D physically unrealisable solutions can also be obtained), while for smaller values of D the system exhibits long transients. An examination of the curves 1 and 2 shows a tendency of the

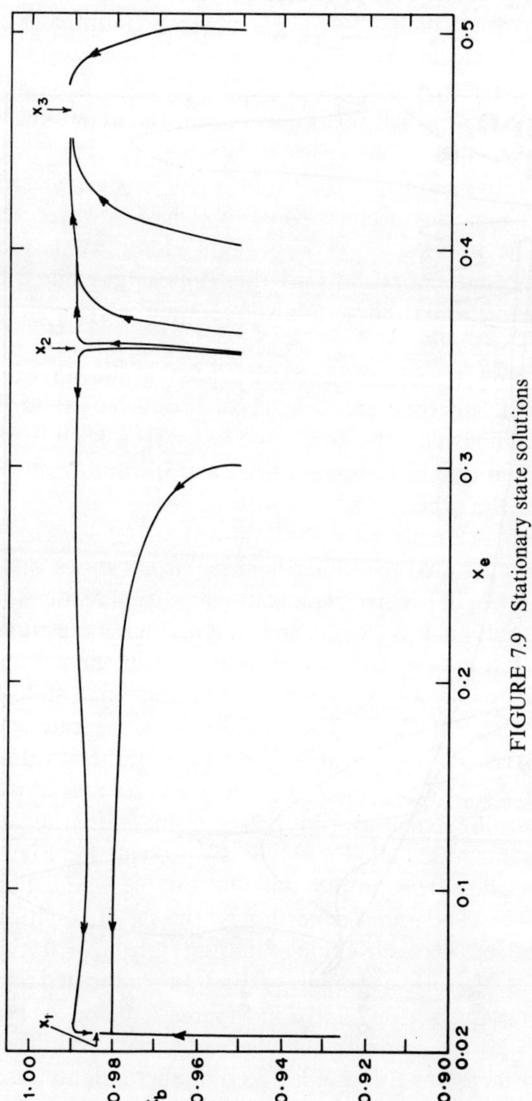

FIGURE 7.9 Stationary state solutions

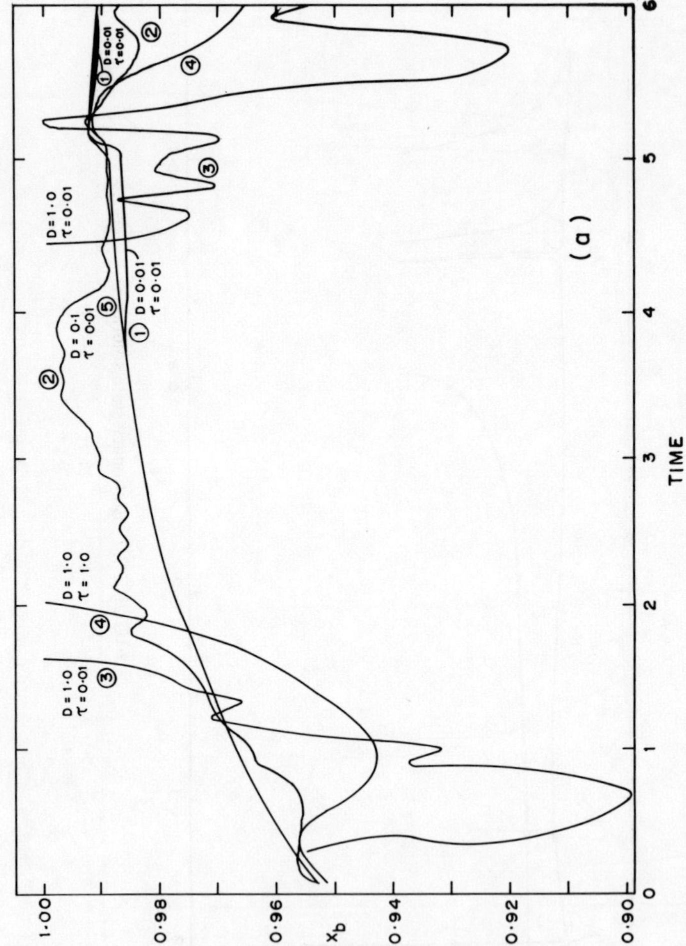

FIGURE 7.10(a) Concentration profiles in the bubble phase for different values of noise intensity and time-correlation effects.

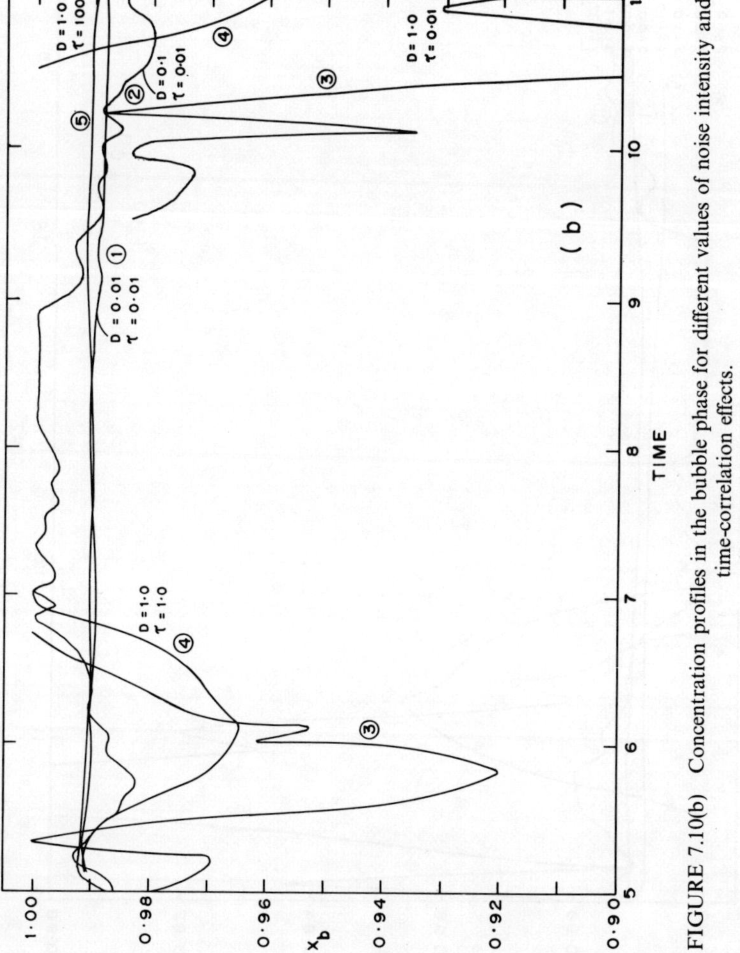

FIGURE 7.10(b) Concentration profiles in the bubble phase for different values of noise intensity and time-correlation effects.

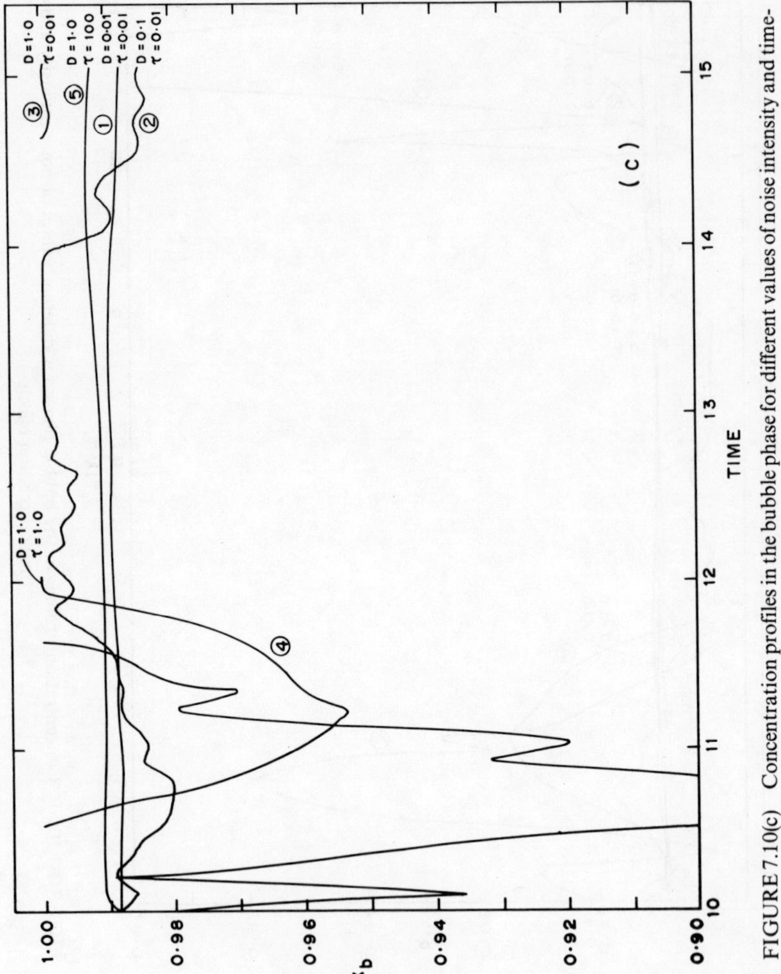

FIGURE 7.10(c) Concentration profiles in the bubble phase for different values of noise intensity and time-correlation effects.

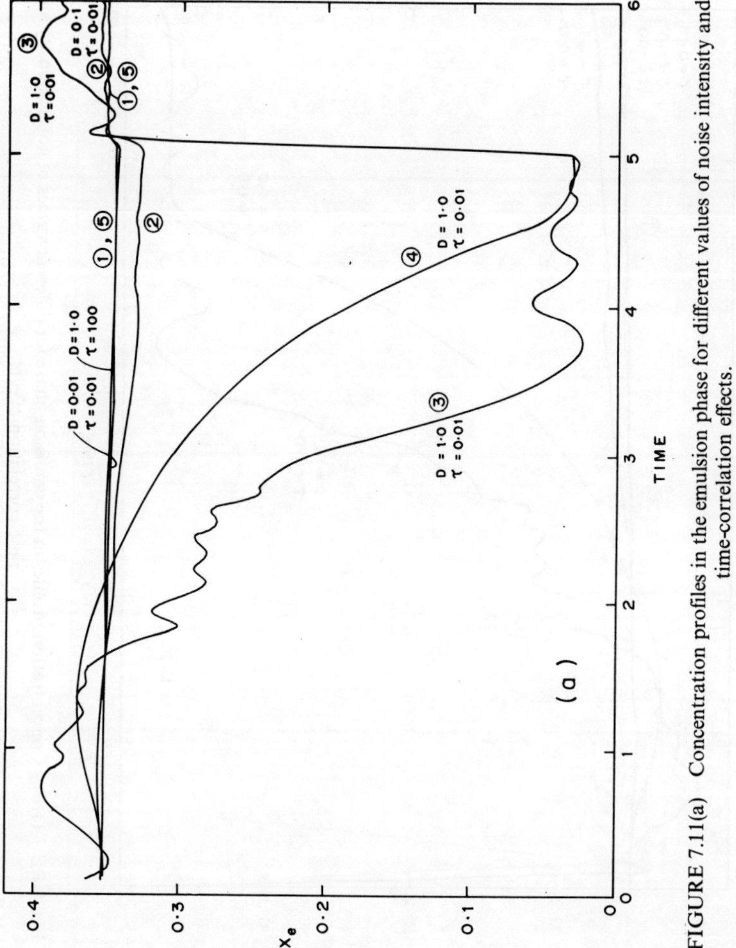

FIGURE 7.11(a) Concentration profiles in the emulsion phase for different values of noise intensity and time-correlation effects.

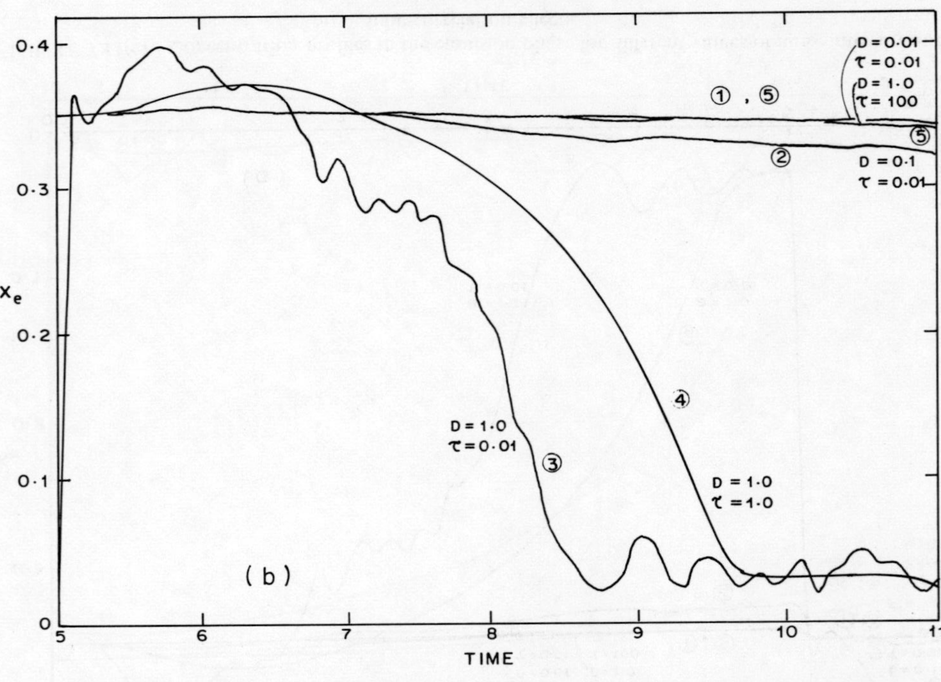

FIGURE 7.11(b) Concentration profiles in the emulsion phase for different values of noise intensity and time-correlation effects.

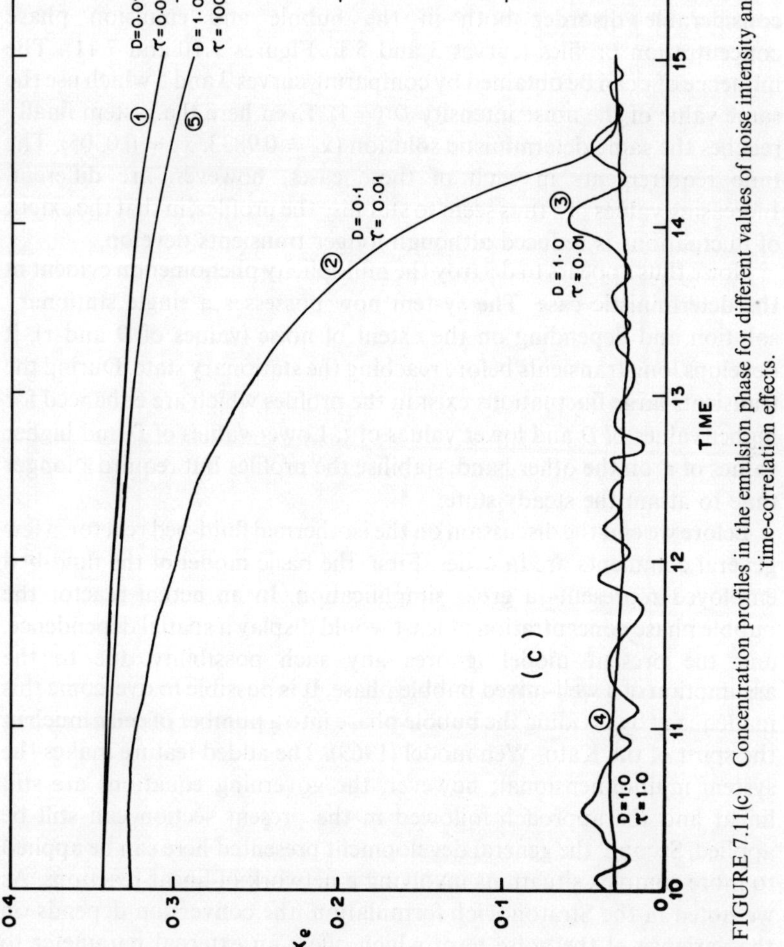

FIGURE 7.11(c) Concentration profiles in the emulsion phase for different values of noise intensity and time-correlation effects.

system to reach the same stationary solution ($x_b = 0.9833$, $x_e = 0.0305$). This is seen more clearly for curve 2, while for curve 1 the system is held localised near the middle unstable solution ($x_b = 0.9888$, $x_e = 0.3533$) for a long time with inclination towards reaching the same stable solution. Of course, during the intermediate times the system shows considerable disorder both in the bubble and emulsion phase concentration profiles (curves 3 and 5 in Figures 7.10 and 7.11). The influence of τ can be obtained by comparing curves 3 and 5 which use the same value of the noise intensity D $(=1)$. Even here the system finally reaches the same deterministic solution ($x_b = 0.9833$, $x_e = 0.0305$). The time requirements in each of these cases, however, are different. Increasing values of τ thus seem to stabilise the profiles, in that the extent of fluctuations is reduced although longer transients develop.

Noise thus appears to destroy the multiplicity phenomenon evident in the deterministic case. The system now possesses a single stationary solution and depending on the extent of noise (values of D and τ), it develops long transients before reaching the stationary state. During the transients large fluctuations exist in the profiles which are enhanced for higher values of D and lower values of τ. Lower values of D and higher values of τ, on the other hand, stabilise the profiles but require a longer time to attain the steady state.

Before we end the discussion on the isothermal fluid-bed reactor, a few general comments are in order. First, the basic model of the fluid-bed employed represents a gross simplification. In an actual reactor the bubble phase concentration at least would display a spatial dependence, and the present model ignores any such possibility due to the assumption of a well-mixed bubble phase. It is possible to overcome this inadequacy by dividing the bubble phase into a number of cells, much in the spirit of the Kato–Wen model (1969). The added feature makes the system multidimensional; however, the governing equations are still linear and the approach followed in the present section can still be applied. Second, the general development presented here can be applied to more complex situations involving a network of linear reactions. As we noted in the Stratonovich formulation, the conversion depends on the variance of the noise term which offers an external parameter to control the selectivity in multistep reactions.

The analysis presented for the isothermal case can be extended to include the nonisothermal effects due to reaction. This requires consideration of additional heat balance equations in the bubble and emulsion phases. Like the mass transfer coefficient, the heat transfer coefficient also now fluctuates and as a first approximation the fluctuation behaviour of both the heat and mass transfer coefficients can be treated as similar. The equations now being nonlinear, there is no hope of obtaining simple solutions. It is expected, however, that the mean results of the isothermal model would describe the situation satisfactorily, since fluctuations in the transport coefficients would not have any significant effect on the dense phase temperature due to its large heat capacity. The result has actually been confirmed by Ligon and Amundson (1981b) who used a hybrid computer technique to solve the problem. The additional results noted during the numerical simulation indicate that, unlike in the case of the isothermal model which always predicts a decrease in conversion with increase in the variance of noise, the conversion may actually increase or decrease depending on the parameters of operation. Also, unlike in the deterministic analysis which may indicate the existence of multiple states in a certain region, the general tendency of the stochastic model is to destroy many features. This feature can be understood as a general consequence of the external disturbance, as noted even for the simpler cases treated in the previous chapter.

7.3 Analysis of Reacting Systems Exhibiting Bistability

7.3.1 Model System with Nonsystemic Autocatalytic Feedback

In this section we shall consider the case of an autocatalytic reaction with exponential acceleration in the rate. This type of autocatalytic feedback, also known as systemic feedback, has been analysed from a macroscopic point of view by Ravikumar, Kulkarni and Doraiswamy (1984), who also cite examples of systems following this rate model. To be specific we shall consider the reaction scheme

$$A \xrightarrow{k_1} A_1 \xrightarrow{k_2} \qquad \text{I}$$

where the rate of consumption of A, according to the model of Ravikumar, Kulkarni and Doraiswamy (1984), follows the rate law

$$r_A = k_1[A] \exp \alpha_1[A_1] \qquad (7.35)$$

As seen in the rate law the product species A_1 exponentially accelerates the rate of its own formation, with α_1 as some arbitrary constant. For a constant input concentration of species A in a CSTR, the macroscopic equation describing the concentration of species A_1 can be written as

$$V\frac{d[A_1]}{dt_1} = k_f([A_{10}] - [A_1]) + Vk_1 A \exp \alpha_1[A_1] - k_2 V[A_1] \tag{7.36}$$

This equation can be put in nondimensional form as

$$\frac{da_1}{dt} = a_{10} - a_1 + Da_1 \exp(\alpha a_1) - Da_2 a_1 \tag{7.37}$$

where

$$t = \frac{k_f[A_{10}]t}{V}, \qquad Da_1 = \frac{k_1 V}{k_f}, \qquad Da_2 = \frac{k_2 V}{k_f[A]}$$
$$a_1 = \frac{[A_1]}{[A]}, \qquad \alpha = \alpha_1[A] \tag{7.38}$$

The stochastic differential equation corresponding to Equation (7.37) (modified to suit the situation of no autocatalytic species in the feed and $k_2 = 0$) can then be written as

$$\frac{da_1}{dt} = -a_1 + \overline{Da}_1 \exp(\alpha a_1) + \exp(\alpha a_1)\xi(t) \tag{7.39}$$

The equation is derived on the assumption that the parameter Da_1 fluctuates around the mean value $Da_1 = \overline{Da}_1 + \xi(t)$. It can be recast in the form

$$\frac{da_1}{dt} = f(a_1) + g(a_1)\xi(t) \tag{7.40}$$

The associated Fokker–Planck equation for the white noise case can be written as

$$\frac{\partial P(a_1, t)}{\partial t} = -\frac{\partial}{\partial a_1} f(a_1)p(a_1, t) + D\frac{\partial}{\partial a_1} g(a_1)\frac{\partial}{\partial a_1} g(a_1)p(a_1, t)$$
$$= -\frac{\partial}{\partial a_1}[-a_1 + \overline{Da}_1 \exp(\alpha a_1)]p(a_1, t)$$
$$+ D\frac{\partial}{\partial a_1} \exp(\alpha a_1)\frac{\partial}{\partial a_1} \exp(\alpha a_1)p(a_1, t) \tag{7.41}$$

The parameter D in this equation, as before, represents the intensity of the noise.

The stationary solution to Equation (7.41) can be obtained only if the model possesses natural, reflecting or periodic boundary conditions (Horsthemke and Lefever, 1984; Gardiner, 1983; Risken, 1984). This can be easily tested by checking whether the process reaches its boundaries or not. The probability flux at the boundaries can be evaluated using the analytical condition

$$J = \int \exp\left\{ - \int_{\delta}^{a_1} \frac{2f(z)}{g^2(z)} dz \right\} da_1 = \infty, \qquad \delta > 0 \tag{7.42}$$

and should remain finite for the equation to possess a solution. Here the limits of integration for the two integrals are $(0, \delta)$ and $(\delta, a_{1\infty})$, and $a_{1\infty}$ represents an upper bound for the variable a_1. Evaluating the inner integral in this equation where $f(z)$ and $g(z)$ are those identified in Equation (7.40), we obtain

$$J = \int \exp\left[\frac{2\overline{Da_1} \exp(-2\alpha z)}{\alpha} - \frac{2\exp(-2\alpha z)}{\alpha} - \frac{\exp(-2\alpha z)}{2\alpha^2} \right]_{\delta}^{a_1} da_1 \tag{7.43}$$

This integral possesses finite values at the boundaries $(0, a_{1\infty})$ and therefore suggests that the system does not possess natural, periodic or reflecting boundaries.

In view of this, the Fokker–Planck Equation (7.41) does not possess a stationary solution and we obtain the simple result that the presence of white noise destroys all the deterministic solutions. The situation, however, can be rectified by creating artificial boundaries—by imposing certain restrictions on the physical model. Thus, for instance, if we suppose that the activating influence (α) vanishes for no autocatalytic species in the system (i.e. $\alpha = 0$ for $a_1 = 0$), we have a natural lower boundary. Likewise, $Da/\alpha \to \infty$ for $a_1 = a_{1\infty}$ creates a natural boundary at the other end. Note that these constraints can be easily applied to the physical model described by Equation (7.37), which now needs to be supplemented by additional equations valid at the boundaries.

The stationary solution to the Fokker–Planck Equation (7.41) in presence of this restriction can be obtained as

$$p_0(a_1,\infty)=\frac{N}{\exp(\alpha a_1)}\exp\left[-\frac{1}{D}\left\{\frac{(1+2\alpha a_1)}{4\alpha}\exp(-2\alpha a_1)+\frac{Da_1}{\alpha}\exp(\alpha a_1)\right\}\right] \tag{7.44}$$

where N represents the normalisation constant to be obtained subject to the condition $\int_0^1 p(a_1,t)\,da_1=1$. In view of $\xi(t)$ being treated as a white noise, the process $a_1(t)$ is Markovian and the solution given by Equation (7.44) therefore also describes the transition probability of the process. This property, however, is lost when $\xi(t)$ is treated as a coloured noise and the simple formulation as above is no longer possible. An approximate Fokker–Planck equation for such situations, and especially for cases where the coloured noise is described by an Ornstein–Uhlenbeck process with small time-correlation effects, has been derived earlier in Chapter 6 and is rewritten here for reasons of clarity:

$$\frac{\partial p(a_1,t)}{\partial t}=-\frac{\partial}{\partial a_1}f(a_1)p(a_1,t)+D\frac{\partial}{\partial g_1}g(a_1)\frac{\partial}{\partial a_1}h(a_1)p(a_1,t) \tag{7.45}$$

where

$$h(a_1)=g(a_1)\left[1+\tau g(a_1)\frac{d}{da_1}[f(a_1)/g(a_1)]\right] \tag{7.46}$$

τ in Equation (7.46) represents the correlation time for the Ornstein–Uhlenbeck noise. The solution to this equation in terms of the white noise solution is given by

$$p_{st}(a_1,\infty)=p_0(a_1,\infty)[1+\tau(E_1-E_2-E_3)] \tag{7.47}$$

where

$$E_1=-\frac{1}{2D}\int_{a_1-}^{a_1+}\frac{f^2(a_1)}{g^2(a_1)}p_0(a_1,\infty)\,da_1$$

$$E_2=g(a_1)\frac{d}{da_1}\left[\frac{f(a_1)}{g(a_1)}\right]\quad\text{and}\quad E_3=\frac{1}{2D}\left[\frac{f(a_1)}{g(a_1)}\right]^2 \tag{7.48}$$

The following stationary solution can now be obtained by using these

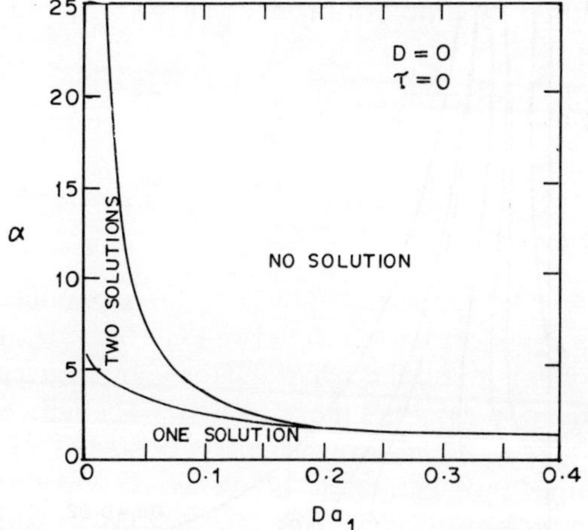

FIGURE 7.12 Bifurcation plot in the α–Da_1 parameter space.

equations:

$$p_{st}(a_1, \infty)$$

$$= p_0(a_1, \infty)\left[1 + \tau\left\{E_1 + 1 - a_1 - \frac{1}{2D}\left[\frac{-a_1 + \overline{Da_1}\exp(\alpha a_1)}{\exp(\alpha a_1)}\right]^2\right\}\right] \tag{7.49}$$

Equations (7.44) and (7.49) give, respectively, the effects of incorporating white and coloured noises in the deterministic system.

The results obtained by solving Equation (7.37) are presented in Figure 7.12 as a bifurcation diagram in the $\alpha - Da_1$ parameter space. The analysis reveals the existence of regions in parameter space with no solution, single solution, and two solutions. The stationary concentration distributions as a function of α for several values of Da_1 are presented in Figure 7.13. The multiplicity pattern seen in Figure 7.12 is also clearly evident in this figure. As may be noted, the region where two solutions exist is spread over a wider $a_1 - \alpha$ space for lower values of Da_1. For instance, for a value of $Da_1 = 0.02$ the region of two solutions lasts up to $\alpha = 18$, while for $Da_1 = 0.4$ the two-solution region is almost nonexistent. The stability of the solutions based on linear bistability analysis is also indicated in the figure.

The stochastic solution to Equation (7.45) has been obtained using Equation (7.49) and the results are shown for two sets of values of D and τ

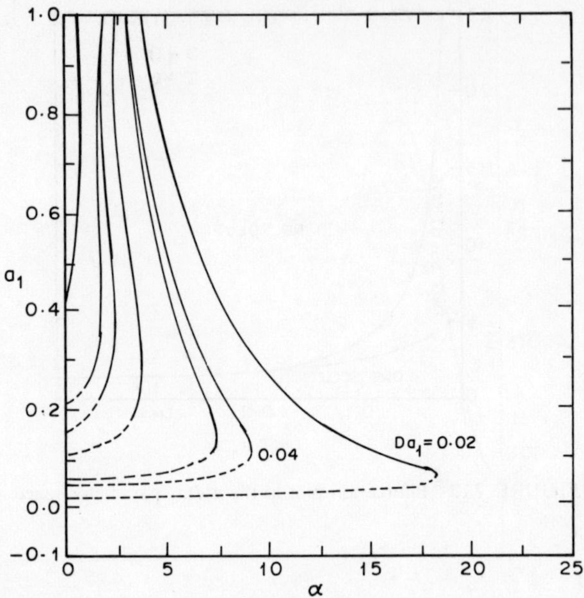

FIGURE 7.13 $a_1 - \alpha$ plots for various Da_1 values (deterministic analysis).

in Figures 7.14 and 7.15. The maximum and minimum in the probability distribution curves correspond, respectively, to the stable and unstable solutions. The region marked in the Figure 7.14 is magnified for the sake of clarity and is shown as an inset. The maximum and the minimum thus obtained for several sets of parameter values have been subsequently used to prepare plots of $a_1 - \alpha$ for a given value of Da_1 ($= 0.2$) and different values of D and τ (see Figures 7.16–7.19). The stability of the solutions is also indicated along the curves. For comparison the deterministic solution is also included. The results indicate that the parameter space wherein solutions exist narrows down in presence of noise.

The figures also indicate that the stochastic solution, like its deterministic counterpart, possesses a stable and an unstable branch. Both the stable and unstable solution branches of the stochastic solution, however, lie below the deterministic solution and suggest a shrinkage of the two-solution region. This trend continues as the value of τ is increased for the same value of D. Figure 7.17 indicates the situation for higher values of τ ($=0.1$) for $D = 0.001$.

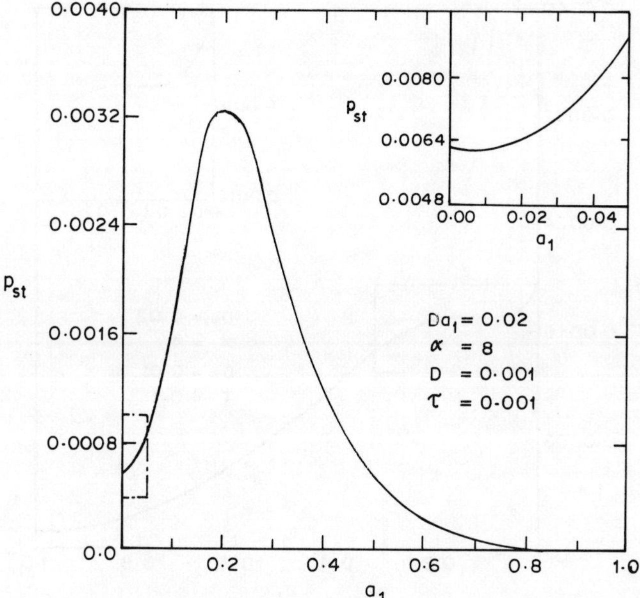

FIGURE 7.14 Stationary probability density distribution.

In order to understand the influence of τ on the system behaviour we shall concentrate on the curve *ABC* marking the stable solution branch in Figure 7.16. The stable solution branch is seen to bend in Figure 7.17 to create a point of maximum when τ increases. The general trends regarding the movements of the points *A*, *B* and *C* on increase in τ are evident in Figures 7.18 and 7.19, where curves are plotted for the same value of D and for $\tau = 0.5$ and 1.0, respectively. The point *A* in general moves towards higher values of α while the point *C* moves to the left on the α-axis. The result suggests that the two-solution region exists for higher values of α over a smaller range. The maximum indicated by point *B* moves down with increase in τ. In fact, with increase in τ, the three points are seen to move towards each other, narrowing the region of parameter space contained within them and would eventually merge into each other signifying total shrinkage. The system would not possess any solution for values of τ greater than this critical value. Detailed calculations indicate that higher values of D require lower values of τ, and vice versa.

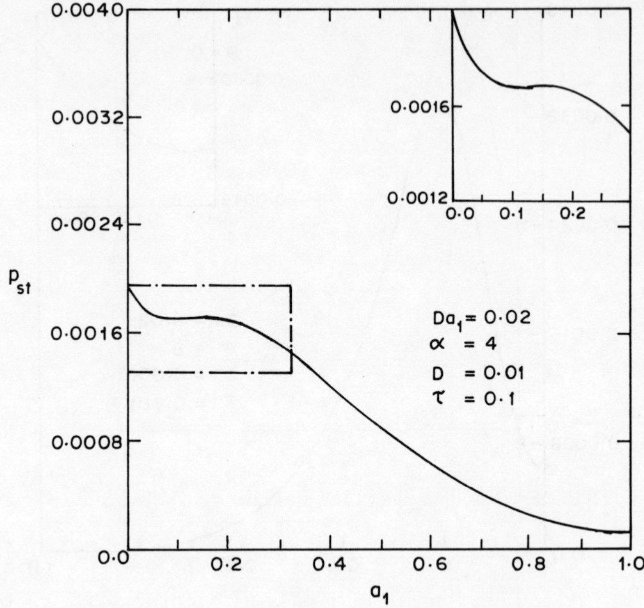

FIGURE 7.15 Stationary probability density distribution.

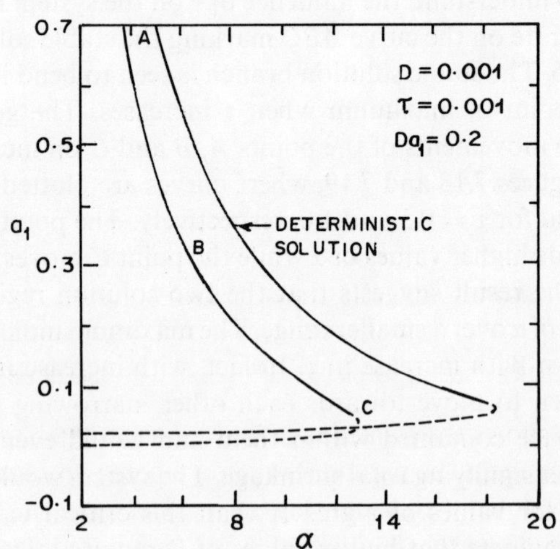

FIGURE 7.16 a_1–α plot in the presence of coloured noise.

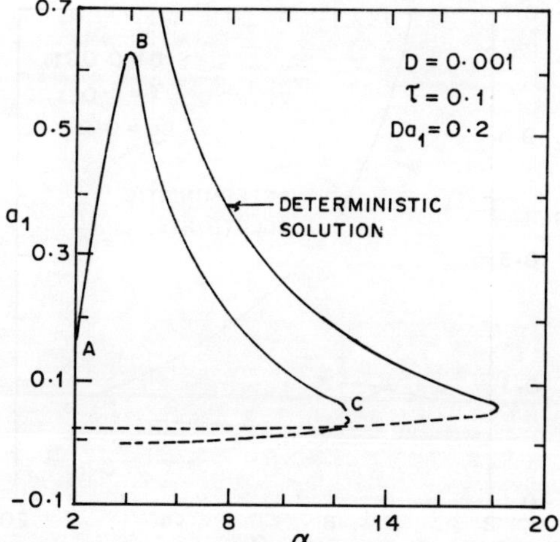

FIGURE 7.17 a_1–α plot in the presence of coloured noise.

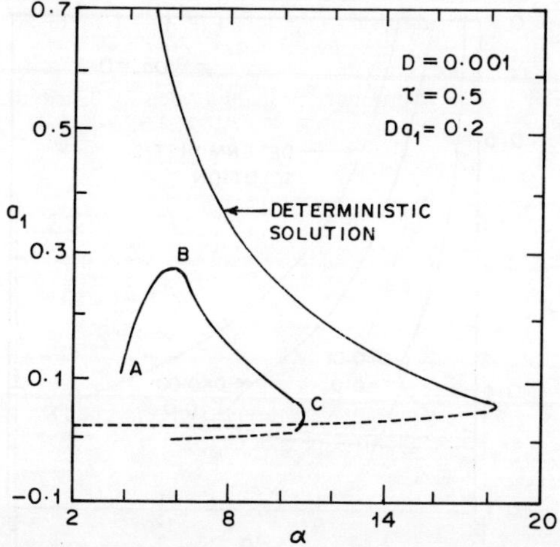

FIGURE 7.18 a_1–α plot in the presence of coloured noise.

Figure 7.20 shows the influence of variation in the intensity of white noise (D). Here again the parameter region where two solutions exist is seen to narrow down with increasing intensity of white noise. The mode of contraction, however, is different when compared with the effects observed in the presence of coloured noise (i.e. both D and τ). As can be

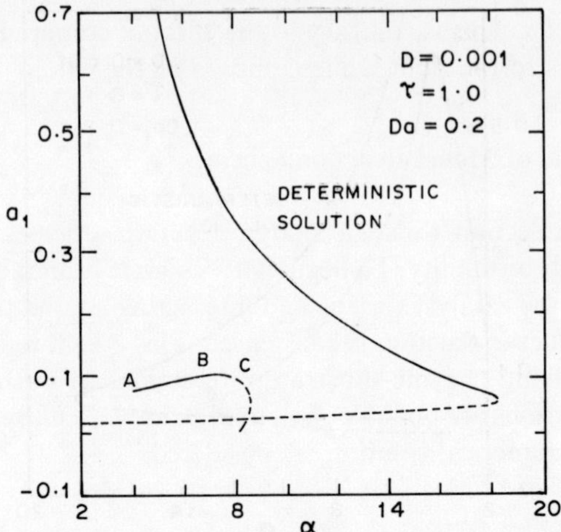

FIGURE 7.19 a_1–α plot in the presence of coloured noise.

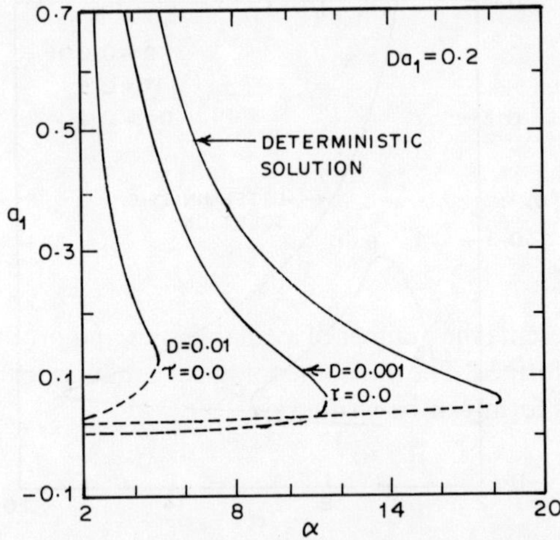

FIGURE 7.20 a_1–α plot for different values of intensity of noise (white noise case).

seen, white noise does not cause any bending of the stable solution branch and in consequence does not produce any point of maximum in it. Both the points *A* and *C* of the stable solution branch move to the left on the α-axis, indicating that the two-solution region now begins at an earlier value of α. This is contrary to the effect of coloured noise where point A moves to the right while C moves to the left.

7.3.2 .Analysis of Model Reaction Schemes

In the present section we shall analyse reaction schemes that exhibit features such as bistability. To begin with a simple example of reactants adsorbing on the catalyst surface and interacting among themselves to yield the product of reaction is considered. This is followed by another example involving enzyme-substrate interaction. In both these cases analytical solutions are possible. As a final example, a more general case that requires numerical solution is considered.

Models Yielding Analytical Solution

We shall first consider a model that exhibits the phenomenon of bistability on a catalyst surface (Bykov and Yablonskii, 1981):

$$X \underset{k_{-1}}{\overset{k_1}{\rightleftharpoons}} AX$$

$$AX + 2X \underset{k_{-2}}{\overset{k_2}{\rightleftharpoons}} 3X \qquad \text{I}$$

$$X \xrightarrow{k_3} Q$$

where X represents the number of available sites and product Q desorbs as fast as it is formed. The phenomenological equation for the scheme in dimensionless form can be written as

$$\frac{dx}{dt} = -\alpha x^3 + \beta x^2 - \gamma x \qquad (7.50)$$

where the variable x denotes the number of available sites and the coefficients α and β involve the influence of environment. The various coefficients can be identified as

$$\alpha = x_0^2(k_2 + k_{-2}), \qquad \beta = k_2 x_0^2 \qquad \text{and} \qquad \gamma = k_1 + k_{-1} + k_3$$

It is assumed that x_0 (the total number of sites available on the surface) fluctuates, as a result of the fluctuating environment, around a mean value. For the parameter values $\alpha = 0.001$, $\gamma = 0.0002$ and $\beta = 0.001$ to 0.0013, the system shows bistability of steady states. As a result of fluctuating x_0, the coefficients α and β get affected and in turn fluctuate around a mean value. Let these coefficients have mean values $\bar{\alpha}, \bar{\beta}$ with variances σ_1, σ_2, respectively. We thus have

$$\alpha = \bar{\alpha} + \sigma_1 \quad \text{and} \quad \beta = \bar{\beta} + \sigma_2 \tag{7.51}$$

The phenomenological Equation (7.50) now can be written as

$$\frac{dx}{dt} = -\bar{\alpha}x^3 + \bar{\beta}x^2 - \gamma x + (\sigma_1 x^3 + \sigma_2 x^2)\frac{dw}{dt} \tag{7.52}$$

where dw/dt denotes Gaussian white noise. Interpreting Equation (7.52) in the Stratonovich sense, the corresponding form of the Ito differential equation is given by

$$\frac{dx}{dt} = [-\bar{\alpha}x^3 + \bar{\beta}x^2 - \gamma x + \tfrac{1}{2}\{x^2(\sigma_1 x + \sigma_2)(3\sigma_1^2 x + 2\sigma_2 x)\}] + x^2(\sigma_1 x + \sigma_2)\frac{dw}{dt} \tag{7.53}$$

The evolution of probability density should satisfy the Fokker–Planck equation

$$\frac{\partial p(x,t)}{\partial t} = -\left[\frac{\partial}{\partial x}\tfrac{3}{2}\sigma_1^2 x^5 + \tfrac{5}{2}\sigma_1\sigma_2 x^4 + (\sigma_1^2 - \bar{\alpha})x^3 + \bar{\beta}x^2 - \gamma x\right] \times p(x,t) + \frac{1}{2}\frac{\partial^2}{\partial x^2}[x^4(\sigma_1 x + \sigma_2)^2]p(x,t) \tag{7.54}$$

Clearly $x = x_0 = 0$ is one of the solutions of this equation since it simultaneously satisfies the condition $F(x_0) = G(x_0) = 0$, where $F(x)$ and $G(x)$ refer, respectively, to the first and second square brackets of Equation (7.54).

A look at the deterministic Equation (7.50) indicates that $x = 0$ is also one of the deterministic solutions. The equation being cubic can possess three solutions in certain parameter ranges which are plotted in Figure 7.21. The figure clearly reveals that $x = 0$ is always one of the solutions and a critical value of β for fixed values of α and γ exists at which bifurcation occurs. The stochastic solution $x = 0$ thus coincides with the deterministic results.

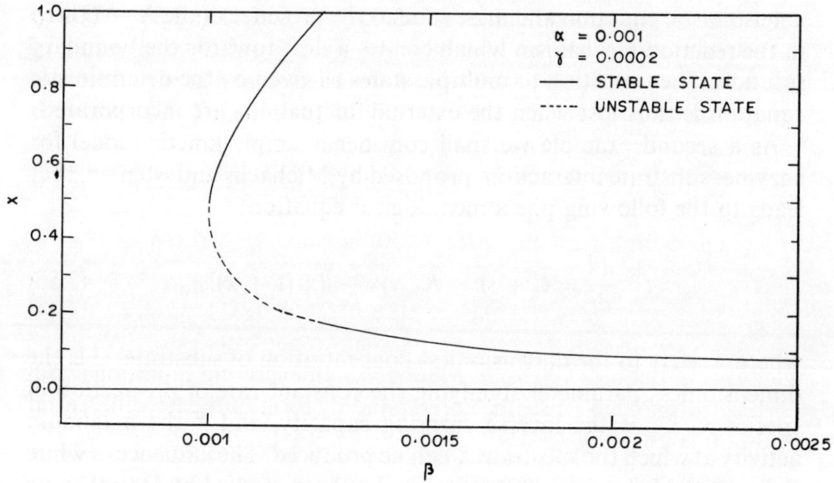

FIGURE 7.21 Stationary solutions for different values of β (scheme I)

Let us now investigate the other possible solutions of the system. For this purpose, the stationary probability distribution can be obtained from Equation (7.54) as

$$p_{st}(x) = \frac{N}{x^4(\sigma x + \sigma)^2} \times \exp \int^x \frac{3\sigma_1^2 x^5 + 5\sigma_1\sigma_2 x^4 + 2(\sigma_1^2 - \bar{\alpha})x^3 + 2\bar{\beta}x^2 - 2\gamma x}{\sigma_1^2 x^6 + 2\sigma_1\sigma_2 x^5 + \sigma_2^2 x^4}\, dx \tag{7.55}$$

where the integral under the exponential sign can be evaluated by breaking it into a number of integrals of standard form. N in this equation represents the normalisation constant.

In order that the stationary solution of the Fokker–Planck equation as given by Equation (7.55) be considered as a probability density, it has to be normalisable, i.e. $\int_0^x p_\infty(x)\, dx < \infty$. Actual integration of Equation (7.55), however, reveals that this condition is not met and in fact because of the strong divergence at $x = 0$, as indicated in Equation (7.55), $\int_0^x p_\infty(x)\, dx \rightarrow \infty$. The stochastic formulation thus admits no other solution.

This result is different from what one obtains by solving the deterministic equation and arises due to the presence of the $X \rightarrow Q$ step in the reaction mechanism which creates a drift towards the boundary solution. The transition to multiple states as given by the deterministic equation is thus lost when the external fluctuations are incorporated.

As a second example we shall consider a simple kinetic model for enzyme–substrate interaction, proposed by Michaelis and Menten, that leads to the following phenomenological equation:

$$\frac{dx}{dt} = C + (1 - K_m x)x - \beta x/(1 + x) \tag{7.56}$$

where x refers to the dimensionless concentration of substrate, C is the dimensionless parameter signifying the constant rate of production of species x, K_m is the inverse carrying capacity, and β the maximum activity at which the substrate X can be produced. The influence of white noise on the behaviour of this system has been studied by Horsthemke and Lefever (1977) who showed that the noise can induce phase transition.

In the present case the noise is assumed to be a coloured noise, described as an Ornstein–Uhlenbeck process, with zero mean and its correlation function given by

$$\langle \xi(t)\xi(t') \rangle = \gamma(t - t') = \frac{D}{\tau}\exp[-(t - t')/\tau] \tag{7.57}$$

For the sake of illustration it is assumed that the parameter β alone fluctuates as $\beta = \bar{\beta} + \xi(t)$. Incorporation of the fluctuations in Equation (7.56) leads to a stochastic differential equation

$$\frac{dx}{dt} = C + (1 - K_m x)x - \bar{\beta}x/(1 + x) + x\xi/(1 + x) \tag{7.58}$$

Treating ξ as a white noise process, the following Fokker–Planck equation can be associated with Equation (7.58):

$$\frac{\partial p(x, t)}{\partial t} = -\frac{\partial}{\partial x}[\{C + (1 - K_m x)x - \bar{\beta}x/(1 + x)\}p(x, t)] + D\frac{\partial^2}{\partial x^2}\left[\left(\frac{x}{1 + x}\right)^2 p(x, t)\right] \tag{7.59}$$

The stationary solution to this equation gives the probability density as

$$p_0(x) = N \exp \frac{1}{D}\left\{-\frac{C}{x} + (C + 2 - K_m - \beta)x + \tfrac{1}{2}(1 - 2K_m)x^2 - \tfrac{1}{3}K_m x^3 + (2C + 1 - \beta - 2D)\ln x + 2D\ln(1 + x)\right\} \tag{7.60}$$

In the event $\xi(t)$ is treated as a coloured noise described by Equation (7.57), the approximate Fokker–Planck equation can be written using the prescription of Sancho *et al.* (1982) as

$$\frac{\partial p(x,t)}{\partial t} = -\frac{\partial}{\partial x}[\{C + (1 - K_m x)x - \beta x/(1 + x)\}p(x,t)] + D\frac{\partial}{\partial x}\frac{x}{1+x}\frac{\partial}{\partial x} \times \left[\frac{x}{1+x} + \tau\left\{\left[1 - 2K_m x - \frac{\bar{\beta}}{(1 + x)^2}\right]\frac{x}{1+x} - \left(\frac{1}{1+x}\right)^2\left[C + x - K_m x^2 - \frac{\bar{\beta}x}{(1 + x)}\right]\right\}\right]p(x,t) \tag{7.61}$$

with the stationary solution

$$p_{st}(x) = p_0(x)\left\{1 + \tau\left[N_1 - \frac{1 + x}{x}\left\{\left[1 - 2K_m x - \frac{\bar{\beta}}{(1 + x)^2}\right]\frac{x}{1 + x} - \left[C + (1 - K_m x)x - \frac{\bar{\beta}x}{1 + x}\right]\frac{1}{(1 + x)^2}\right\} - \frac{1}{2D}\left[\frac{C + (1 - K_m x)x - \bar{\beta}x/(1 + x)}{x/(1 + x)}\right]^2\right]\right\} \tag{7.62}$$

The constant N_1 appearing in this equation can be estimated from

$$N_1 = -\frac{1}{2D}\int_0^\infty \left[\frac{C + (1 - K_m x)x - \bar{\beta}x/(1 + x)}{x/(1 + x)}\right]^2 p_0(x)\,dx \tag{7.63}$$

The stationary probability distribution given by Equation (7.62) is plotted in Figure 7.22 for a set of parameter values ($C = 4.5$, $K_m = 0.1$, $\bar{\beta} = 7.5$ and $D = 6.5$). The choice of $D = 6.5$ is made based on Figure 7.23 which shows a plot of the stationary solution of Equation (7.58) for

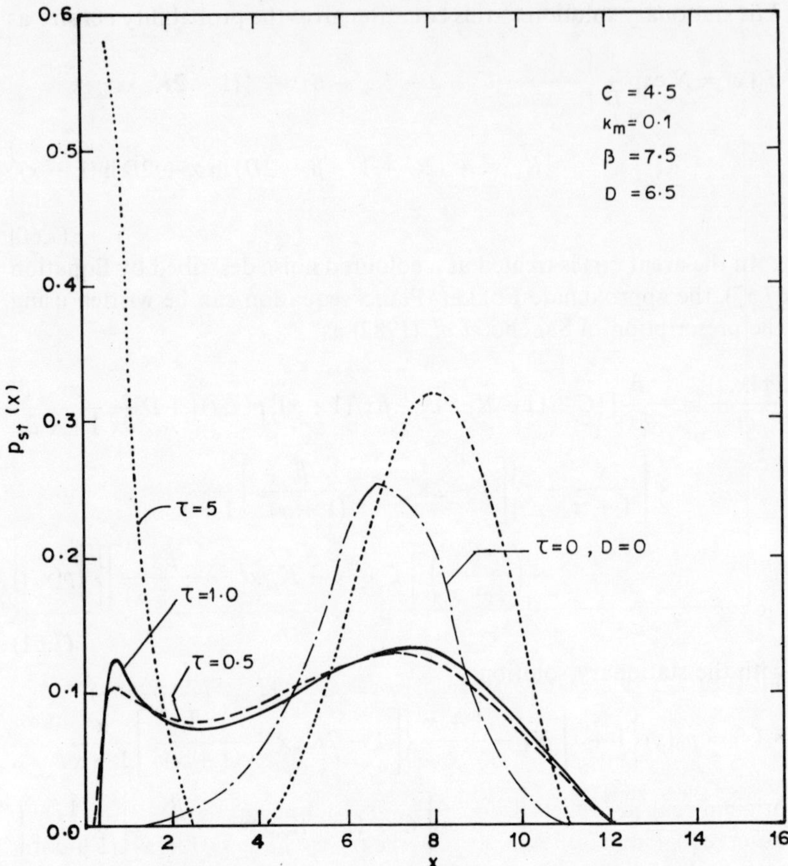

FIGURE 7.22 Stationary probability distribution of x for $D = 6.5$.

several values of D. It may be noted that $D = 6.5$ lies near the transition point where the system changes over from a unique state to the three-state region. The stationary probability distribution of this equation also shows only one peak corresponding to $x = 6.7$ ($\tau = 0$) for this value of D.

Figure 7.22 clearly brings out the effect of τ. It is seen that the original probability distribution for $\tau = 0$ (which shows only one peak corresponding to the stable steady state at $x = 6.7$) now develops an additional peak around $x = 0.7$ and a minimum around $x = 2.7$. The

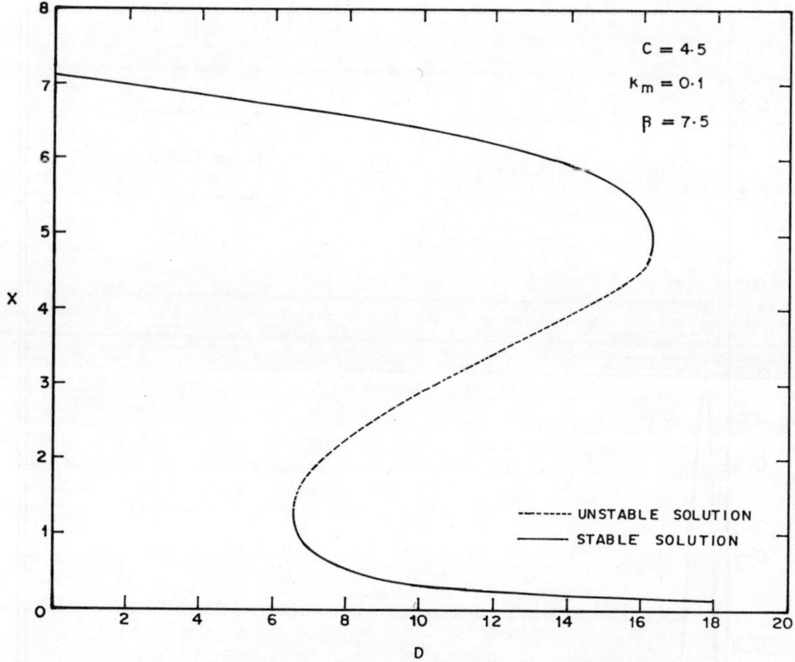

FIGURE 7.23 Plot of stationary solutions of Equation (7.56) for several values of D (Michaelis–Menten kinetics).

original peak at 6.7 gets dislocated and now appears at $x = 7.3$. We thus observe that in presence of τ (=0.5) the system moves over from a unique state to the three-state region. Coloured noise of the same intensity as the white noise can therefore induce transition where there was none with the white noise. Figure 7.24 sketches the results when D is sufficiently large (=8.5) to induce such a transition by itself. The role of τ now is to sharpen these peaks further. The higher steady state solution also moves to the right on incorporation of τ.

Analysis of a scheme showing bistability on a catalyst surface indicates that the incorporation of external noise destroys the multiplicity feature and the system eventually attains a state of no conversion (i.e. $x = 0$ is the only attainable steady state solution). The results obtained with enzymatic reactions indicate that incorporation of coloured noise can induce multiplicity when it is absent in presence of white noise. The

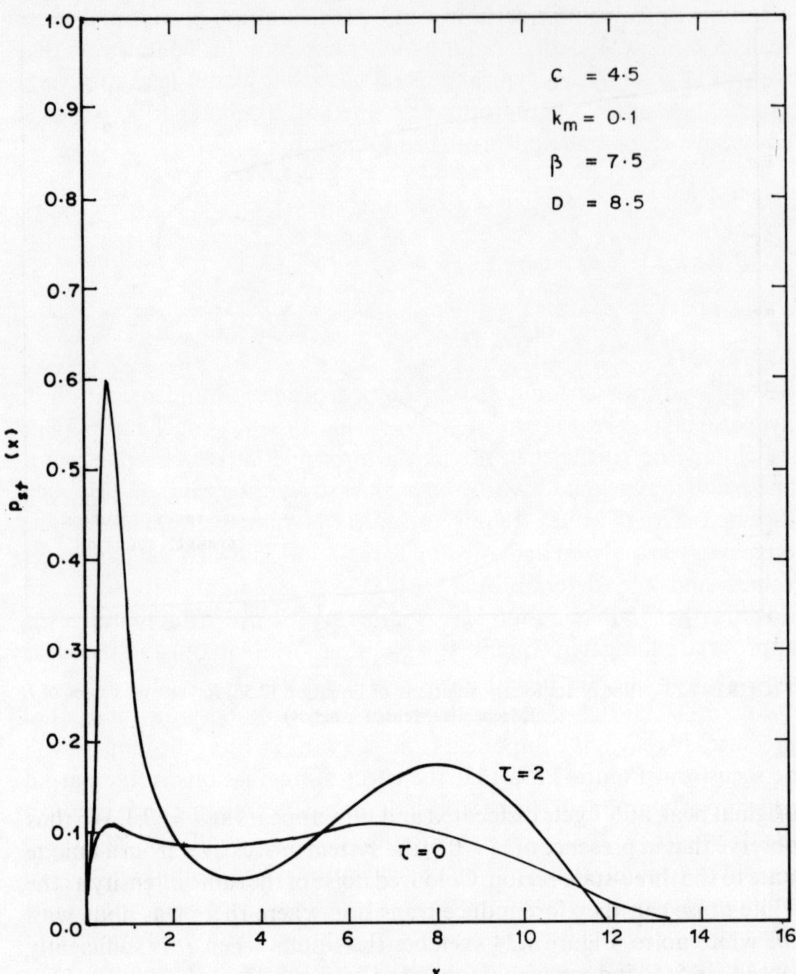

FIGURE 7.24 Stationary probability distribution of x for $D = 8.5$.

implications of these results in the proper interpretation of the reactions are evident.

Models Requiring Numerical Solution

In the examples analysed above it has been possible to obtain analytical solutions. This would not, however, be the situation always and more realistic cases would often require numerical solution. To illustrate this, we will now analyse a typical general case that would lead to several practical schemes on appropriate reduction. To be specific we shall consider a three-stage adsorption mechanism

$$A_n + nZ \rightleftharpoons nAZ$$

$$B_m + mZ \rightleftharpoons mBZ$$

$$pAZ + qBZ \rightleftharpoons (p+q)Z + A_pB_q$$

Several examples encountered in the hydrogenation and oxidation of hydrocarbons are known to follow this type of mechanism. The stoichiometric coefficients, the partial pressures of the components A and B and the various forward and backward reaction rate constants can assume different values leading to several possible combinations. The typical cases analysed are listed in Table 7.1 along with the parameter values and the deterministic steady state solutions. In order to investigate the behaviour of these systems in presence of fluctuations, the appropriate Langevin equations have been formulated using the basis described in Chapter 6, and have been numerically solved.

Figures 7.25–7.28 describe the effects of fluctuations in the partial pressures of reactant components and the various rate constants. It will be seen from Figure 7.25a that the effect of fluctuations in the partial pressure of component A varies with the intensity and time correlation of the fluctuations. The stochastic solution takes an oscillatory form and oscillates around the deterministic solution. The amplitude of the solution clearly depends on the noise intensity and increases with it. The effect of time correlation of noise is to dampen these oscillations and restore the solution close to the deterministic solution. Figure 7.25b shows the effect of fluctuations in the partial pressure of component B. The general features seen in this figure are similar to the ones seen in Figure 7.25a. The stochastic solutions in this case, however, lie above those of the deterministic solutions. Figures 7.26–7.28 show, respectively, the effect of fluctuations in the rate constants for the three steps. As is evident the effects for the particular case analysed are not very pronounced.

Similar conclusions can be reached for the various other cases listed in

TABLE 7.1
Some of the reaction schemes analysed and their stationary solutions

Reaction scheme		Stoichiometric coefficients		Parameters	Deterministic solutions	
					X	Y
$A_n + nZ \rightleftarrows nAZ$	I	$m=1$	$n=2$	$k_{-1}=k_{-2}=k_{-3}=0$	0	1
$B_m + mZ \rightleftarrows mBZ$		$p=1$	$q=1$	$k_1=10,\ k_2=2,\ k_3=50$	0.13189	0.3618
$pAZ + qBZ \rightleftarrows (p+q)Z + A_pB_q$				$p_A=0.1,\ p_B=0.5$	0.3618	0.13189
					1	0
	II	$m=1$	$n=2$	$k_{-1}=k_{-2}=k_{-3}=0$	0	1
		$p=2$	$q=1$	$k_1=10,\ k_2=1,\ k_3=10$	0.1137	0.3862
				$p_A=0.5,\ p_B=0.1$	0.4781	0.02187
					1	0
	III	$m=1$	$n=2$	$k_{-1}=k_{-2}=k_{-3}=0$	0	1
		$p_1=1$	$q=2$	$k_1=10,\ k_2=5,\ k_3=10$	0.2144	0.8535
				$p_A=0.5,\ p_B=0.5$	0.7285	0.1464
					1	0
	IV	$m=2$	$n=2$	$k_{-1}=k_{-2}=k_{-3}=0$	0	1
		$p=1$	$q=1$	$k_1=1,\ k_2=1,\ k_3=5$	0.0101	0.8648
				$p_A=0.5,\ p_B=0.5$	0.7898	0.08513

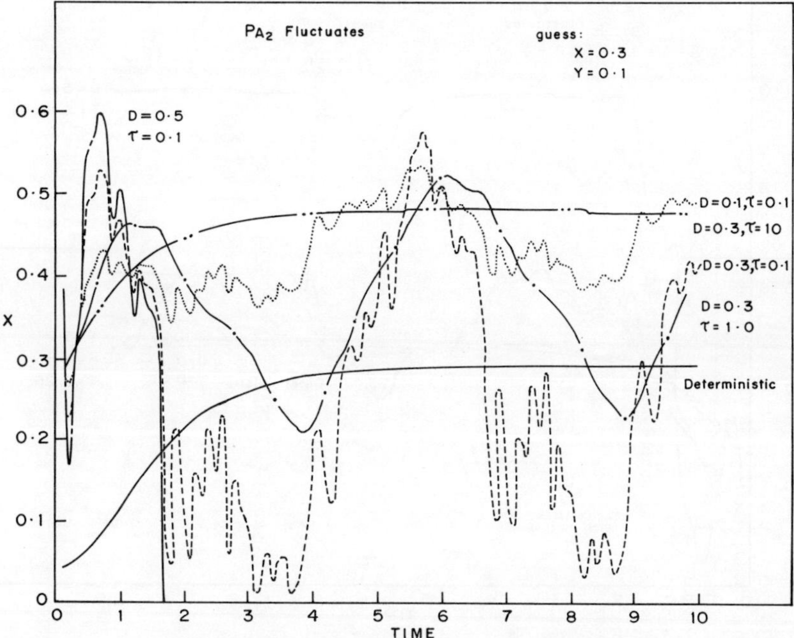

FIGURE 7.25a Concentration profiles for varying values of noise parameters (fluctuations in the partial pressure of component A)

Table 7.1 (see Dabke, Kulkarni and Doraiswamy, 1986b, for a detailed analysis).

7.3.3 Effect of Temperature Fluctuations on the Kinetic Behaviour of Reacting Systems

In the present section we consider a continuous stirred tank reactor maintained at a mean temperature of T with an input–output flow rate of F. It is assumed that the control bath temperature varies randomly about the mean value $\bar{T}$ and the actual temperature of the bath can be written as

$$T = \bar{T} + \xi(t) \tag{7.64}$$

where $\xi(t)$ represents the random variation or the external noise. Let us

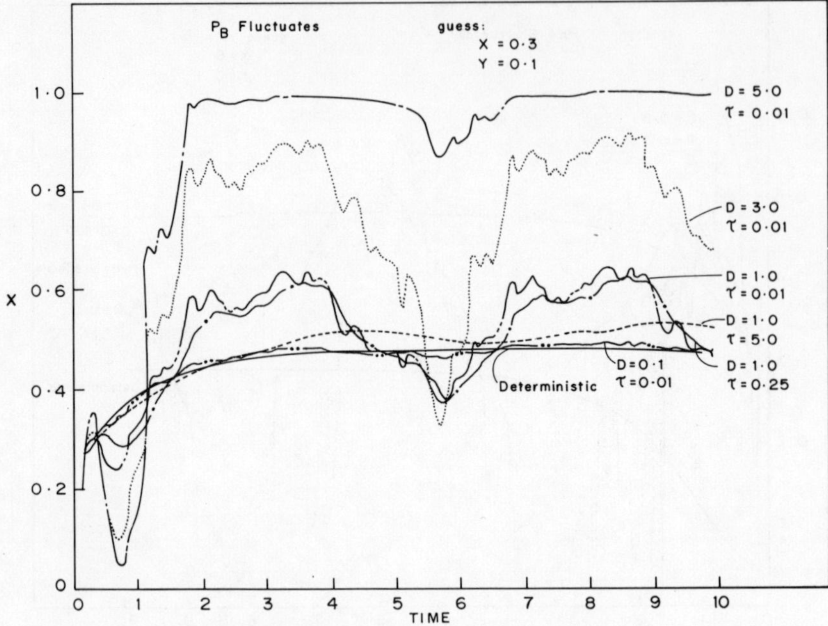

FIGURE 7.25b Concentration profiles for varying values of noise parameters (fluctuations in the partial pressure of component B)

now consider a reaction with rate $r(x)$ taking place in this reactor. The macroscopic equation describing the behaviour of the system can then be written as

$$\frac{dx}{dt} = \frac{F}{V}(x_0 - x) - r(x) \tag{7.65}$$

$$r(x) = k_0 \exp(\alpha) f(x), \qquad \alpha = -E/RT \tag{7.66}$$

where the temperature dependence of the rate is written explicitly and $f(x)$ represents the concentration dependent part. In view of the variation in the temperature T of the controlled bath, the parameter α also varies randomly, which for the sake of simplicity can be written likewise as $\alpha = \bar{\alpha} + \xi(t)$.

The statistical properties of the noise are supposed to be described by a stationary Gaussian process although one could consider any other distribution. To keep closer to the reality, we allow for the time-

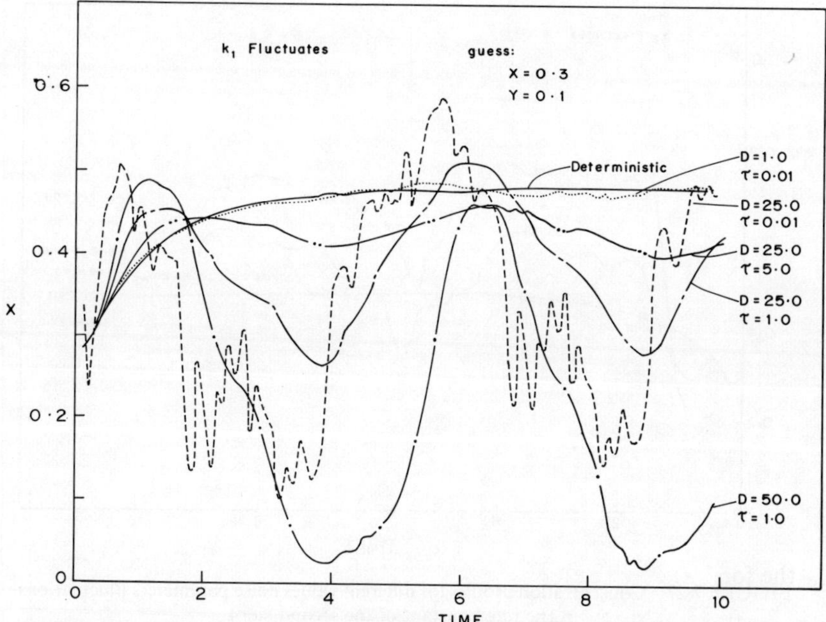

FIGURE 7.26 Concentration profiles for different values of noise parameters (fluctuations in the rate constant of the first step)

correlation effects of the noise. The noise considered is thus a nonwhite Gaussian noise that can be described as an Ornstein–Uhlenbeck process. Mathematically the noise is thus represented by

$$\frac{d\xi}{dt} = \tau^{-2}\,\xi + \tau^{-1} D\,dW \tag{7.67}$$

with zero mean ($\langle \xi(t) \rangle = 0$) and time correlation given by $\langle \xi(t)\xi(t') \rangle = (D/\tau)\exp(-(t-t')/\tau)$. D and τ in this equation represent the intensity and the correlation time of the noise, and dW the Wiener process.

The macroscopic Equation (7.65) in presence of noise can now be written as

$$\frac{dx}{dt} = \frac{F}{V}(x_0 - x) - k_0\, e^{\bar{\alpha}}\, e^{\xi(t)} f(x) \tag{7.68}$$

which can be recognised as a standard Langevin equation with multiplicative coloured noise. One can use an ordered cumulant

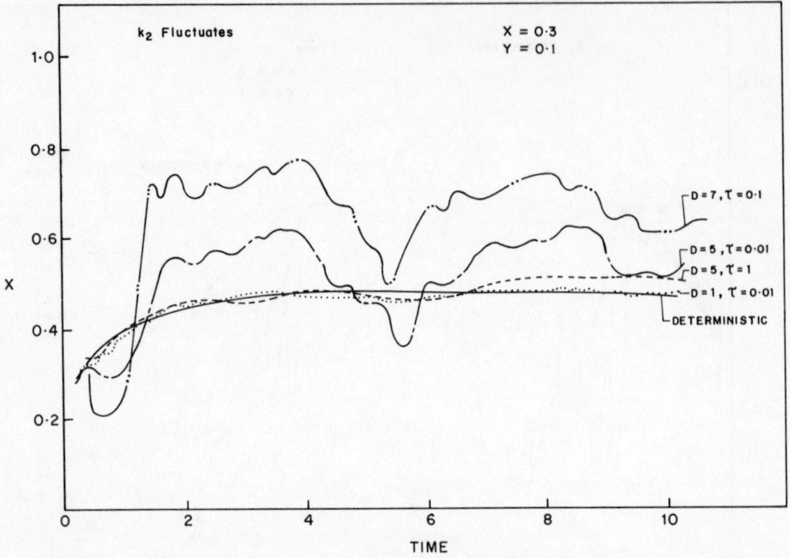

FIGURE 7.27 Concentration profiles for different values noise parameters (fluctuations in the rate constant of the second step)

technique (van Kampen, 1976; Sagues, 1984) to write the appropriate Fokker–Planck equation, which for small D and τ (and finite D/τ) takes the form

$$\frac{\partial p(x,t)}{\partial t} = -\frac{\partial}{\partial x}\left[\frac{F}{V}(x_0 - x) - e^{D/2\tau}\, e^{\bar{\alpha}} k_0 f(x)\right] p(x,t)$$
$$+ \tau\, e^{D/\tau}\left[\sum_{n=1}^{\infty} (1/nn!)(D/\tau)^n\right] \frac{\partial}{\partial x}[k_0\, e^{\bar{\alpha}} f(x)]\, \frac{\partial}{\partial x}[k_0\, e^{\bar{\alpha}} f(x)]\, p(x,t) \tag{7.69}$$

Equation (7.69) embodies the effect of noise as evident from the correction to the drift term (first term). We shall now investigate the solution to this equation for different functional forms of $f(x)$.

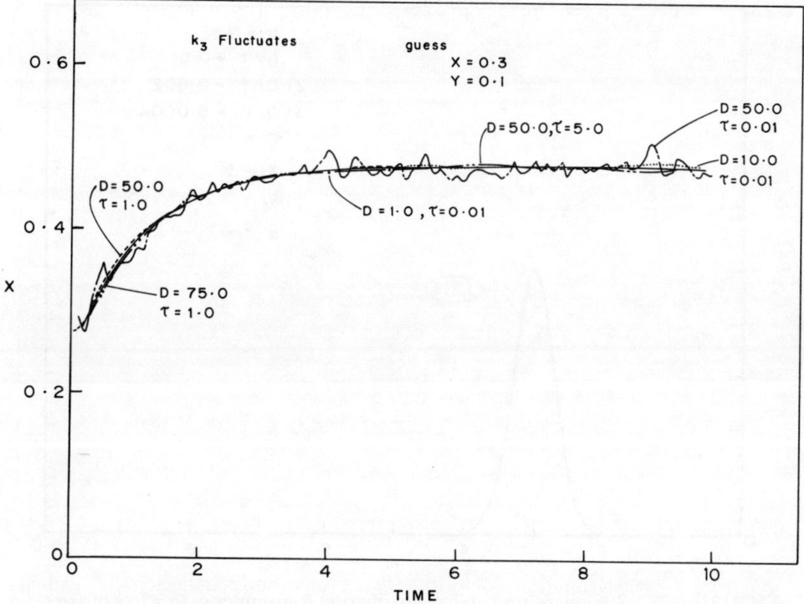

FIGURE 7.28 Concentration profiles for different values of noise parameters (fluctuations in the rate constant of the third step)

Zero-Order Case

Here the functional form $f(x)$ takes the value equal to unity. The formal stationary solution to Equation (7.69) is then

$$p_{st}(x) = \frac{N}{e^{\bar{\alpha}}k_0} \exp\left[\left(A + B + \frac{Ax}{2x_0}\right)x\right] \tag{7.70}$$

where

$$A = \frac{Fx_0}{Vk_0^2 D \exp(2\bar{\alpha} + D/\tau)\lambda}, \qquad B = \frac{\exp(\alpha + D/2\tau)}{k_0 D\lambda \exp(2\bar{\alpha} + D/\tau)}$$

and

$$\lambda = 1 + \frac{1}{4}\frac{D}{\tau} + \frac{1}{18}\left(\frac{D}{\tau}\right)^2 \tag{7.71}$$

N in this equation refers to the normalisation constant. Equation (7.70) for the stationary probability distribution is plotted in Figure 7.29 for a

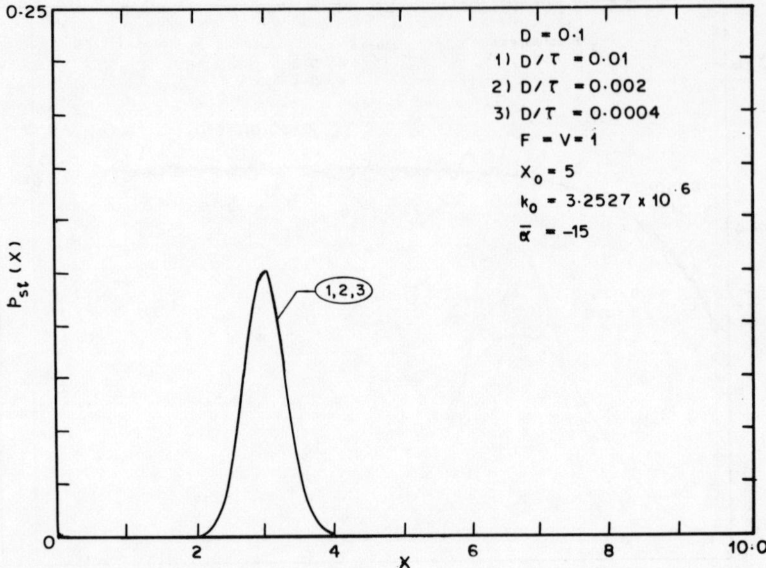

FIGURE 7.29 Stationary probability distribution (zero-order case, effect of D/τ).

set of parameter values. The maximum of the distribution corresponds to the stationary value for the variable $x(x_s)$. It is evident from the figure that the stationary solution ($x_s = 3$) as obtained in presence of noise is considerably different from the macroscopic stationary solution ($x_s = 4$). The other parameter values are listed under legend in the figure.

It is evident from the stationary solution given by Equation (7.70) that the final steady state value can get affected as a result of variations in the noise intensity D or in the ratio D/τ. Figures 7.29 and 7.30 show the effects of these variations. It is clear that the variations in D or in the ratio D/τ have only marginal effects on the stationary solution reached. The solution, however, is considerably different from the macroscopic solution.

To elucidate this point further let us return to the Fokker–Planck Equation (7.69) which for the case of zero-order reaction takes the Ito form. One could obtain the point at which the maximum occurs explicitly for this case by equating the derivative dp/dx to zero. The

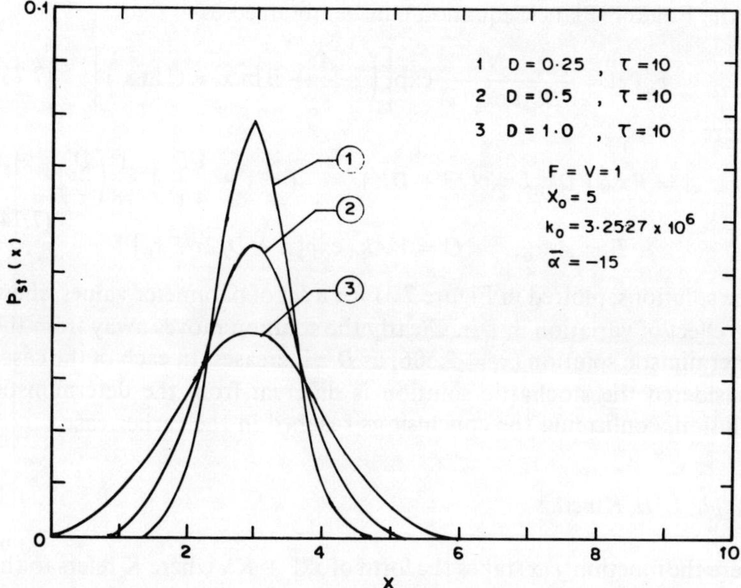

FIGURE 7.30 Stationary probability distribution (zero-order case, effect of D).

simple exercise leads to the equation

$$\frac{F}{V}(x_0 - x_s) - k_0 \exp(\bar{\alpha} + D/2\tau) = 0 \tag{7.72}$$

Clearly the point of maximum depends on D/τ, and the presence of noise would always affect the stationary solution. This simple example of zero-order reaction therefore shows that nonlinear noise such as the one considered here would always affect the final solution reached. Ignorance of the nature and extent of noise can lead to wrong estimates when such noise infected data are used without due accounting of the noise. This is especially serious since such noise would always be present in any practical system.

First-Order Case

Here the function $f(x)$ takes a value equal to x. The stationary solution to the Fokker–Planck equation can be obtained as

$$p_{st}(x) = -\frac{N}{k_0 x \exp(\bar{\alpha})} \exp\left[\left(-\frac{A}{x} + B \ln x + C \ln x\right)\right] \tag{7.73}$$

where

$$A = F x_0 / V D k_0^2 \lambda \exp(2\bar{\alpha} + D/\tau), \qquad \lambda = 1 + \frac{1}{4}\frac{D}{\tau} + \frac{1}{18}\left(\frac{D}{\tau}\right)$$
$$B = A/x_0, \qquad C = VAk_0 \exp[\bar{\alpha} + D/2\tau/Fx_0] \tag{7.74}$$

The solutions, plotted in Figure 7.31 for a set of parameter values, show the effect of variation in D/τ. Clearly the solution moves away from the deterministic solution ($x_s = 2.506$) as D/τ increases. In each of the cases considered the stochastic solution is different from the deterministic solution, confirming the conclusions reached in the earlier case.

Simple L–H Kinetics

Here the function $f(x)$ takes the form of $x/1 + Kx$ where K refers to the

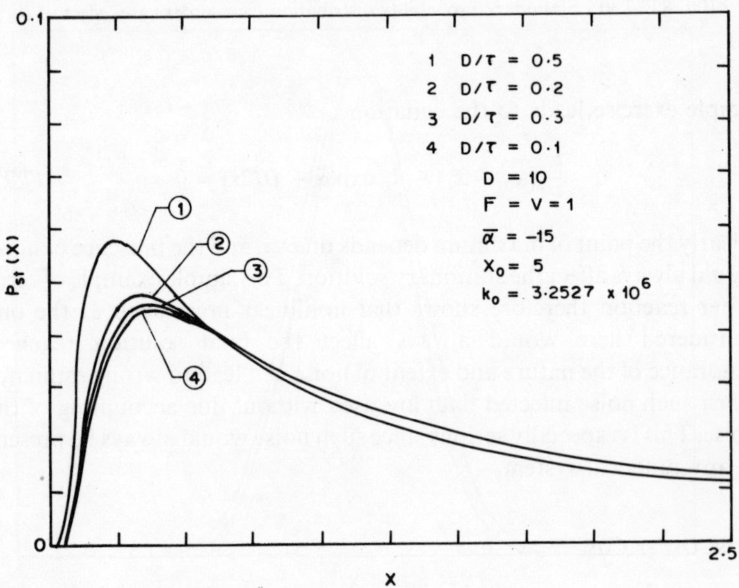

FIGURE 7.31 Stationary probability distribution (first-order case, effect of D/τ).

adsorption equilibrium constant. The stationary solution for this case can be obtained as

$$p_{st}(x) = -\frac{n(1+Kx)}{k_0 x \exp(\bar{\alpha})}$$
$$\times \exp\left[\left(-\frac{A}{x} + AK^2 x + 2AK\ln x\right) - B\left(\ln x + \frac{K^2x^2}{2} + 2Kx\right)\right.$$
$$\left. - C\left(\ln x + \frac{K^2x^2}{2} + 2Kx\right)\right] \tag{7.75}$$

where the parameters A, B and C are defined as in the previous case.

The stationary solution is plotted in Figure 7.32 for different values of D/τ but the same values of D and other parameters. It is evident from a comparison of the curves that the numerical magnitude of D/τ has only a slight influence on the stationary solution which is now localised around $x_s = 2.44$. Note that the macroscopic stationary solution for this case

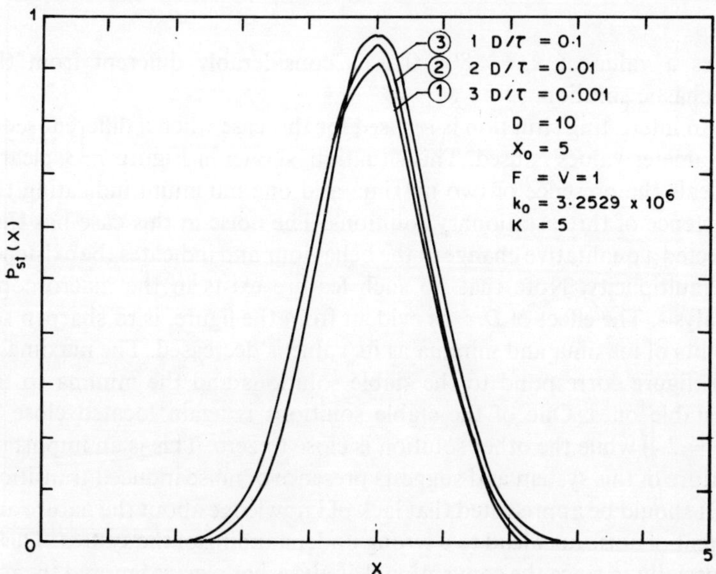

FIGURE 7.32 Stationary probability distribution (L–H kinetics, effect of D/τ).

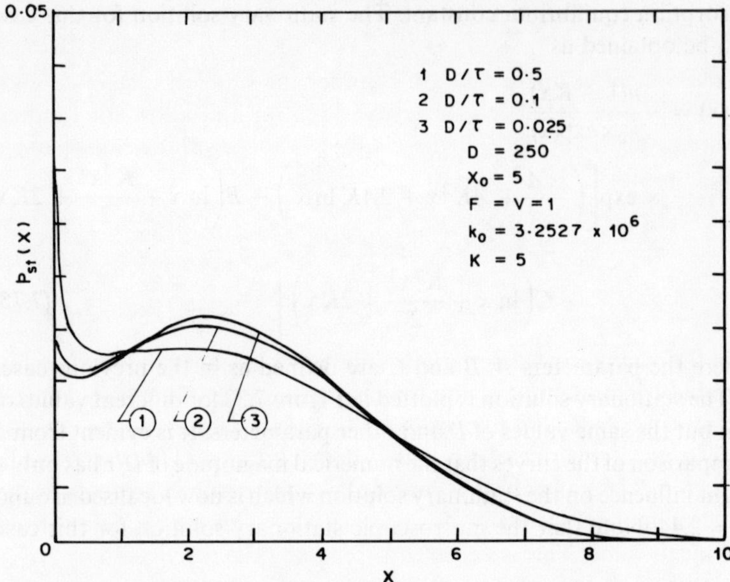

FIGURE 7.33 Stationary probability distribution (L–H kinetics, effect of D/τ).

gives a value of $x_s = 4.806$ that is considerably different from the stochastic solution.

An interesting situation is realised for this case when a different set of parameter values is used. This situation, shown in Figure 7.33, clearly reveals the presence of two maxima and one minimum indicating the existence of three stationary solutions. The noise in this case has thus effected a qualitative change in the behaviour and indicates the existence of multiplicity. Note that no such feature exists in the macroscopic analysis. The effect of D/τ, as evident from the figure, is to sharpen the points of maxima and minima as its value is decreased. The maxima in this figure correspond to the stable solutions and the minima to the unstable ones. One of the stable solutions is again located close to ($x_s = 2.4$) while the other solution is close to zero. This is an important feature of this system and suggests presence of noise induced transition.

It should be appreciated that lack of knowledge about the nature and extent of noise may lead to a wrong understanding of the system. This is especially so since the conventional analysis has always ignored the role of noise.

7.3.4 Catalyst Deactivation in Presence of External Diffusion

In the present example we shall consider the simplest case of catalyst deactivation in presence of external diffusion and allow for random variation in the activity of the catalyst. The governing macroscopic equations are

$$-\frac{dC_R}{dt} = k_m a_s(C - C_R) - aC_R, \qquad -\frac{da}{dt} = k_d a$$

$$a = 1, \quad t = 0 \qquad (7.76)$$

In the macroscopic description as above it is assumed that the activity variable a represents a value that is constant for the entire catalyst surface and depends only on time. In reality, however, the catalyst surface has sites with varying activity distributions and the variable a in Equation (7.76) can at best be regarded as a mean value. To describe the situation more realistically, it is necessary to allow for deviations from this mean value. This is achieved by defining the activity by

$$a = \bar{a} + \xi(t) \qquad (7.77)$$

where $\xi(t)$ represents the random deviation from the mean value a. This definition of the activity variable necessitates a consideration of the stochastic differential equation associated with Equation (7.76). It is in general difficult to solve this stochastic equation, although the corresponding macroscopic equation can be solved analytically. Dabke, Kulkarni and Doraiswamy (1986c) have used a numerical method to solve this case and their results are presented below.

Numerical solutions of the set of equations in presence of noise have been obtained for varying values of the noise parameter D, deactivation kinetics k_d and the Damköhler number $Da = k/k_M a_s$. The Damköhler number can take the limiting values of 0 and ∞ corresponding, respectively, to the kinetically controlled and external transfer controlled regimes.

Figure 7.34 shows the dimensionless concentration profiles of the reactant on the surface for different values of the noise intensity D, while Figure 7.35 depicts the effect of variation in the deactivation rate constant at a constant level of noise intensity. It is clear from Figure 7.34 that increase in D implies increased deviation from the macroscopic results. The reactant concentrations, now higher on the catalyst surface,

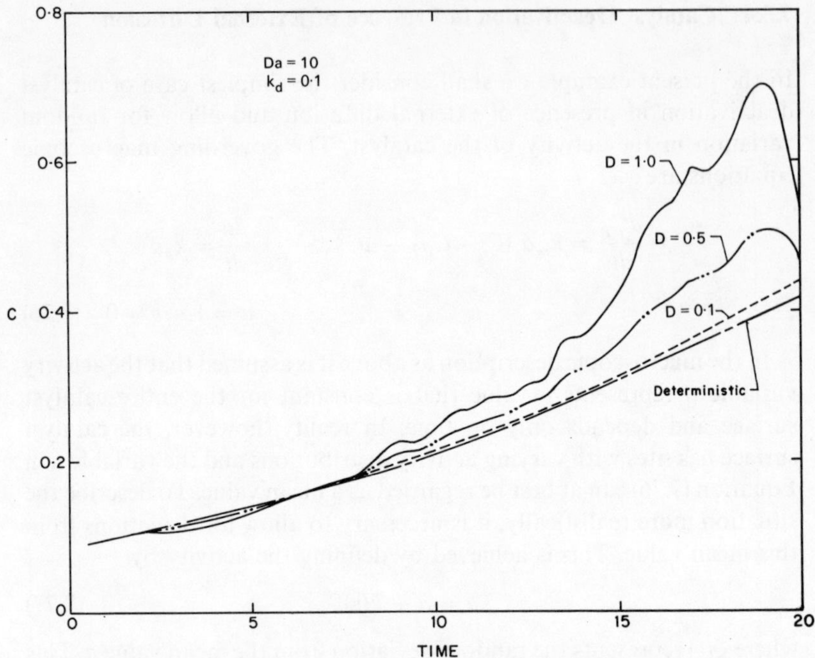

FIGURE 7.34 Concentration profiles for different values of noise intensity D.

indicate that the reactant processing ability of the surface decreases. This can be attributed to higher dispersion about the mean value of activity when D increases.

Figure 7.35 showing the effect of the deactivation kinetic constant suggests that, for a fixed intensity of noise, the effect of noise is greater for catalysts with higher values of the deactivation constant. In other words, rapidly deactivating catalysts are more susceptible to the effects of noise than the slowly deactivating catalysts. The general trends shown in Figures 7.34 and 7.35 for a specific value of Da (=10) also hold for other values of Da.

It is also convenient to express the results of the numerical analysis in the form of effectiveness factors, which can be defined as the ratio of the global rate at any instant of time to the purely chemical reaction controlled rate at zero time. Figures 7.36 and 7.37 show the deterministic results and the ratio of the effectiveness factor in the absence of noise to

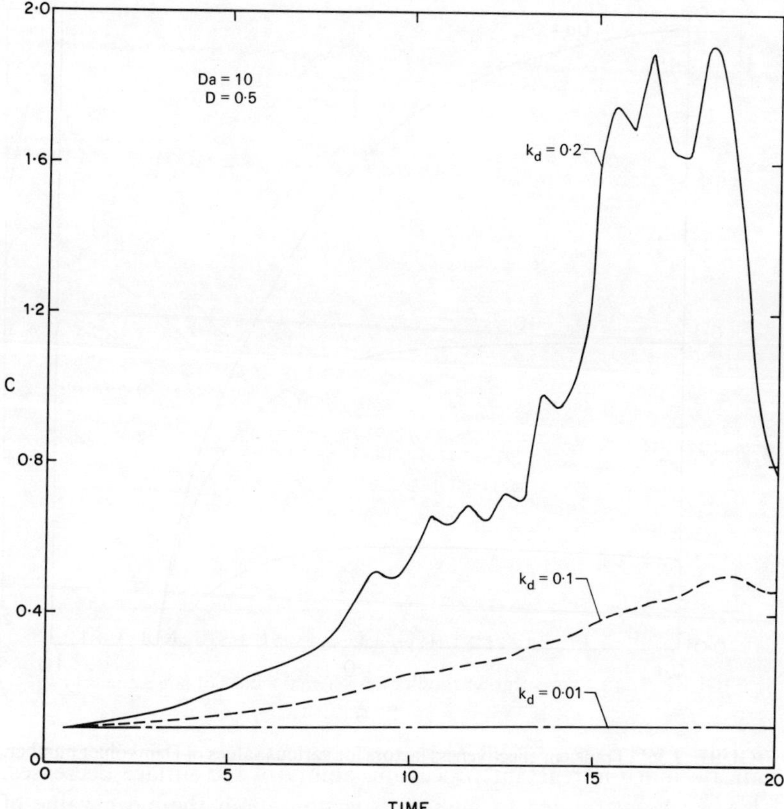

FIGURE 7.35 Concentration profiles as functions of the variation in the deactivation rate constant.

that in its presence as a function of Damköhler number. When the noise has no effect one would expect this ratio to be unity. The analysis of Figure 7.37 reveals that the effect of noise is more severely felt by systems with low values of *Da* than by systems with higher *Da* values. In other words, systems operating in or closer to the kinetic controlled regime feel the severity of noise more than those operating in or closer to the external transfer controlled regime. As seen in the figure, in the long time limit the ratio exceeds unity suggesting that the actual value of the effectiveness factor (η_{noise}) is smaller than the deterministic value.

We shall now quantitatively illustrate one more feature of the results

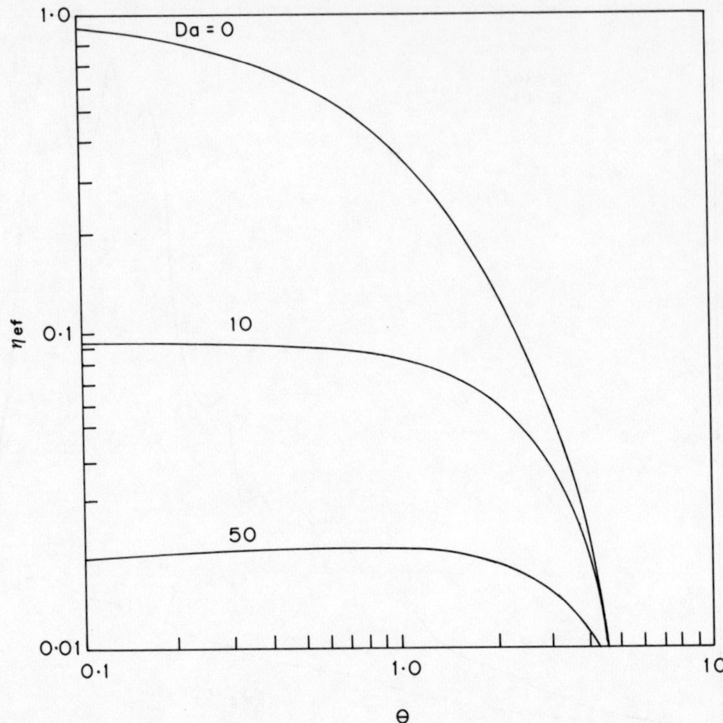

FIGURE 7.36 Transient effectiveness factors for various values of Damköhler number.

shown in Figure 7.37, which is not clear at first glance. Thus let us take some arbitrary value of time ($k_d t = 1.7$). The value of the effectiveness factors ratio can be read from Figure 7.37 as 2.66 for $Da = 0$. The deterministic value of η from Figure 7.36 for these values of time and Da is 0.22. The actual value of the effectiveness factor in presence of noise with $D = 1$ can then be obtained as equal to 0.0827. Note that this represents the actual value that one would observe if an experiment were conducted.

Often, the intention of conducting an experiment is to determine the values of the main and deactivation rate constants, and to begin with one is not aware of the extent of noise in the system. If a macroscopic theory is followed, as is customarily done, one would associate a value of Da ($\simeq 9$ in the present case) corresponding to this observed value of η

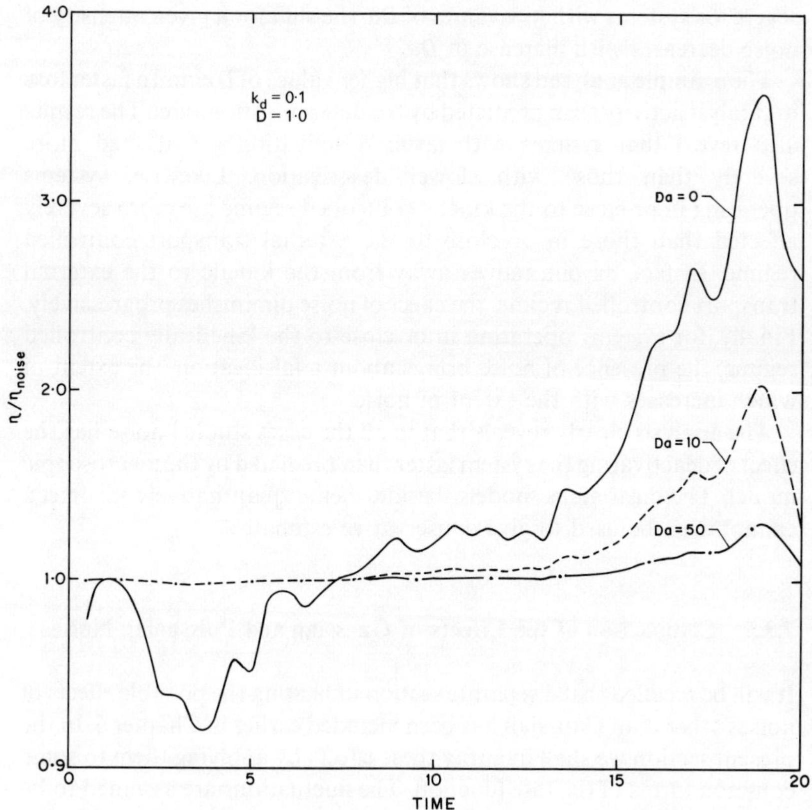

FIGURE 7.37 Ratio of effectiveness factors in absence and presence of noise for different Damkohler numbers.

(=0.0827) from Figure 7.36. We therefore note that what is actually a kinetically controlled process ($Da = 0$) starts appearing as a process with higher values of Da. In other words, the presence of noise shifts the system from a kinetically controlled region towards the external transfer controlled region. The presence of noise therefore brings about a falsification and the implications of the results are clear. Of course the extent of shift depends on the extent of noise, and, as would be expected (based on the results shown in Figure 7.34), the shift becomes more pronounced with increase in D. Also, since the effect of noise is more severe for systems with low values of Da, the shift for a given intensity of noise decreases with increase in Da.

The example analysed shows that higher values of D entail a faster loss in catalyst activity than predicted by the deterministic model. The results also reveal that systems with faster deactivation get affected more severely than those with slower deactivation. Likewise, systems operating in or close to the kinetic controlled regime are more severely affected than those in or close to the external transport controlled regime. In fact, as one moves away from the kinetic to the external transport controlled regime, the effect of noise diminishes progressively. Finally, for systems operating in or close to the kinetically controlled regime, the presence of noise brings about a falsification, the extent of which increases with the extent of noise.

The analysis clearly reveals that in all the cases studied noise has the effect of deactivating the system faster than predicted by the macroscopic model. The customary models, besides being quantitatively incorrect, cannot even be used to give conservative estimates.

7.3.5 Comparison of the Effects of Gaussian and Poissonian Noises

It will be recalled that a separate section indicating the possible effects of noises other than Gaussian has been included earlier in Chapter 6. In the present section we shall quantify these effects by applying them to some common forms of the rate function. The fluctuations are assumed to be white or nonwhite Poissonian. The results of the cases analysed afford a comparison with the corresponding results for the case of Gaussian noise and bring out the difference between the two types of noises.

As a first example we shall analyse the case of a quadratic nonlinearity that is modelled by the phenomenological equation

$$\frac{dx}{dt} = \frac{1}{2} - x + \beta x(1 - x) \tag{7.78}$$

Such an equation also provides an example of a common model for certain chemical reactions, and the corresponding stochastic differential equation when β is allowed to fluctuate takes the form

$$\frac{dx}{dt} = \frac{1}{2} - x + \bar{\beta}x(1 - x) + \xi x(1 - x) \tag{7.79}$$

This equation has been analysed by Horsthemke and Lefever (1980) and Lindenberg and West (1983) who have considered $\xi(t)$ as a dichotomous or Gaussian coloured noise. Dabke, Kulkarni and Doraiswamy (1986d) have also analysed this case by considering $\xi(t)$ as a Poisson coloured process. As noted earlier, the results for Gaussian white and nonwhite cases can be recovered as limiting cases of this more general analysis providing a basis for comparison.

Equation (6.143) gives the stationary probability distribution for this case which can be differentiated to obtain its points of maxima:

$$F(x) - Dg(x)[g'(x) - \tau G'(x)] + \frac{sg(x)}{3}F'(x) = 0 \qquad (7.80)$$

This equation can be written more explicitly as

$$x^3\left(\frac{2\bar{\beta}s}{3} - 2D\tau - 2D\right) + x^2\left(3D - \bar{\beta} + 3D + \frac{s}{3} - \bar{\beta}s\right) + x\left(\bar{\beta} - D - D\tau + \frac{s\bar{\beta}}{3} - 1 - \frac{s}{3}\right) + \frac{1}{2} = 0 \quad (7.81)$$

Equation (7.81) is a cubic equation and can possess more than one root for certain sets of parameter values. The limiting cases of this equation can be obtained by letting s and/or τ tend to zero. The results for several cases are presented in Figures 7.38–7.42.

Figure 7.38 shows the maxima of the probability density distribution for different values of D for the case of white Gaussian noise ($s \to 0$, $\tau \to 0$). The stochastic and deterministic solutions ($x = 0.5$ for $\bar{\beta} = 0$) coincide for smaller values of D, while for higher values of noise intensity the figure indicates a transition from one state to three states. For $D > 3$ we thus observe that the system can exist in three different stationary states. One of these solutions now corresponds to the original deterministic state ($x = 0.5$) which, however, is now unstable. The two other solutions owe their existence to noise.

The influence of the time correlation of the noise is shown in Figure 7.39, where we begin with a one-solution region in Figure 7.38 ($D = 1$). The points of maxima of the distribution for the case of $D = 1.0$ ($s = 0$, $\beta = 0$) are shown. The figure reveals that incorporation of time-correlation effects can generate additional solutions in the region where white noise intensity alone is insufficient. Thus for $\tau > 1.5$, even at $D = 1$ we observe the existence of three states. The effect of τ thus seems to aid

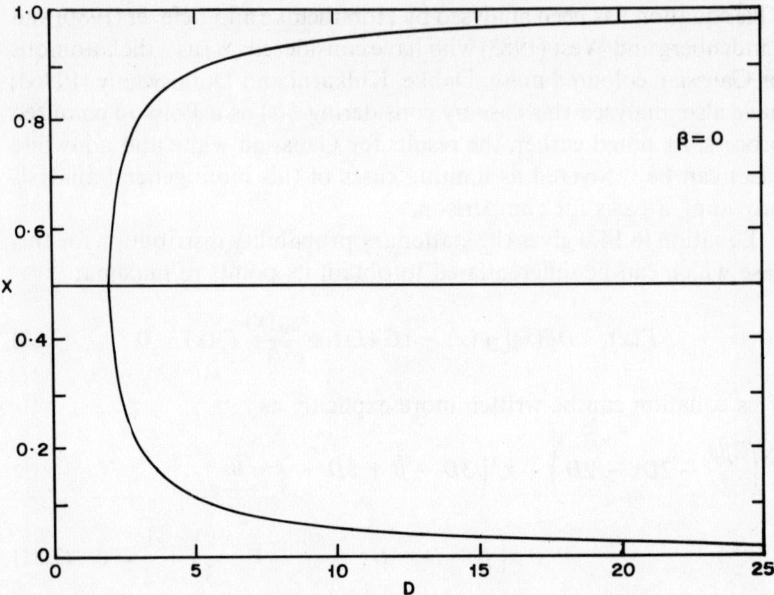

FIGURE 7.38 Effect of white (Gaussian) noise intensity on the stationary solutions.

generation of new transitions and also to supplement the influence of D. The original deterministic solution now represents one of the three solutions and as before is rendered unstable.

Figure 7.40a shows the results for the case of Poisson white noise with varying intensity. We begin with a case of white Gaussian noise of intensity $D = 3$. Reference to Figure 7.38 clearly indicates the existence of three states for this case. Incorporation of s of increasing intensity reveals that the upper and middle states are brought progressively closer until at around $s = 0.75$ the system possesses only one solution. A new transition from three states to one state is therefore evident when white Poisson noise of a certain intensity exists. Note that we have $\lambda s^2 = 2D = 6$ in this case. Variation in s therefore entails variation in λ that measures the mean number of δ pulses. The results reveal that a lesser number of pulses of higher intensity can cause this transition from three states to one state. A larger number of pulses but of weaker intensity is not adequate.

Figure 7.40b suggests that, if we were to begin with a smaller value of D, while Figure 7.38 indicates only one solution, increasing s has the

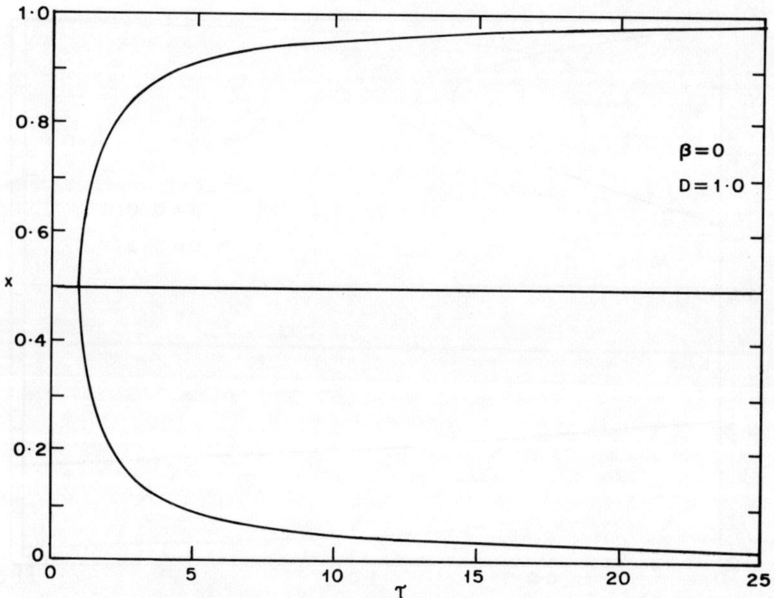

FIGURE 7.39 Effect of coloured Gaussian noise on the stationary solutions.

effect of decreasing this solution, and no new transitions can be created.

We shall now examine the influence of time-correlation of noise. Figure 7.41 shows this effect where we begin with the case of $s = 0.9$. Reference to Figure 7.40a shows that for this value of intensity of white Poisson noise the system has only one solution. Incorporation of τ seems to steer the system back to the region of three states. The influence of τ therefore seems to be equivalent to reducing the intensity of noise s or increasing the number of pulses λ. The two additional solutions created move away from each other with increase in τ, s and τ are thus seen to have opposing influences.

The opposing influences of s and τ are also evident in Figure 7.42 where, for fixed values of D and τ, the points of maxima of the stationary distribution are plotted as a function of the white noise intensity s. To begin with we have a system possessing three stationary solutions. Progressive increase of s leads to the transition from three states to a single state.

The example analysed brings out certain interesting conclusions. D and τ seem to supplement each other for Gaussian noise, while for

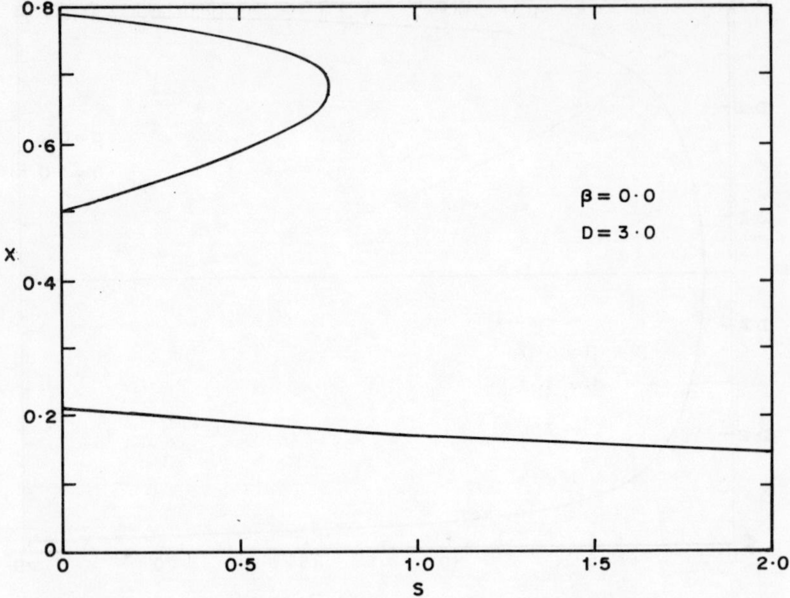

FIGURE 7.40a Stationary solutions in presence of white (Poisson) noise ($D = 3$).

Poisson noise (for a fixed value of D) the two act in opposing directions. The example also suggests new transitions that have their origin solely in the noise. It is important to note that information regarding these transitions cannot be obtained from the macroscopic description. In the instance when additional solutions are generated the original (deterministic) solution also exists and represents the unstable middle solution. Increase in s, on the other hand, tends to destroy this middle solution (see Figure 7.40a).

The second example we shall consider is the case of cubic nonlinearity (see Sancho and San Miguel, 1982b; Lindenberg and West, 1983), modelled by the equation

$$\frac{dx}{dt} = \alpha x - \beta x^3 \tag{7.82}$$

where α and β are positive parameters. For brevity we shall allow α to fluctuate around a mean value of $\bar{\alpha}$ leading to the following stochastic

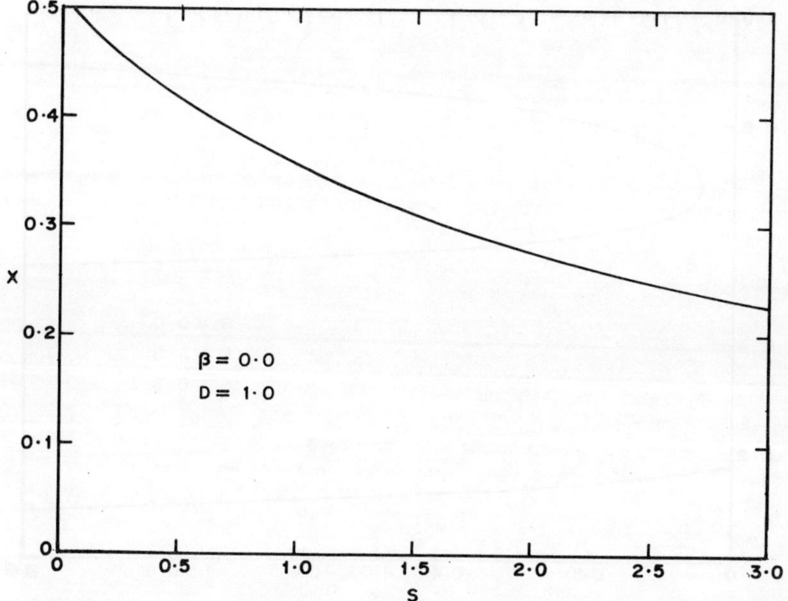

FIGURE 7.40b Stationary solutions in presence of white (Poisson) noise ($D = 1$).

description:

$$\frac{dx}{dt} = \bar{\alpha}x - \beta x^3 + x\xi(t) \tag{7.83}$$

Treating $\xi(t)$ as a coloured Poisson noise, we can obtain the stationary probability distribution as before and the maxima of the distribution:

$$x = 0 \tag{7.84}$$

$$x = \pm\left(\frac{\alpha + s\alpha/3 - D}{\beta + s\beta - 6D\tau\beta}\right)^{1/2} = \left(\frac{\alpha}{\beta}\right)^{1/2}\left[\frac{1 + s/3 - D/\alpha}{1 + s - 6D\tau}\right]^{1/2} \tag{7.85}$$

Note that in the absence of $\xi(t)$, $x = 0$ (unstable) and $x = \pm(\alpha/\beta)^{1/2}$ (stable) are the deterministic solutions of the system. According to the stochastic analysis, $x = 0$ is still a solution of the system, while the second solution is modified by a factor that involves the noise parameter. The second solution clearly exists only if certain conditions are met and is dominant if

$$\alpha(1 + s/3) > D \qquad \text{and} \qquad (1 + s) > 6D\tau \tag{7.86}$$

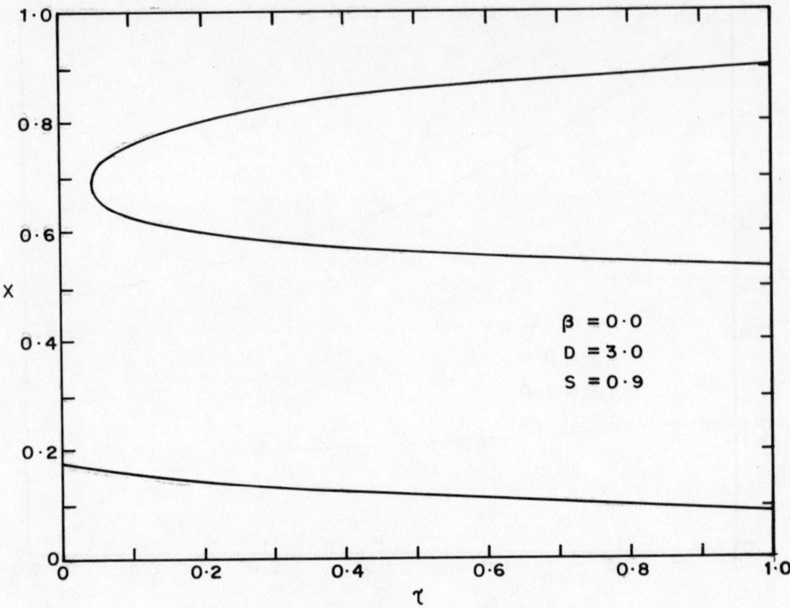

FIGURE 7.41 Time-correlation's influence on the stationary solutions (Poisson noise).

or

$$\alpha(1 + s/3) < D \qquad \text{and} \qquad (1 + s) < 6D\tau \tag{7.87}$$

In the event none of these conditions are obeyed the most dominant solution corresponds to $x = 0$.

There are some additional features contained in Equation (7.87) which we shall discuss by considering specific types of noises. Thus for the case of white Gaussian noise, $x = 0$ is the dominant solution until $\alpha < D$. The solution $x = 0$ is stable if $\alpha > 0$ and becomes unstable for $\alpha > 0$. This can be easily verified by taking the first derivative of Equation (7.82) and examining its sign. Since in the present example α is treated as a positive parameter the solution $x = 0$ is unstable. In the range of $0 < \alpha < D$ calculation of the stationary probability density shows a divergence at $x = 0$. It should be noted that though $x = 0$ is no longer a stable stationary point, it remains the most probable value. We then note the following features: (1) $\alpha = 0$ is a transition point in the sense that the stable solution $x = 0$ changes over to an unstable solution for positive α, (2) in the region $0 < \alpha < D$, $x = 0$ still remains the most probable solution, and (3) at $\alpha > D$ a new transition corresponding to

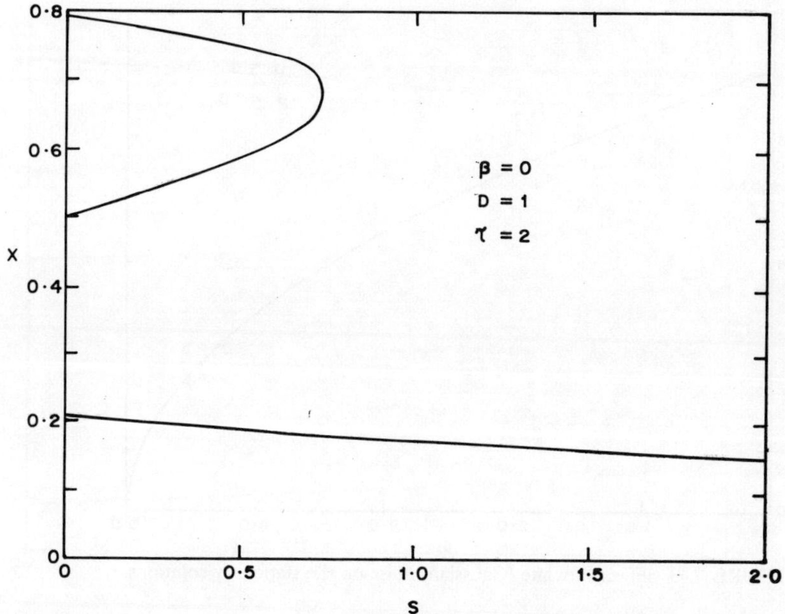

FIGURE 7.42 Influence of Poisson noise intensity on the stationary solutions.

$x = [\alpha - D/\beta]^{1/2}$ is created. The typical case for a set of α and β values is shown in Figure 7.43.

Let us now analyse the situation where a coloured Gaussian noise exists. We note from Equation (7.85) that the second solution exists even when α is less than D. The situation is indicated in Figure 7.44. Note that this transition does not exist in the white noise limit and is created purely due to time-correlation effects. The necessary requirement for this transition to occur is: if $\alpha < D$ then $6D\tau > 1$. If this condition is not met then again $x = 0$ is the only solution that the system would possess. The transition also exists when α exceeds D. However, in this case the necessary requirement is that $6D\tau < 1$. The case corresponding to this situation is presented in Figure 7.45.

Same features are noted when we consider the white Poisson noise. Even when $\alpha \leqslant D$, if the condition that $\alpha(1 + s/3) > D$ is met the system would exhibit the transition. Again, if the condition is not satisfied, $x = 0$ would represent the only solution (Figure 7.46). For a situation where white Poisson noise is not enough to bring about a transition,

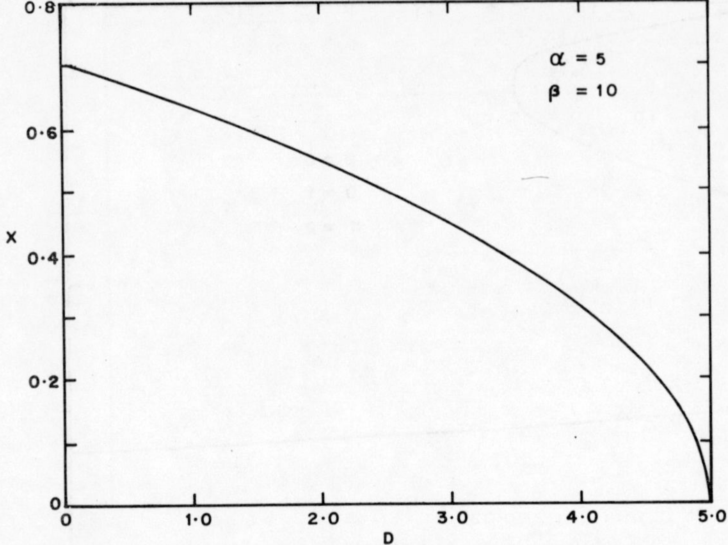

FIGURE 7.43 Effect of white (Gaussian) noise on the stationary solutions.

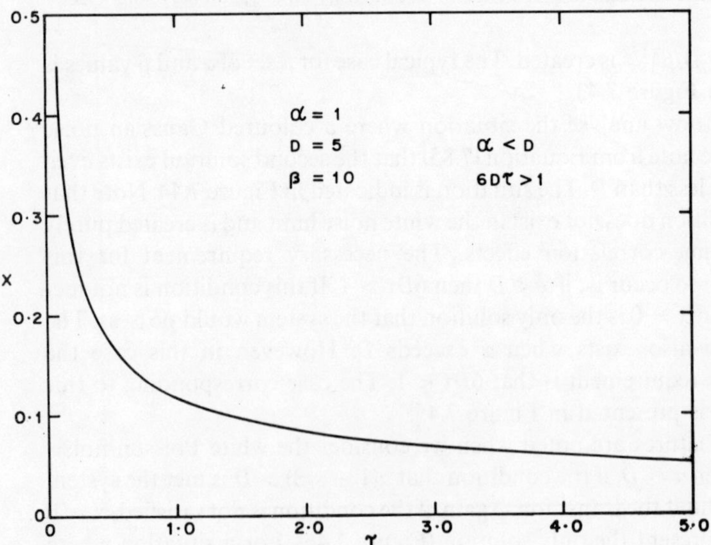

FIGURE 7.44 Effect of coloured (Gaussian) noise on the stationary solutions.

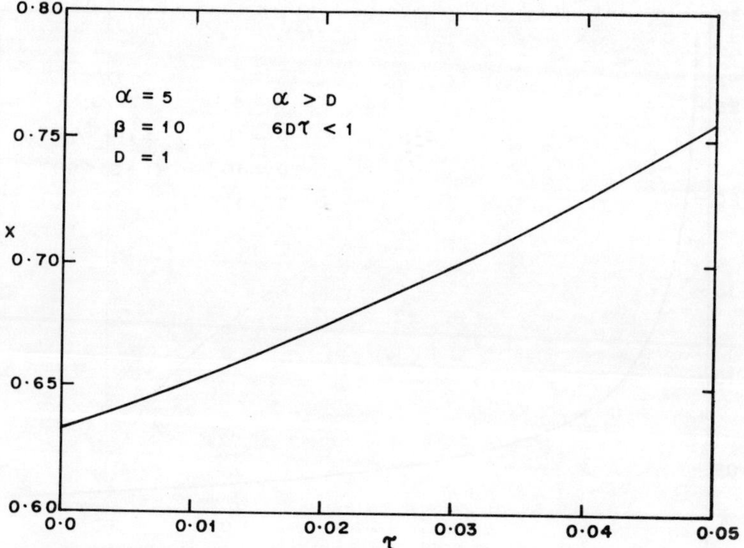

FIGURE 7.45 Effect of varying correlation-time on stationary solutions.

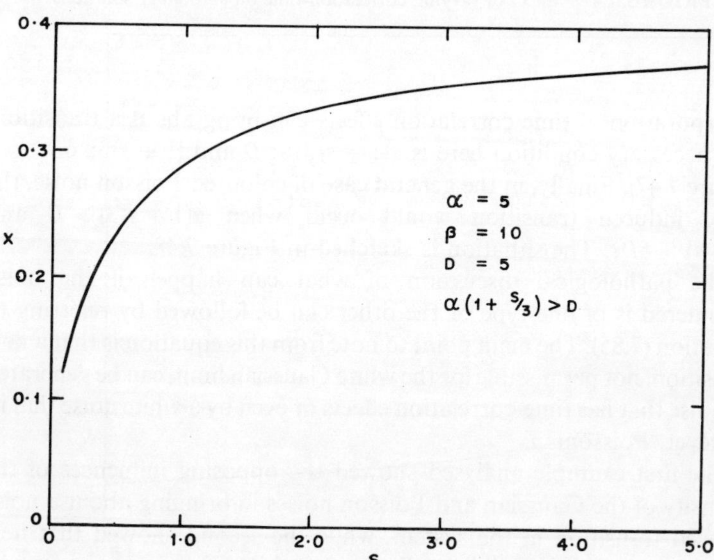

FIGURE 7.46 Influence of (Poisson) white noise on stationary solutions.

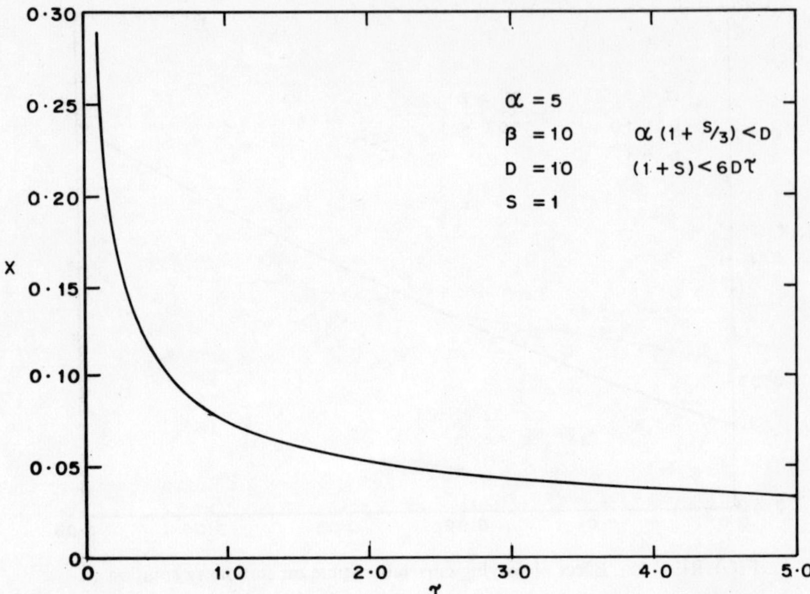

FIGURE 7.47 Effect of varying correlation-time on stationary solutions.

incorporation of time-correlation effects can bring about a transition. The necessary condition here is $\alpha(1 + s/3) < D$ and $(1 + s) < 6D\tau$ (see Figure 7.47). Finally, in the general case of coloured Poisson noise, the noise induced transition would occur when $\alpha(1 + s/3) > D$ and $(1 + s) > 6D\tau$. The situation is sketched in Figure 7.48.

The pathological discussion of what can happen if the noise considered is of one type or the other can be followed by referring to Equation (7.85). The main point to note from this equation is that a new transition, not permissible for the white Gaussian limit, can be generated by noise that has time-correlation effects or even by a white noise that is, however, Poissonian.

The first example analysed showed the opposing influences of the intensity of the Gaussian and Poisson noises in bringing about a noise induced transition in the system, while the second showed that new transitions, not permissible in the case of white Gaussian noise, can be induced due to Poisson noise.

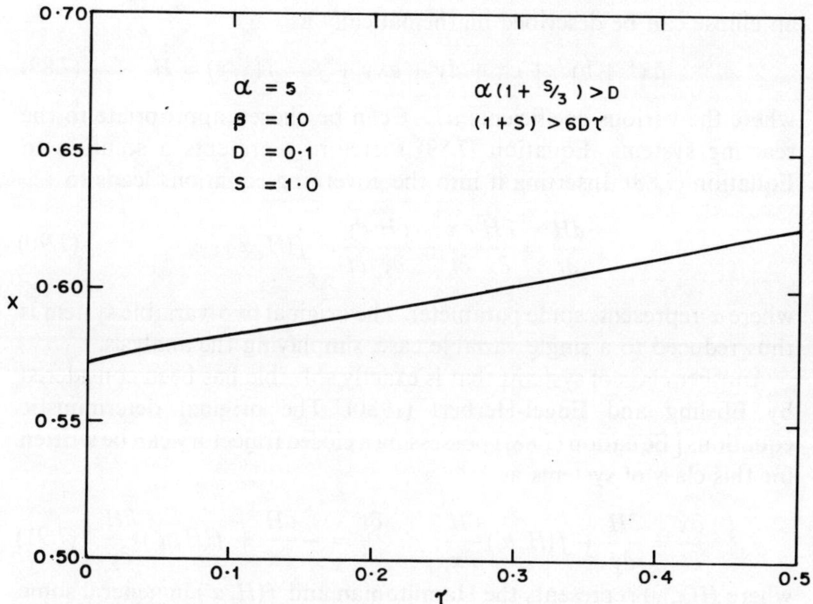

FIGURE 7.48 Effect of varying correlation-time on stationary solutions.

7.4 Analysis of Reacting systems exhibiting Oscillatory Behaviour

7.4.1 Analytical Evaluation of the Role of Coloured Noise

We start with an exactly solvable case of a limit cycle and study the influence of external (Gaussian) noise on the system behaviour. This simple case helps to obtain a clear insight into the behaviour of the system in the presence of noise. Behaviour of more complex models can then be understood more easily. We consider a two-variable system governed by the deterministic equations

$$\frac{dx}{dt} = f_1(x, y), \qquad \frac{dy}{dt} = f_2(x, y) \tag{7.88}$$

which possess the limit cycle behaviour for certain parameter values. A general criterion for the system of Equations (7.88) to possess an exactly solvable elliptical limit cycle has been presented by Escher (1979). Such an ellipse can be described mathematically as

$$ax^2 + by^2 + cx + dy + exy + f = f(x, y) = H \tag{7.89}$$

where the various coefficients $a \dots f$ can be chosen appropriate to the reacting systems. Equation (7.89) therefore represents a solution to Equation (7.88). Inserting it into the governing equations leads to

$$\frac{dH}{dt} = \frac{\partial H}{\partial x}\frac{\partial x}{\partial t} + \frac{\partial H}{\partial y}\frac{\partial y}{\partial t} = f(H, \alpha') \tag{7.90}$$

where α' represents some parameter. The original two-variable system is thus reduced to a single variable case, simplifying the analysis.

Another class of systems that is exactly solvable has been considered by Ebeling and Engel-Herbert (1980). The original deterministic equations [Equation (7.88)] possessing a closed trajectory can be written for this class of systems as

$$\frac{\partial x}{\partial t} = \frac{\partial H}{\partial y} + f(H, \alpha')\frac{\partial H}{\partial x}, \qquad \frac{\partial y}{\partial t} = -\frac{\partial H}{\partial x} + f(H, \alpha')\frac{\partial H}{\partial y} \tag{7.91}$$

where $H(x, y)$ represents the Hamiltonian and $f(H, \alpha')$ in general some nonlinear function of H and external parameter α'. Incorporating the effect of variations in the parameter α', Equation (7.91) becomes

$$\frac{\partial x}{\partial t} = \frac{\partial H}{\partial y} + f(H, \bar{\alpha}')\frac{\partial H}{\partial x} + g(H)\xi(t) \tag{7.92}$$

$$\frac{\partial y}{\partial t} = -\frac{\partial H}{\partial x} + f(H, \bar{\alpha}')\frac{\partial H}{\partial y} + g(H)\xi(t) \tag{7.93}$$

where $g(H)$ represents some function of H that is associated with the parameter α' and depends on the specific form of $f(H, \bar{\alpha}')$, and $\xi(t)$ represents the noise.

Equations (7.92) and (7.93) can again be converted into a single equation in H as

$$\frac{dH}{dt} = h(H)[f(H, \bar{\alpha}') + g(H)\xi(t)] \tag{7.94}$$

using the relation

$$h(H) = (\partial H/\partial x)^2 + (\partial H/\partial y)^2$$

Ebeling and Engel-Herbert have solved this stochastic differential equation and the stationary probability distribution is obtained as

$$p_0(H,\infty)=\frac{N}{h(H)g(H)}\exp\int_{H_1}^{H_2}\frac{1}{D}\frac{f(H,\bar{\alpha}')}{h(H)g^2(H)}\,dH \tag{7.95}$$

N in this equation represents the normalisation constant that can be obtained subject to the condition $\int_{H_1}^{H_2} p_0(H,\infty)\,dH=1$, and D as before is the intensity of white noise. The limits of integration of H_1 and H_2 represent the lower and upper bounds for the values of the variable H.

The simple treatment as above becomes invalid if $\xi(t)$ represents a coloured noise where the nonvanishing correlation time of the noise (τ) makes the temporal evolution of the system a non-Markovian process. An approximate Fokker–Planck equation valid for small correlation times developed in Chapter 6 can now be used, and leads to

$$\frac{\partial p(H,t)}{\partial t}=-\frac{\partial}{\partial H}h(H)f(H,\bar{\alpha}')p(H,t)+D\frac{\partial}{\partial H}h(H)g(H)\frac{\partial}{\partial H}l(H)p(H,t) \tag{7.96}$$

Function $l(H)$ in Equation (7.96) is defined as

$$l(H)=h(H)g(H)\left[1+\tau h(H)g(H)\frac{d}{dH}\left(\frac{f(H,\bar{\alpha}')}{g(H)}\right)\right] \tag{7.97}$$

The stationary solution to Equation (7.96) can be obtained as

$$p_{\rm st}(H,\infty)=p_0(H,\infty)\{1+\tau[E_1-E_2-E_3]\} \tag{7.98}$$

where

$$E_1=-\frac{1}{2D}\int_{H_1}^{H_2}\frac{f^2(H,\bar{\alpha}')}{g^2(H)}p_0(H,\infty)\,dH$$

$$E_2=h(H)g(H)\frac{d}{dH}\frac{f(H,\bar{\alpha}')}{g(H)},\qquad E_3=\frac{1}{2D}\frac{f^2(H,\bar{\alpha}')}{g^2(H)} \tag{7.99}$$

Equation (7.88) gives the stationary probability distribution of H in presence of Ornstein–Uhlenbeck noise.

Equation (7.95) has been used by Ebeling and Engel-Herbert (1980) to analytically show the influence of white noise on the oscillatory system. The effect of coloured noise as contained in Equation (7.98), however, has been reported by Ravikumar and Kulkarni (1985). We shall now quantitatively evaluate the effect of noise by considering some case examples.

As a first example, we shall take the several functions in Equation (7.94) as $h(H) = 2H$, $g(H) = (2H)^{-1/2}$, $f(H, \bar{\alpha}') = (\bar{\alpha}' - 2H)(2H)^{-1/2}$. The specific forms of the functions correspond to a set of deterministic equations of the form of Equation (7.88) and possess a stable limit cycle described by $2H = x^2 + y^2 = \bar{\alpha}'$. Clearly, $H = \bar{\alpha}'/2$ then represents the deterministic solution. In order to obtain a stochastic solution to this case, we first obtain the stationary probability distribution using Equation (7.95) as

$$p_0(H, \infty) = N(2H)^{-1/2} \exp\left\{\frac{1}{D}[\alpha'(2H)^{1/2} - \tfrac{1}{3}(2H)^{3/2}]\right\} \quad (7.100)$$

The extremal points of the probability distribution are obtained as

$$\frac{dp_0(H, \infty)}{dH} = 0 = (2H)^{1/2}(\bar{\alpha}' - 2H) - D \quad (7.101)$$

The analysis of this equation reveals that for the case of no noise ($D = 0$) the system possesses the deterministic solution at $2H = \bar{\alpha}'$. For finite values of D, the equation possesses another root that is generated due to white noise. The additional solution can be identified as an unstable limit cycle and lies inside the deterministic stable orbit. The unstable limit cycle grows with the extent of noise and eventually merges with the deterministic solution annihilating it. The system does not possess any solution beyond this extent of noise ($D = 0.4$ in this case).

The effect of coloured noise on this system can be obtained by using Equation (7.98) where $\bar{\alpha}'$ has been assigned a value of one. The results obtained are displayed in Figures 7.49 and 7.50 for varying values of correlation time (τ) as the value of the intensity of noise (D) is increased. Let us first concentrate on the results shown in Figure 7.49 for the case of $D = 0.1$. It is evident that at low values of τ ($\tau \to 0$) the system possesses two solutions, one of which is stable. The case of $\tau = 0$ of course corresponds to the white noise case. As the value of τ is increased, the stable solution that begins at $H = 0.445$ shifts towards the deterministic solution ($H = 0.5$). The unstable solution branch shows a decreasing tendency. The branch, however, disappears abruptly at around $\tau = 0.15$. In this region ($\tau = 0$ to 0.15) we therefore have two stationary solutions to the system. One of these corresponds to a stable limit cycle and the other to the unstable limit cycle. Between $\tau = 0.15$ and 0.3 the system possesses only one solution that corresponds to the stable limit cycle. At $\tau = 0.3$ the system shows bifurcation to two other solutions. For $\tau > 0.3$,

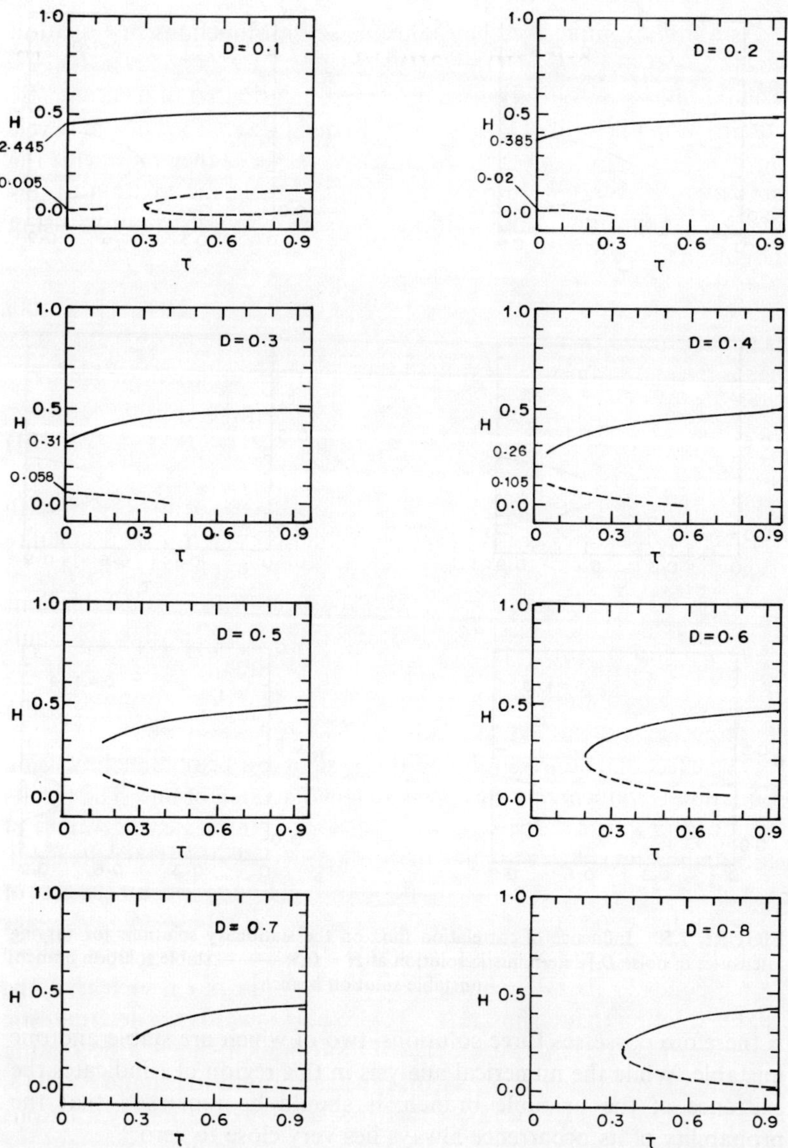

FIGURE 7.49 Influence of correlation time on the stationary solutions for varying intensities of noise D [deterministic solution at $H = 0.5$; ——— stable solution branch, ----- unstable solution branch].

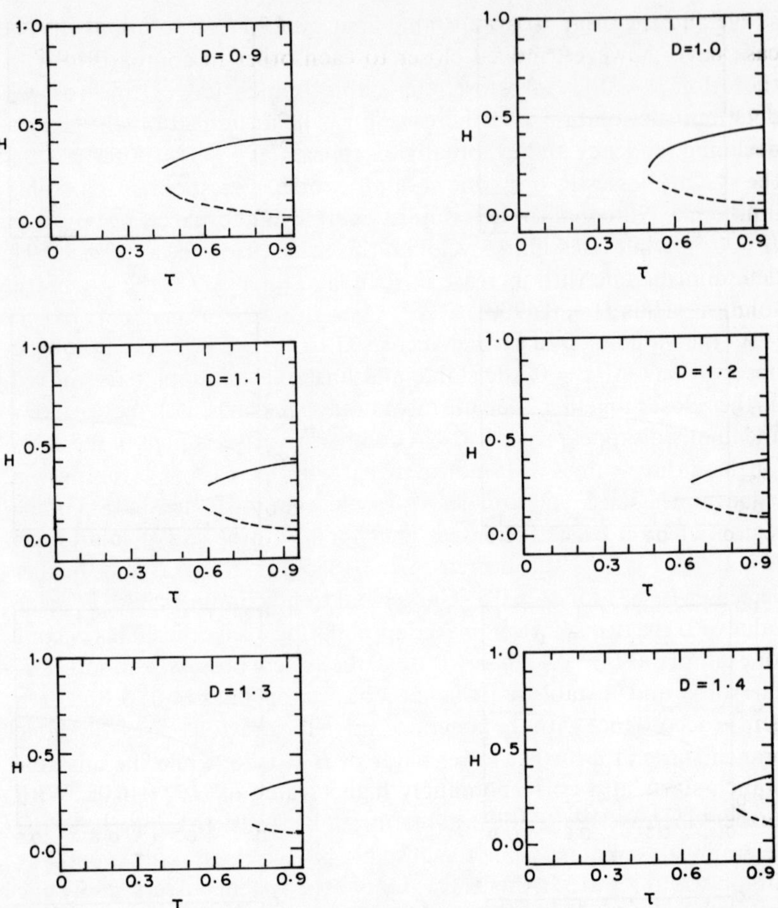

FIGURE 7.50 Influence of correlation time on the stationary solutions for varying intensities of noise D [deterministic solution at $H = 0.5$; ——— stable solution branch, ----- unstable solution branch].

it therefore possesses three solutions, two of which are stable and one unstable. While the numerical analysis in this region of τ indicates the existence of the unstable branch, it should be remarked that the probability of its occurrence always lies very close to zero.

Let us now examine the influence of increase in the intensity of noise on the behaviour of this system. As noted in the figure for $D = 0.2$, at $\tau = 0$ the system again possesses two solutions, one corresponding to a stable and the other to an unstable limit cycle. These solutions in this case have, however, moved closer to each other in comparison with those for $D = 0.1$. As before, the stable branch moves towards the deterministic solution with increase in τ. The unstable branch shows a declining tendency and eventually disappears at $\tau = 0.3$. After $\tau = 0.3$ the system possesses only one solution which corresponds to a stable limit cycle. Note that the deterministic stable limit cycle corresponds to $H = 0.5$, while the limit cycle in this case tries to approach the deterministic one with increase in τ. In fact, for $\tau > 0.8$, the stochastic solution yields $H = 0.49001$.

As the value of D is further increased to 0.3 we note the following observations. At $\tau = 0$, the stable and unstable solutions have moved further closer together. The unstable branch declines with increase in τ and finally disappears at $\tau = 0.4$. A comparative assessment of the cases for these three values of D indicates that with increase in D for a given value of τ the stable solution decreases while the unstable solution in the region where it exists increases. Thus for $\tau = 0$ the stable solution of $H = 0.445$ for $D = 0.1$ decreases to 0.385 for $D = 0.2$ and further decreases to 0.31 for $D = 0.3$. The general trend continues until at some value of D the two solutions merge annihilating each other. This occurs, for example, at $D = 0.4$ where for $\tau = 0$ the system possesses no solution. The stable and unstable branches now appear only at $\tau = 0.05$, and even here in accordance with the general trend in the previous cases the stable branch starts at a further lower value of $H = 0.26$, while the unstable branch starts at a correspondingly higher value of $H = 0.0105$. With further increase in τ the stable branch tends to approach the deterministic solution, while the unstable branch decays and eventually disappears at $\tau = 0.55$. After this value of τ the system possesses only one stable limit cycle as its solution.

The remaining figures for progressively higher values of D essentially bring forth the same pattern. As the value of D is increased the initial region where no solutions exist increases. Thus for $D = 0.6$ the no-solution region extends from $\tau = 0$ to 0.2 as against $\tau = 0$ to 0.05 for $D = 0.4$. The figures also indicate that the stable and unstable solution branches begin at higher values of τ as D is increased. The unstable solution branch that decays consequently requires larger values of τ with increase in D to completely disappear. The stable branch that tends to approach the deterministic solutions also requires higher values of τ with increase in D to approach the deterministic limit cycle.

The results of the present case indicate conflicting roles of D and τ. For a deterministic system with one stable limit cycle as its solution, incorporation of D (i.e. white noise alone) brings about an additional solution that corresponds to the unstable limit cycle. An increase in the value of D expands the unstable solution while the stable solution is brought down. Eventually for some value of D the two solutions merge, annihilating each other and the system possesses no solution after this critical value of D. The incorporation of coloured noise (both D and τ) for a given value of D acts the other way. Increasing the correlation time of the noise for a fixed value of its intensity (D) favours the stable solution and helps its rise towards the deterministic solution. The unstable solution decays with increase of correlation time.

In the second example we consider the several functions in Equations (7.94): $f = (\bar{\alpha}' - 2H)(2H)^{-1}$, $g(H) = (2H)^{-1}$ and $h(H) = 2H$. These functions again correspond to a deterministic set of equations that possess the limit cycle described by $2H = x^2 + y^2 = \bar{\alpha}'$. Here, too, $H = \bar{\alpha}'/2$ would represent the deterministic solution. To obtain a stochastic solution we write the probability distribution in presence of white noise according to Equation (7.95) as

$$p_0(H, \infty) = N \exp\left[\frac{H}{D}(\bar{\alpha}' - H)\right] \tag{7.102}$$

where N represents the normalisation constant. The points of maxima of this equation indicate that they always occur at $H = \bar{\alpha}'/2$, which corresponds to the deterministic solution. The analysis, therefore, indicates that presence of white noise has no influence on the deterministic solution.

Let us now investigate the role of coloured noise. According to Equation (7.98) the stationary probability distribution is

$$p_{st}(H, \infty) = p_0(H, \infty)[1 + \tau(E_1 - E_2 - E_3)] \tag{7.103}$$

where

$$E_1 = -\frac{1}{2D}\int_0^1 (\bar{\alpha}' - 2H)^2 p_0(H_1, \infty)\, dH, \qquad E_2 = 2$$
$$E_3 = \frac{1}{2D}(\bar{\alpha}' - 2H)^2 \tag{7.104}$$

The stationary probability distribution calculated using Equations (7.103) and (7.104) also indicates that the maxima always occur at

$H = \bar{\alpha}'/2$ for any combination of D and τ. The coloured noise also therefore has no influence on the deterministic solution. In other words, the original deterministic system remains unaltered in the presence of both white and coloured noises.

7.4.2 Noise Induced Transitions in Reacting Systems

In the cases analysed thus far we began with simple model systems which, under appropriate conditions, exhibited a unique stable, bistable or limit cycle behaviour. The present section considers a model system that exhibits *all these types of behaviour* on variation in some parameter. The Brusselator scheme is known to exhibit all these features when an external parameter such as flow rate changes. We shall analyse this situation here in presence of noise and compare the results with those known for the case of no noise in the system.

We begin with a reaction scheme

$$A + 2X \rightleftharpoons 3X$$
$$X \rightleftharpoons B$$
$$C + X \longrightarrow A$$

and allow it to occur in a CSTR where the reactants enter at a certain mean rate $k_0 = F/V$. The dimensionless equations describing the evolution of concentration profiles can be written as

$$\frac{dx}{d\tau'} = ax^2 + x^3 - x + b - cx + j(x_0 - x) \tag{7.105}$$

$$\frac{da}{d\tau'} = -\delta ax^2 + \delta x^3 + cx + j(a_0 - a) \tag{7.106}$$

$$\frac{db}{d\tau'} = \eta x - \eta b + j(b_0 - b) \tag{7.107}$$

$$\frac{dc}{d\tau'} = -\psi cx + j(c_0 - c) \tag{7.108}$$

where the following parameter definitions are employed:

$$a = \frac{k_1 A}{\sqrt{k_{-1}k_2}}, \qquad b = \frac{k_{-2}}{k_2}\sqrt{\frac{k_{-1}}{k_2}}B, \qquad x = \sqrt{\frac{k_{-1}}{k_2}}x, \qquad \tau' = \frac{t}{k_2}$$

$$c = \frac{k_3}{k_2}C, \qquad k_0 = \frac{F}{V}, \qquad j = \frac{k_0}{k_2} \tag{7.109}$$

$$\delta = \frac{k_1}{k_{-1}}, \qquad \eta = \frac{k_{-2}}{k_2}, \qquad \psi = \frac{k_3}{\sqrt{k_{-1}k_2}}$$

The stationary solution to these equations can be obtained by equating them to zero. We thus have

$$b_s = \frac{\eta}{\eta + j}x_s + \frac{j}{\eta + j}b_0 \tag{7.110}$$

and x_s is given by a cubic equation

$$-x^3(1 + R) + x^2\left(\psi + Q - \frac{\psi R}{\delta} + \frac{\psi M}{\eta}\right)$$

$$-x\left(1 - \frac{\psi Q}{\delta} - \frac{\psi N b_0}{\eta} + \psi_0 \kappa_0 + j - M\right) + Nb_0 + jx_0 = 0 \tag{7.111}$$

where

$$R = \frac{\delta}{j}\left[1 + j - \frac{\eta}{\eta + j}\right], \qquad Q = \delta x_0 + a_0 + \frac{\delta b_0}{\eta + j}$$

$$M = \frac{\eta}{\eta + j} \quad \text{and} \quad N = \frac{j}{\eta + j} \tag{7.112}$$

The stationary solution for a_s can likewise be obtained.

A typical plot showing the stationary solutions on variation in j is shown in Figure 7.51. It is clear from this figure that as the flow rate j increases from zero, we first enter a region where the system possesses a unique unstable solution. Increase in j destroys this feature and the system possesses a unique stable solution. Further increase in j leads to bistability, and finally to a single stable state. We shall now investigate the behaviour of this system in presence of noise which enters due to variation in the flow to the reactor. This can be defined as $k_0 = \bar{k}_0 + \xi(t)$.

Incorporation of this variation in the phenomenological equations

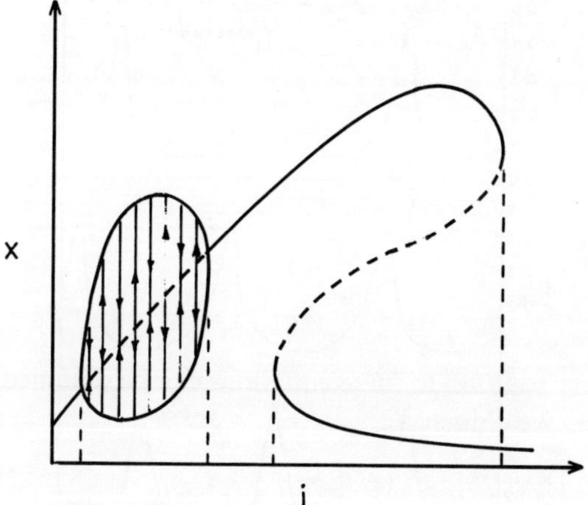

FIGURE 7.51 Stationary solutions and their stability.

leads to the addition of the term $\xi(x_0 - x)$ to Equation (7.105), $\xi(a_0 - a)$ to Equation (7.106), and $\xi(b_0 - b)$ to Equation (7.107). We shall regard ξ in these equations as a general coloured noise with finite time-correlation effects and represent it as an Ornstein–Uhlenbeck process:

$$\frac{d\xi}{dt} = -\varepsilon^{-2}\xi + \varepsilon^{-1} D\, dW \tag{7.113}$$

where D and τ have the usual meanings. The set of Equations (7.105)–(7.108) along with Equation (7.113) have been numerically integrated and the results obtained are discussed below.

The deterministic solution for a set of parameter values $\delta = 10$, $\eta = 0.01$, $b_0 = 0.02$, $a_0 = 16.65$ and $x_0 = c_0 = \psi = 0$ for varying values of the dimensionless input flow j is plotted in Figure 7.51 where we see a limit cycle region for the range $j = 0.01$ to 0.164. For j lying in the range 0.165–1.0 the system exhibits three states. Between $j = 0.164$ (limit cycle region) and $j = 0.165$ (unstable region) there exists a narrow range of j values over which the system exhibits a unique stable state. The numerical computations for the stochastic models have been carried out in each of these three regions to examine the influence of noise, and the general results obtained are discussed below.

We first begin with the region where deterministic solutions show oscillatory behaviour. A specific value of $j = 0.1$ which lies within the

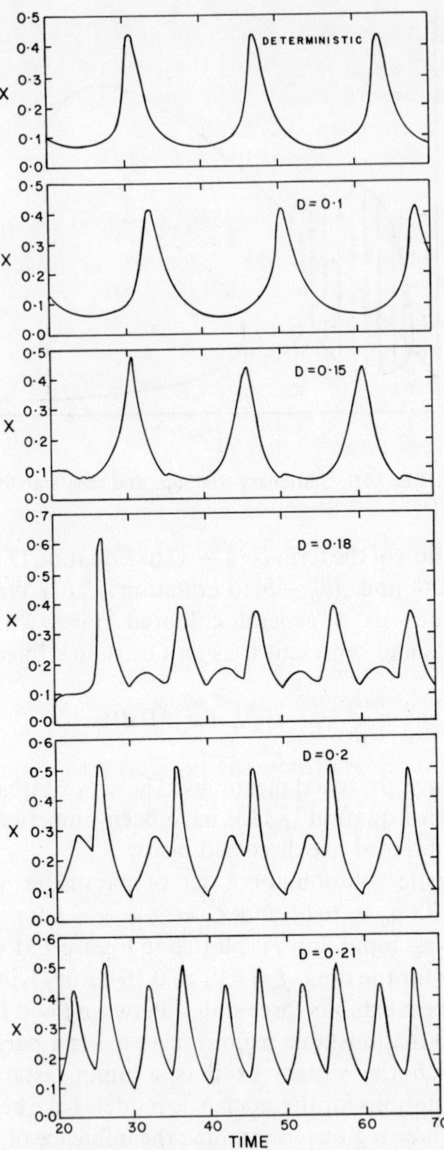

FIGURE 7.52 Effect of increase in noise intensities for systems in oscillatory mode.

limit cycle region has been chosen for the analysis. Figure 7.52 shows the influence of the intensity of noise on the oscillatory behaviour; it is seen that the oscillations are displaced towards the right on the time scale. The amplitude of the oscillations falls slightly while the time period of oscillations remains unaltered. Increase in the value of D ($D = 0.15$) causes a small aberration in the profile indicating that perhaps a second peak is under development. This aberration grows with the intensity of noise and the original single peak oscillating system breaks into a two-peak oscillating one ($D = 0.18, 0.2, 0.11$). Further, the amplitude of the first peak is dependent on the extent of noise and grows with it while the second peak is always slightly shorter than the deterministic peak. The time period of oscillations is also seen to be nearly halved, and between the two peaks of the deterministic case two doubly-peaked oscillating cycles have been accommodated.

The effects of time correlation of the noise for an arbitrary value of its noise intensity ($D = 0.20$) are shown in Figure 7.53 where we observe that the effects of D and τ are in opposing directions. In fact, for sufficiently higher values of correlation time, the double-peak cycle is transformed into the original single-peak cycle. The amplitude and time period of oscillations now increase with increase in time correlation and the original deterministic case is reproduced (although a slight displacement towards higher values of time still persists).

A comparative study of Figures 7.52 and 7.53 reveals that the effect of increase in D is to generate a double-peak cycle from the original single peak cycle, while the effect of increasing time correlation tends to undo the effect of D. Also, the time period, which decreases with increase in D and becomes in fact nearly half of the original cycle, is restored with increase in time correlation. The two parameters of noise, its intensity and time correlation, thus show opposing effects for a system exhibiting limit cycle behaviour.

Let us now turn to the monostable region (of Figure 7.51). For the parameter values chosen this situation exists for $j = 0.165$. The numerical results for the stochastic case are shown in Figure 7.54, wherein the deterministic case is also included for the sake of comparison. It appears from this figure that the monostable character of the system in this region is susceptible to noise even at a very low level ($D = 0.1$) and is transformed into an unstable region showing limit cycle behaviour. Increase in D tends to enhance the amplitude of these oscillations while their time periods remain practically unaltered. Figure

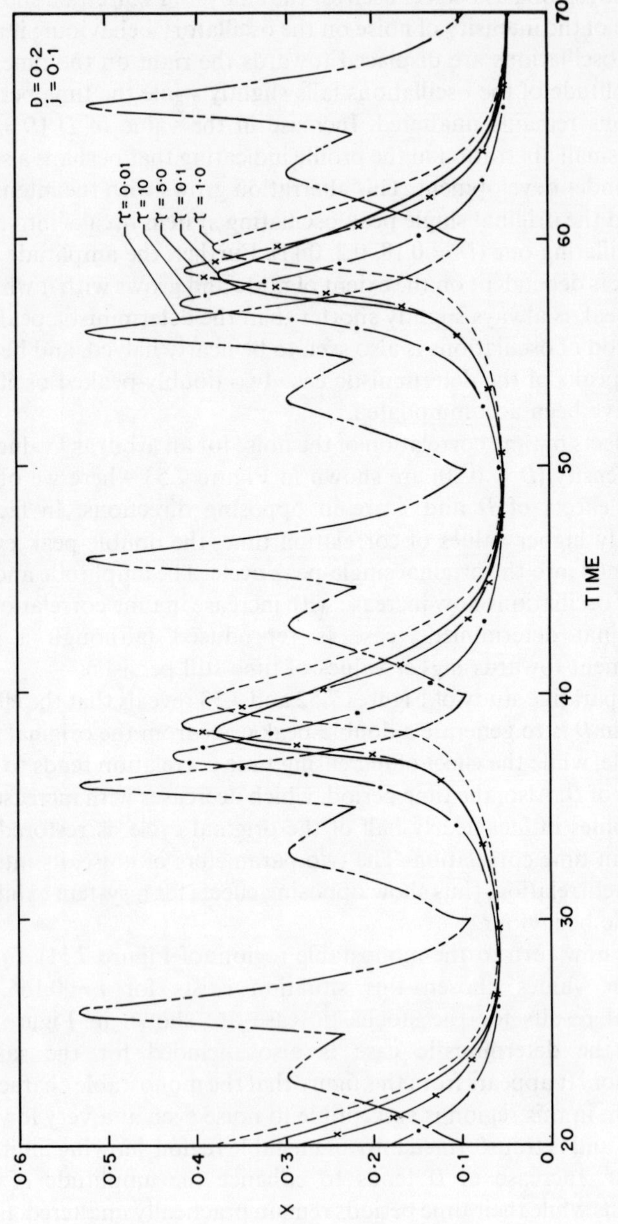

FIGURE 7.53 Time-correlation effects for systems in oscillatory mode

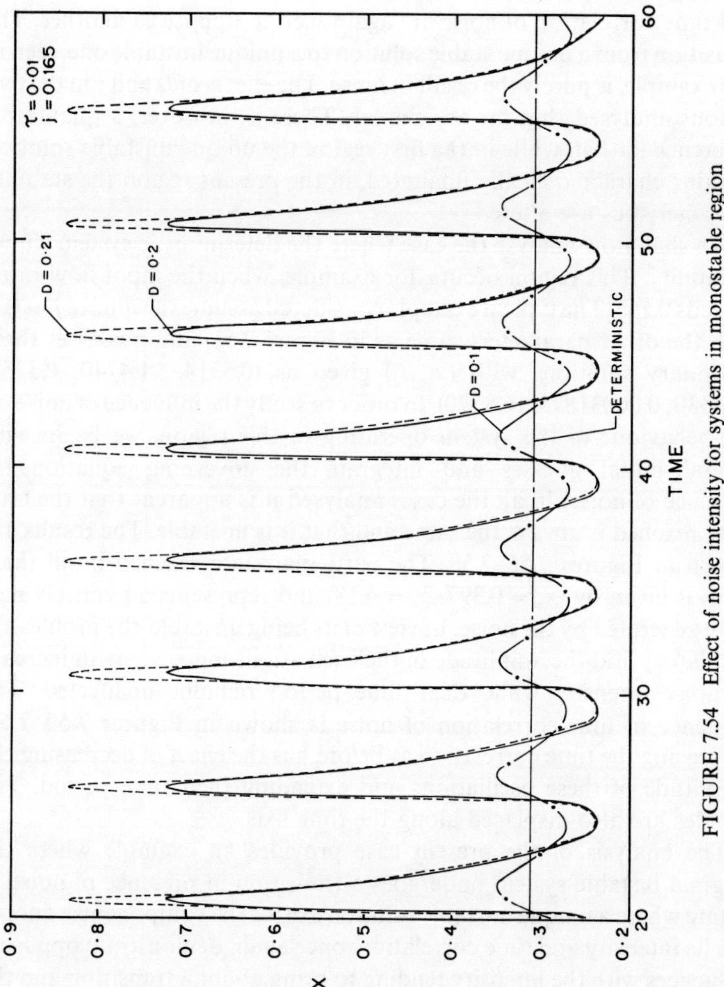

FIGURE 7.54 Effect of noise intensity for systems in monostable region

7.55 shows the effect of time correlation of noise of a certain intensity ($D = 0.1$) on the behaviour of the system operating in this monostable region. The figure clearly shows that, with increasing time-correlation effects, the system tends towards deterministic behaviour. The intensity and time correlation of noise are again seen to oppose each other. The transition from a unique stable solution to a unique unstable one, seen in this example, is purely the result of noise. The effects of D and τ in the two regions analysed thus far are similar. There is, however, a qualitative difference in that, while in the first region the unique unstable solution remains characteristically unaltered, in the present region the stability characteristics are altered.

We shall now analyse the case where the deterministic system shows bistability. This region occurs, for example, when the input flow rate j exceeds 0.165. The typical example considered assumes a value of $j = 0.2$, with the other parameters same as in Figure 7.51, and possesses three stationary solutions with (x_s, a_s) given as (0.5314, 11.4740; 0.3359, 16.4530; 0.00031875, 16.8570). In order to study the influence of noise on the behaviour of the system operating in this region, we begin with several initial guesses and integrate the governing equations in presence of noise. In all the cases analysed it is apparent that the final state reached is always the same and that it is unstable. The results are shown in Figures 7.56–7.58. The particular state reached in all these cases is given by ($x_s = 0.397$, $a_s = 4.18$) and represents an entirely new state generated by the noise. In view of its being unstable, the profiles are oscillatory and the amplitude of the oscillations increases with increase in noise intensity while their time period remains unaffected. The influence of time-correlation of noise is shown in Figures 7.59–7.61. Increasing the time correlation as before has the effect of decreasing the amplitude of these oscillations and extending their time period. The profiles are also displaced along the time axis.

The analysis of the present case provides an example where the original bistable system undergoes a transition in presence of noise to a state where a unique unstable solution exists. The components of noise, viz. its intensity and time correlation, once again demonstrate opposing influences with the intensity tending to bring about a transition and the time correlation resisting this change.

The present section analysed the influence of noise on a model system which exhibits limit cycle, monostability and bistability behaviour when some external parameter such as flow rate changes. In each of these

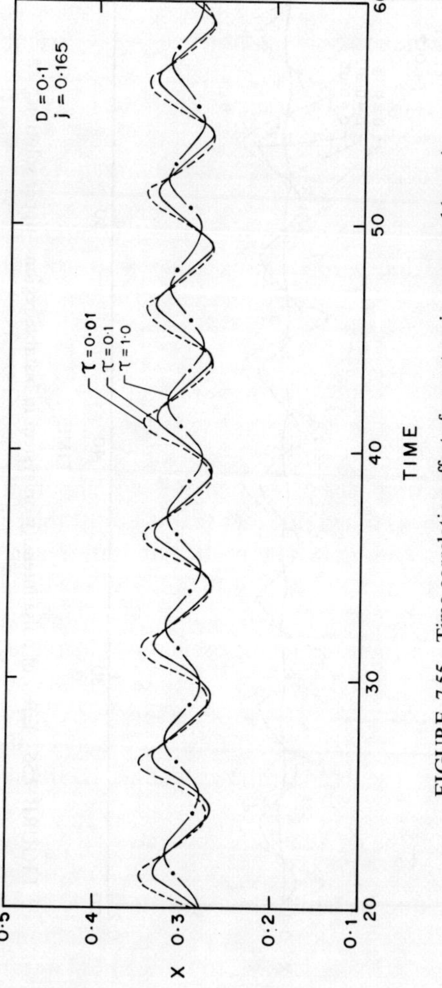

FIGURE 7.55 Time-correlation effects for system in monostable region.

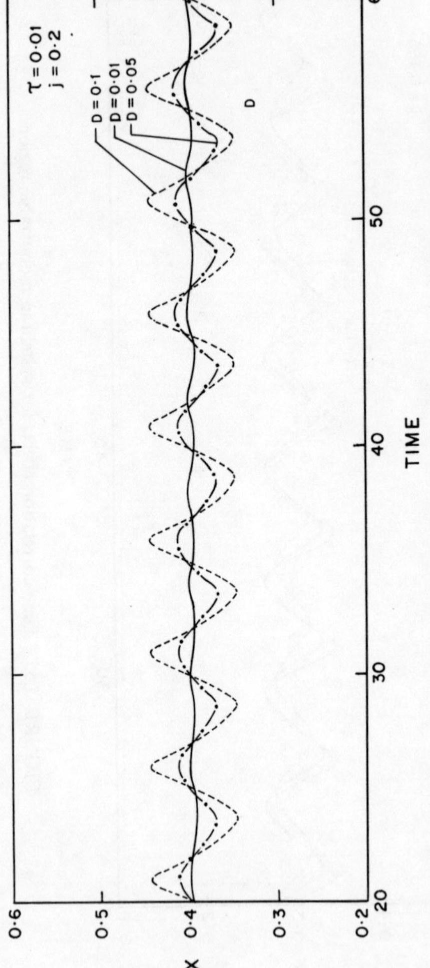

FIGURE 7.56 Effect of noise intensity on system in bistable region—upper solution.

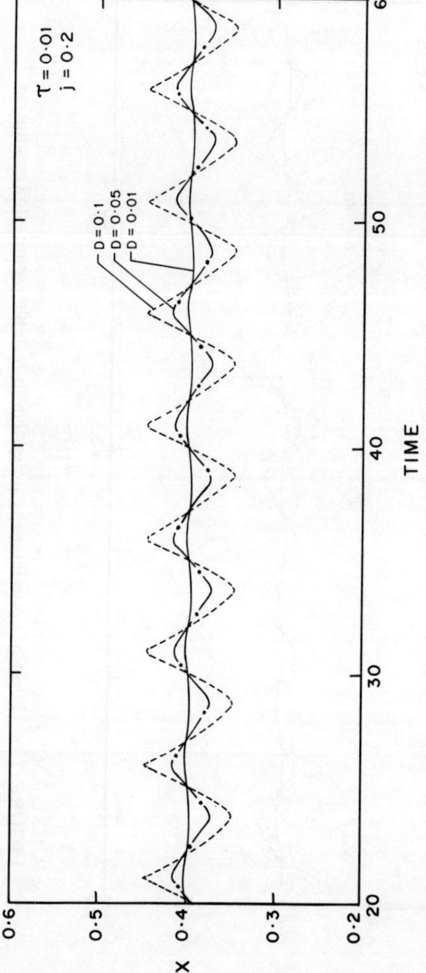

FIGURE 7.57 Effect of noise intensity on system in bistable region—middle solution.

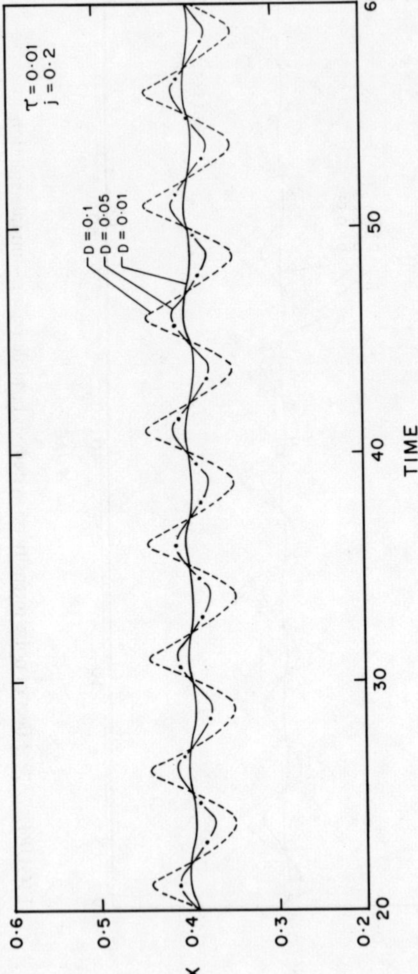

FIGURE 7.58 Effect of noise intensity on system in bistable region—lower solution.

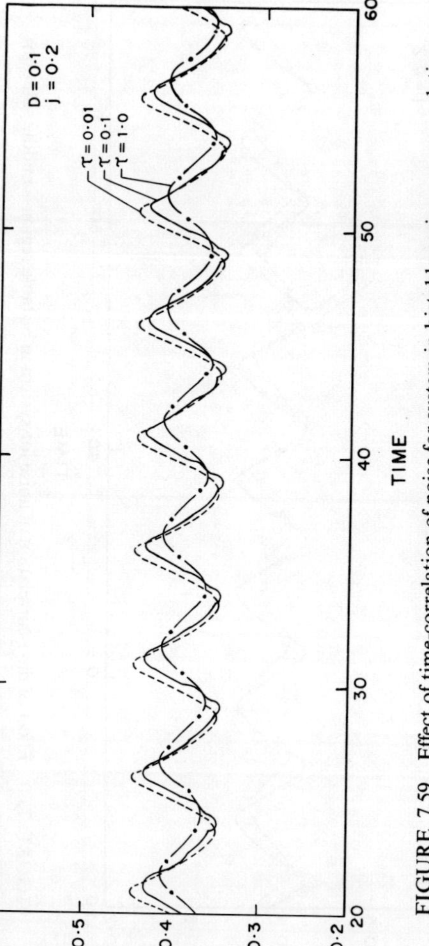

FIGURE 7.59 Effect of time-correlation of noise for system in bistable region—upper solution.

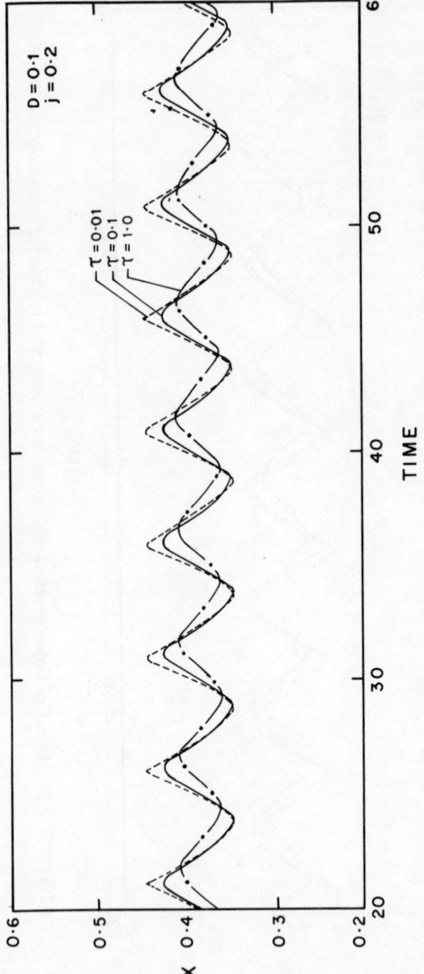

FIGURE 7.60 Effect of time-correlation of noise for system in bistable region—middle solution.

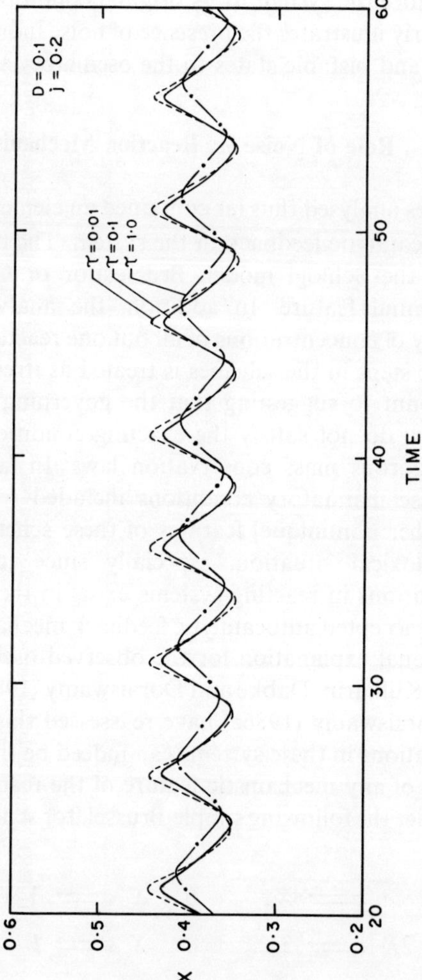

FIGURE 7.61 Effect of time-correlation of noise for system in bistable region—lower solution.

regions, the role of noise has been to create a unique unstable state around which the system oscillates. In general, the intensity of noise tends to increase the amplitude of oscillations while the time-correlation effects tend to restore the system to its original condition. This simple model system clearly illustrates the presence of noise induced transition from monostable and bistable states to the oscillatory state.

7.4.3 Oscillations: Role of Noise vs. Reaction Mechanisms

The model schemes analysed thus far contained an elementary step that generates an autocatalytic feedback in the system. The model reaction schemes such as the Schlogl model, Brusselator or Oregonator all contain this essential feature. In addition, the analysis invariably assumes constancy of concentrations of all but one reactant species and at least one of the steps in the schemes is treated as irreversible. These assumptions amount to suggesting that the governing equations for these schemes now do not satisfy the exacting conditions of detailed balancing and rigorous mass conservation laws. In fact, a rigorous analysis with these mandatory conditions included reveals that the oscillatory (or other nonunique) features of these schemes disappear. This is a paradoxical situation, especially since incontrovertible evidence of oscillations in reacting systems exists in the literature.

The commonly accepted autocatalytic feedback mechanism thus fails to provide a rational explanation for the observed oscillations in the reacting systems. Kulkarni, Dabke and Doraiswamy (1985) and Dabke, Kulkarni and Doraiswamy (1986e) have reassessed this situation and shown that oscillations in these systems can indeed be the consequence of noise, and not of any mechanistic feature of the reaction scheme.

We shall consider the following simple Brusselator scheme as a model system:

$$A \rightleftharpoons X, \qquad B + X \rightleftharpoons Y + F$$

$$Y + 2X \rightleftharpoons 3X, \qquad X \rightleftharpoons E$$

Unlike the conventional scheme that ignores the reversible steps (Nicolis and Prigogine, 1967) the present scheme assumes that all the steps are reversible (Kulkarni, Dabke and Doraiswamy, 1985; Dabke, Kulkarni and Doraiswamy, 1986e). Also, the common assumption, viz. the concentrations of the several reactants and product species are constant, is done away with. These assumptions in the analysis thus far have essentially eliminated the less important variables and consequently led to a reduced description in terms of a simple two-variable system that is known to exhibit limit cycle behaviour. The present analysis relaxes these common assumptions, so that the full system now ceases to exhibit any oscillatory behaviour. In fact, in a closed system the model leads to the following relation:

$$\left(\frac{y}{x}\right)_e = \frac{k_2}{k_{-2}} = \frac{k_{-3}}{k_3} \tag{7.114}$$

suggesting that, for the detailed balancing condition to hold, the individual reactions in a closed system at equilibrium must balance. We now argue that no system is, however, a truly closed system and that one can only hope to approach this ideal situation. An infinitesimally small extent of coupling of the system with the outside surroundings would always exist in practice and can induce a small variation in the parameter values of the system. These parameters, as before, can be described in terms of their mean values plus fluctuations that are random. The fluctuations may or may not be inconsequential in deciding the behaviour of the system, and their possible role would evidently depend on the nature of the underlying system. Especially for systems operating near the borderline of the regions of stability, these fluctuations can acquire significant importance and may dictate the evolution of the system.

The macroscopic conservation equations, taking full account of all the species and reversibility of the reaction steps, can be written as

$$\frac{dx}{dt} = k_{-4}M_2 + (k_1 - k_{-4})a - (k_{-1} + k_4 + k_{-4})x - (k_{-4} - M_3k_{-2})y$$
$$- k_2b_x - k_{-2}by + k_3x^2y - k_{-3}x^3 \tag{7.115}$$

$$\frac{dy}{dt} = k_{-4}M_2y + k_{-2}by + k_2bx - k_3x^2y + k_{-3}x^2 \tag{7.116}$$

$$\frac{da}{dt} = k_{-1}x + k_{-1}a \tag{7.117}$$

$$\frac{db}{dt} = k_{-5}M_3 - (k_5 + k_{-5})b + k_{-2}M_3y - k_{-2}by - k_2bx \tag{7.118}$$

The lower-case letters in these equations denote the concentrations of the respective species, the various k's the rate constants for the forward and the backward steps, and M_1, M_2 and M_3 constants defined by

$$M_1 = M_2 + M_3, \qquad M_2 = a + e + x + y, \qquad M_3 = b + d \tag{7.119}$$

In view of the closed nature of the system the solutions to these equations for given initial conditions proceed to unique stable equilibrium point. We shall now investigate the behaviour of the system in presence of fluctuations in the parameter values.

The noise, as before, has been treated as an Ornstein–Uhlenbeck process:

$$\frac{d\xi}{dt} = -\varepsilon^{-2}\xi(t) + D\varepsilon^{-1}\,dW \tag{7.120}$$

with $\langle \xi \rangle = 0$ and $\langle \xi(t)\xi(t') \rangle = D\exp(-\varepsilon^{-2}|t - t'|)$.

To illustrate the effect of noise we assume that the rate constant k_3 fluctuates about the mean value with $\xi(t)$ defined by Equation (7.120). The incorporation of noise in the governing equations leads to an additional term $\xi x^2 y$ on the rhs. The set of equations has been solved numerically. The influence of the intensity and time-correlation effects of noise is presented in Figures 7.62 and 7.63. Clearly, the forbidden oscillatory behaviour of the Brusselator is restored in presence of noise. The important result is that even when the mandatory conditions of detailed balancing and mass conservation relations are satisfied, oscillations can still occur purely due to noise. These oscillations are therefore not the intrinsic feature of the system but are noise induced.

It is further seen from these figures that increasing the intensity of noise (D) increases the amplitude of the oscillations. The time period, however, remains unaltered. Figure 7.63 reveals the influence of time-correlation, which is seen to be opposite to that of D. For higher values, the original deterministic solution is recovered.

We shall now investigate the case when the system satisfies the detailed balance condition but where rigorous mass conservation laws are relaxed. In other words, the concentrations of the species such as A, B, etc. are now treated as constant. Note that the macroscopic analysis for this situation still fails to give oscillations. The reduced system of equations has been solved in presence of noise and the results are displayed in Figures 7.64 and 7.65. The plots show exactly the same trends as the original rigorous model.

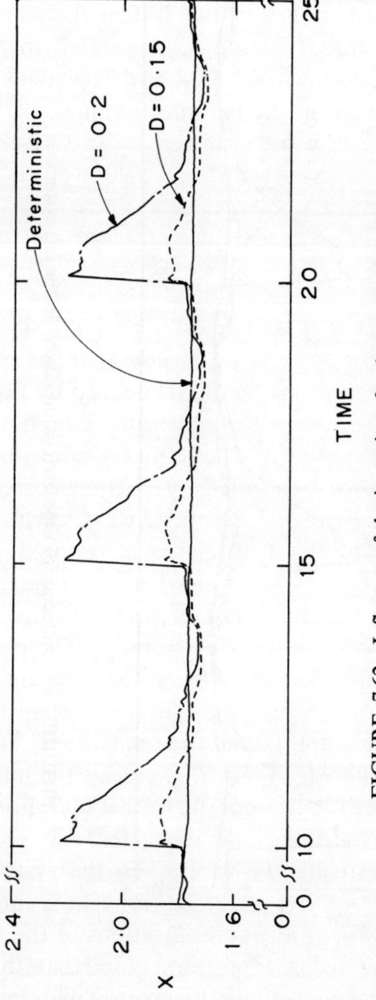

FIGURE 7.62 Influence of the intensity of noise on the full system.

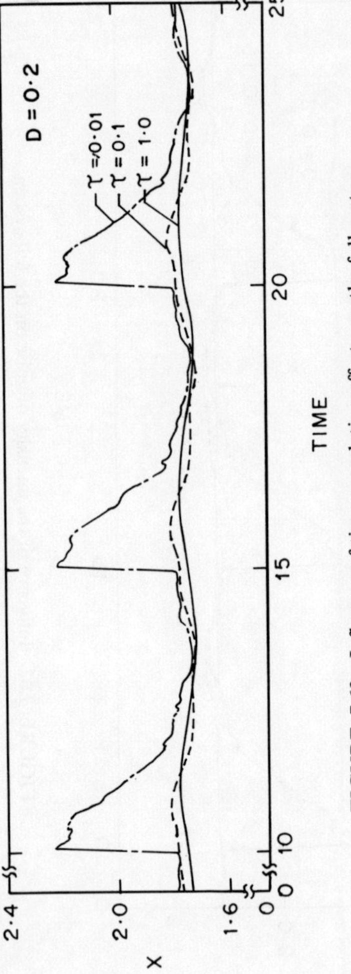

FIGURE 7.63 Influence of time-correlation effects on the full system.

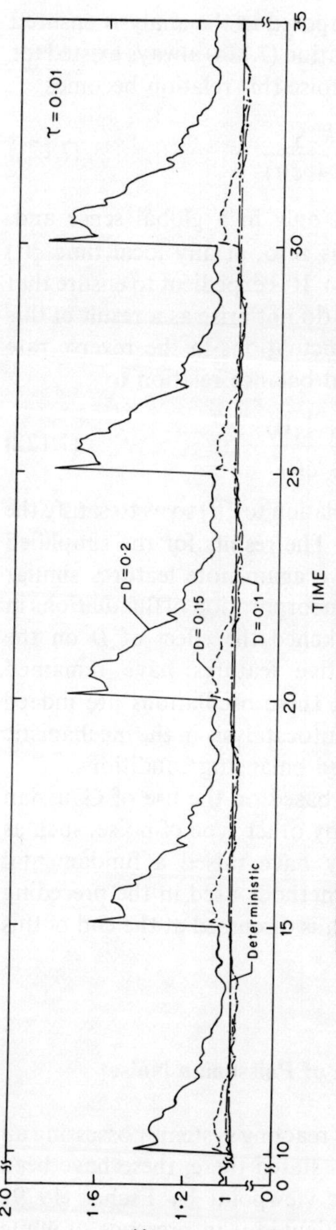

FIGURE 7.64 Effect of noise intensity on the reduced variable system

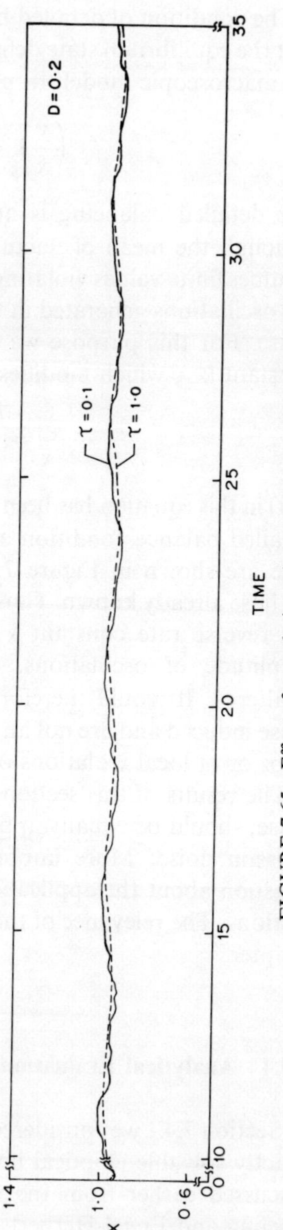

FIGURE 7.65 Effect of correlation-time on the reduced system

The condition of detailed balancing imposed in the analysis ensured that the equilibrium state defined by Equation (7.114) always existed for the macroscopic model. In presence of noise this relation becomes

$$\left(\frac{y}{x}\right)_e = \frac{k_2}{k_{-2}} = \frac{k_{-3}}{\bar{k}_3 + \xi(t)} \tag{7.121}$$

The detailed balancing is now satisfied only in a global sense and, although the mean of fluctuations $\xi(t)$ is zero, at any local time $\xi(t)$ assumes finite values violating the relation. It is expedient to ensure that the oscillations generated in the analysis do not arise as a result of this cause. For this purpose we allow for fluctuations in the reverse rate constant k_{-3} which modifies the detailed balance relation to

$$\frac{y}{x} = \frac{k_{-2}}{k_2} = \frac{\bar{k}_{-3} + \xi_1(t)}{\bar{k}_3 + \xi(t)} \tag{7.122}$$

$\xi_1(t)$ in this equation has been chosen in relation to $\xi(t)$ so as to satisfy the detailed balance condition at any time. The results for the simplified case are shown in Figure 7.66 where we again note features similar to those already known. Thus, while the incorporation of fluctuations in the reverse rate constant k_{-3} has weakened the effect of D on the amplitude of oscillations, the qualitative features have remained unaltered. It would therefore seem that these oscillations are indeed noise induced and are not an artifact of autocatalysis in the mechanistic steps or of local violations of the detailed balancing conditions.

The results of this section, which are based on the use of Gaussian noise, should be equally applicable to any other type of noise, such as Poisson noise. More importantly, they have raised a fundamental question about the applicability of the methods used in the preceding sections. The relevance of this conclusion is discussed at the end of this chapter.

7.4.4 Analytical Evaluation of the Role of Poissonian Noise

In Section 7.4.1 we considered the case of reacting systems possessing an exactly solvable elliptical limit cycle. As stated there, these have been discussed earlier from the macroscopic viewpoint by Escher (1979). Ebeling and Engel-Herbert (1980) investigated it in presence of white

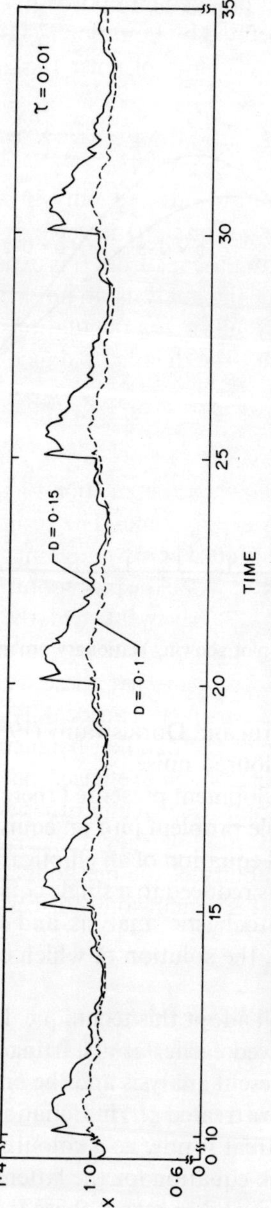

FIGURE 7.66 Effect of noise intensity when detailed balancing is satisfied, even at local times

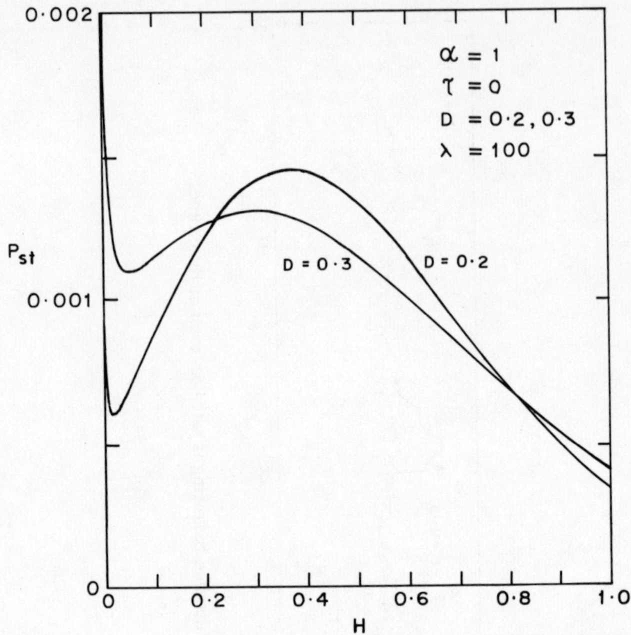

FIGURE 7.67 Typical plot showing stationary probability distribution.

noise while Dabke, Kulkarni and Doraiswamy (1986d) considered these systems in presence of coloured noise.

The key idea in the development presented there was to convert what is essentially a two-variable problem into an equivalent single-variable problem using the general equation of an elliptical limit cycle. Once the description of the system is reduced to a single equation, we can employ the known methods of stochastic analysis and derive an equivalent Fokker–Planck equation, the solution of which contains the effects of noise.

In what follows we shall adopt this technique. In fact, we shall begin with Equation (7.94) derived earlier as a starting point. The difference that exists between the present analysis and the one reported in Section 7.4.1 is that, while earlier we treated $\xi(t)$ in Equation (7.94) as a Gaussian coloured noise, we shall treat it now as a coloured Poisson noise. The associated Fokker–Planck equation for the latter situation has already been derived in Chapter 6, and as stated there the results of this more general case reduce to those of several simplifying cases such as Gaussian white noise, Gaussian coloured noise or Poisson white noise under appropriate conditions.

Let us thus begin with Equation (7.94) rewritten as

$$\frac{dH}{dt} = h(H)[f(H,\bar{\alpha}) + g(H)\xi(t)] \tag{7.123}$$

where, as is known, the transformation $h(H) = (\partial H/\partial x)^2 + (\partial H/\partial y)^2$ converts the original two-variable system into an equivalent single variable problem. The associated Fokker–Planck equation as derived in Chapter 6 is

$$\begin{aligned}\frac{\partial p(H,t)}{\partial t} = &-\frac{\partial}{\partial H}F(H)p(H,t) + D\frac{\partial}{\partial H}g(H)\frac{\partial}{\partial H}[g(H) - \tau G(H)]p(H,t)\\ &-\frac{Ds}{3}\frac{\partial}{\partial H}g(H)\frac{\partial}{\partial H}[g(H) - \tau G(H)]\frac{\partial}{\partial H}g(H)p(H,t)\\ &+ O(Ds^2, \tau^2)\end{aligned} \tag{7.124}$$

where $G(H) = F(H)g'(H) - F'(H)g(H)$.

The equation properly reduces to the limiting forms and provides a more general representation. Of special interest in this equation is that $(s = \tau = 0)$ corresponds to the case of white Gaussian noise, $(s = 0)$ to the case of coloured Gaussian noise, and $(\tau = 0)$ yields the results for the white Poisson noise. The general solution to this equation under stationary conditions can be written as

$$p_{st}(x) = N\exp[I_1][1 + I_2 + I_3] \tag{7.125}$$

where N represents the normalisation constant, and I_1, I_2 and I_3 are integrals defined in Equation (6.144).

We shall now consider a few cases and compare their results with those known for the Gaussian noise. First, we shall take the several functions in Equation (7.94), viz.

$$f(H,\bar{\alpha}) = (\bar{\alpha} - 2H)(2H)^{-1/2}$$

$$g(H,\bar{\alpha}) = (2H)^{-1/2}$$

$$h(H,\bar{\alpha}) = (2H)$$

These specific forms of the functions have already been considered

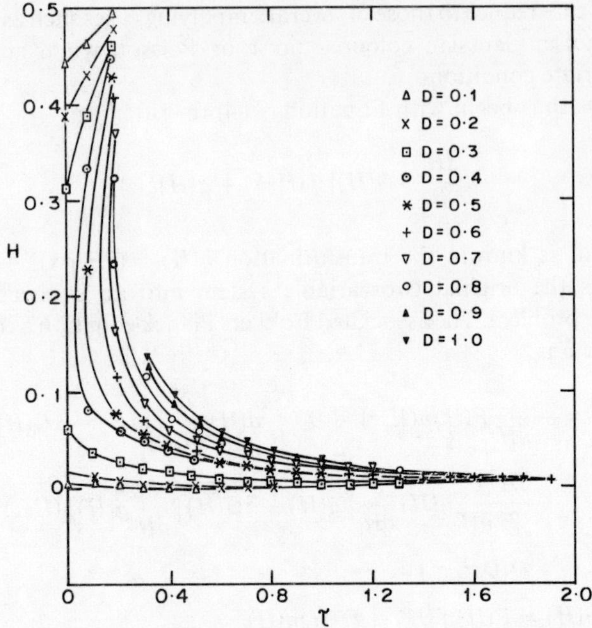

FIGURE 7.68 Stationary solutions for varying strengths of coloured Gaussian noise.

earlier and correspond to a set of deterministic equations which possess a stable limit cycle described by $2H = x^2 + y^2 = \bar{\alpha}$. The deterministic solution is given by $H = \bar{\alpha}/2$, while the stochastic solution can be obtained using Equation (7.125).

The stationary probability distributions for the two sets of parameter values are indicated in Figure 7.67. The chief observation from this figure is that the deterministic solution $H = \bar{\alpha}/2 = 0.5$ no longer represents the stochastic solution and the stationary point has been shifted to some other value of H. Clearly, the extent of shift depends on the extent of noise. In addition, the noise generates a new unstable solution.

The stationary solutions which correspond to the extremal points of the probability distribution can be more easily obtained by differentiating Equation (7.125) with respect to H and equating it to zero. For the specific case of white Gaussian noise ($s \to 0$, $\tau \to 0$), this procedure yields the following equation for the maximum:

$$\frac{\partial P(H, \infty)}{\partial H} = 0 = (2H)^{1/2}(\bar{\alpha} - 2H) - D \tag{7.126}$$

The equation is the same as that obtained earlier [Equation (7.101)] and possesses an additional solution for H. A more general equation for obtaining the extremal points of the probability distribution can likewise be written as

$$\bar{\alpha}(2H)^{1/2} - (2H)^{3/2} - D(2H)^{1/2}[(2H)^{-1/2} - 4\tau] + \frac{s}{3}(2H)^{1/2}[\bar{\alpha}(2H)^{-1/2} - 3(2H)^{1/2}] = 0 \tag{7.127}$$

Several limiting cases can be deduced from this equation by appropriately letting s or τ tend to zero. We shall first take the case when $s \to 0$ which corresponds to the Gaussian coloured noise. Equation (7.127) now becomes

$$\bar{\alpha}(2H)^{1/2} - (2H)^{3/2} - D + 4D\tau(2H)^{1/2} = 0 \tag{7.128}$$

Comparison of Equation (7.126) for white Gaussian noise with Equation (7.128) for coloured noise indicates that an additional term containing τ is now involved. Also, inspection of the last two terms in Equation (7.128) clearly reveals that D and τ have opposing influences. This is confirmed by the numerical calculations shown in Figure 7.68 where the stationary solutions showing time-correlation effects for various values of noise intensity D are presented.

The general Equation (7.127) for the case when the noise is assumed to be white and Poissonian reduces to

$$\bar{\alpha}(2H)^{1/2} - (2H)^{3/2} - D + \frac{s\bar{\alpha}}{3} - s(2H)^{1/2} = 0 \tag{7.129}$$

Note that in view of the opposite signs of the last two terms the net contribution is decided by the values of s and $\bar{\alpha}$. For a given value of $\bar{\alpha}$, changing s would eventually bring about a change in the total contribution of these terms—from supplementing the effect of D to nullifying it.

Equation (7.129) is plotted in Figure 7.69, from which the effect of intensity of the Poisson noise for several values of the Gaussian intensity D is clearly evident. We note from this figure that in the no-solution

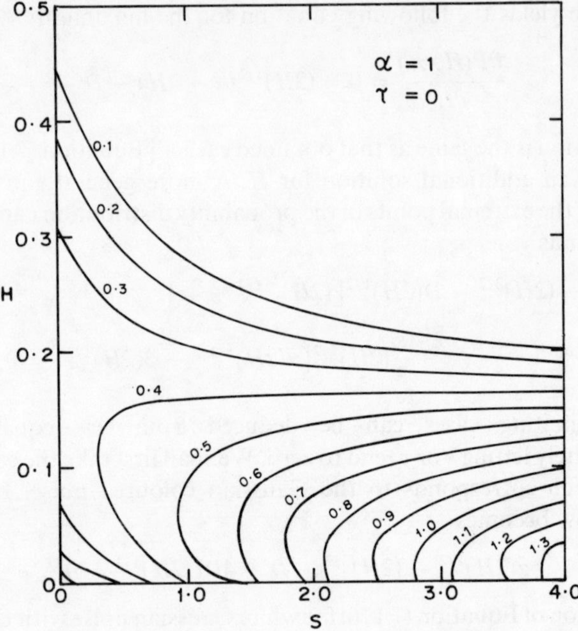

FIGURE 7.69 Stationary solutions for varying intensities of white Poisson noise.

region in presence of white Gaussian noise ($D > 0.4$), introducing s helps to generate the solutions. Transitions not permissible for certain intensities of Gaussian noise can therefore be effected by inducing Poisson noise of a certain intensity. To illustrate this point let us consider the case of $D = 0.4$. Clearly, no stationary solution exists for this case in the Gaussian limit ($s \rightarrow 0$). As s is increased we reach a point (say $s = 0.5$) when we move from the no-solution region to the two-solution region. The region of two solutions lasts over a certain interval of the intensity of Poisson noise and disappears finally to give only one solution. In presence of Poisson noise we therefore have the transition: no solution $\rightarrow$ two solution $\rightarrow$ one solution, with increase in s. As D increases the interval over which the solution region lasts is shifted to higher values of s and also becomes narrower. In fact, for a sufficiently high value of D, we notice only the transition: no solution $\rightarrow$ one solution.

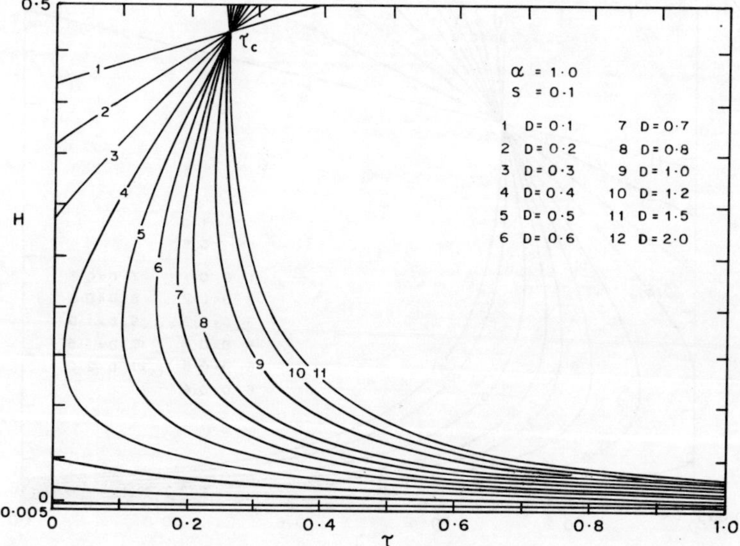

FIGURE 7.70 Stationary solutions for varying strengths of coloured Gaussian noise (*D* and τ) for a fixed intensity (*s*) of Poisson noise.

The effect of coloured Poisson noise on the system behaviour is indicated in Figures 7.70–7.75. Each of these figures shows the stationary solutions for varying time-correlation effects with different values of the white noise intensity *D*. The white Poisson noise intensity (*s*) is varied as we move from Figure 7.70 to 7.75. A typical feature present in all these cases is the existence of a point through which all the branches of stationary solutions pass. The point is located around $\tau = 0.4$ and is marked as τ_c in the figures. It moves to the left of the τ-axis and up on the *H*-axis as the value of noise intensity decreases. In fact, for a given noise intensity *s*, the existence of such a point suggests the invariance of the stationary solution for this value of τ to variations in the intensity of the Gaussian noise *D*. We note that *s* is related to *D* through the relation $\lambda s^2 = 2D$. For fixed *s*, varying *D* thus implies varying λ and one could generalise the existence of the invariant point at this value of τ for varying values λ. As noted earlier, λ monitors the mean number of disturbance pulses in the system.

An examination of the figures reveals that for time-correlation effects

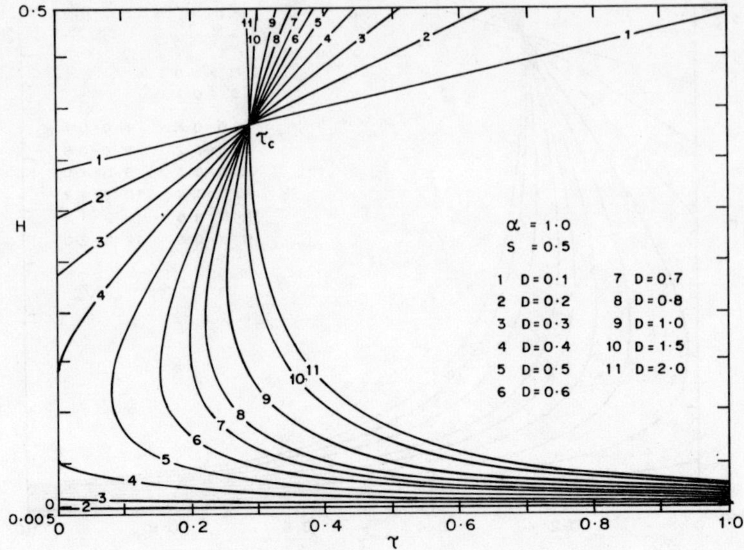

FIGURE 7.71 Stationary solutions for varying strengths of coloured Gaussian noise (D and τ) for a fixed intensity (s) of Poisson noise.

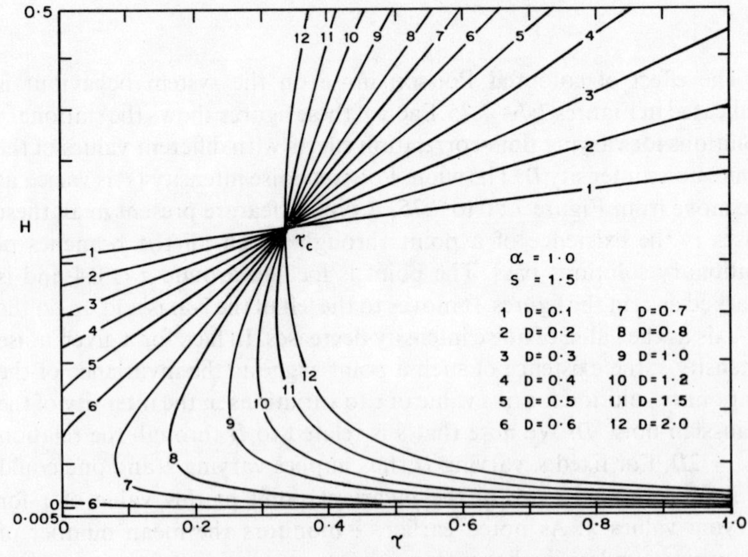

FIGURE 7.72 Stationary solutions for varying strengths of coloured Gaussian noise (D and τ) for a fixed intensity (s) of Poisson noise.

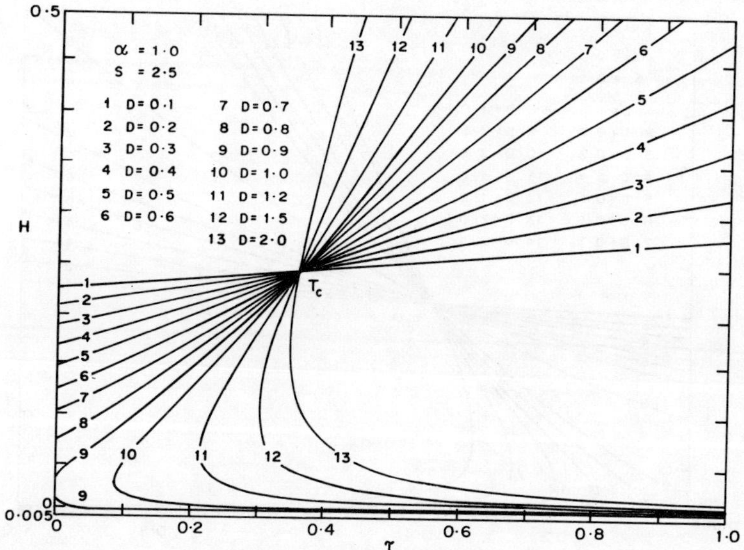

FIGURE 7.73 Stationary solutions for varying strengths of coloured Gaussian noise (D and τ) for a fixed intensity (s) of Poisson noise.

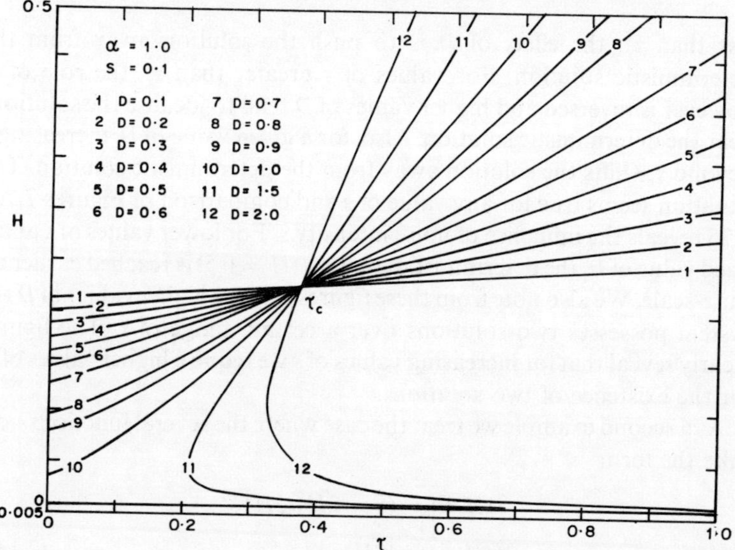

FIGURE 7.74 Stationary solutions for varying strengths of coloured Gaussian noise (D and τ) for a fixed intensity (s) of Poisson noise.

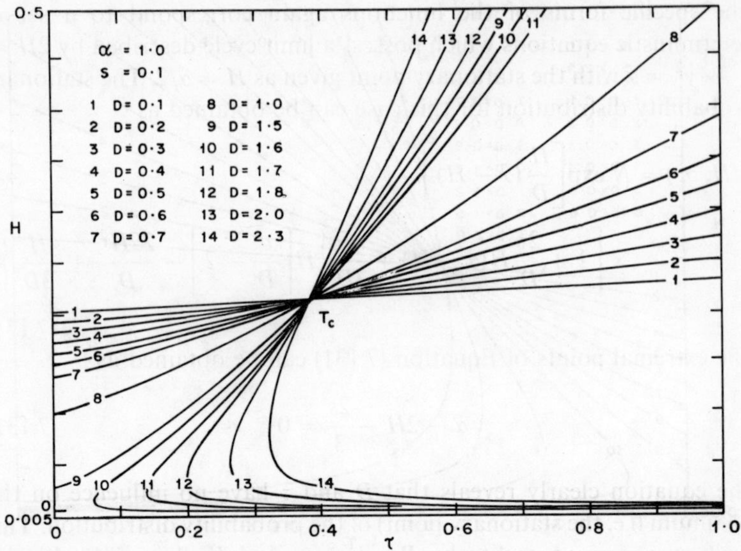

FIGURE 7.75 **Stationary solutions for varying strengths of coloured Gaussian noise (*D* and τ) for a fixed intensity (*s*) of Poisson noise.**

less than τ_c, the effect of *D* is to push the solution away from the deterministic solution. For values of τ greater than τ_c, the role of *D* however is reversed and higher values of *D* tend to localise the solutions near the deterministic solution. Also, for a given value of *D*, increasing τ beyond τ_c shifts the solution away from the deterministic solution. The situation seems true for any value of *s* and comparison of Figures 7.70–7.75 reveals the influence of noise intensity *s*. For lower values of *s* and a fixed value of *D*, the deterministic solution ($H = 0.5$) is reached earlier on the τ-scale. We also note from these figures that for higher values of *D* the system possesses two solutions over a certain range of τ. The figures clearly reveal that for increasing values of *s* we require higher values of *D* for the existence of two solutions.

As a second example we treat the case where the several functions now take the form

$$
\begin{aligned}
f(H, \bar{\alpha}) &= (\bar{\alpha} - 2H)(2H)^{-1} \\
f(H, \bar{\alpha}) &= (2H)^{-1} \qquad (7.130) \\
h(H, \bar{\alpha}) &= (2H)
\end{aligned}
$$

The specific forms of the functions again correspond to a set of deterministic equations which possess a limit cycle described by $2H = x^2 + y^2 = \bar{\alpha}$ with the stationary point given as $H = \bar{\alpha}/2$. The stationary probability distribution for this case can be obtained as

$$P(H, \infty) = N \exp\left[\frac{H}{D}(\bar{\alpha} - H)\right]$$
$$\times \left[1 + \frac{2\tau}{D} H(\bar{\alpha} - H) + \frac{s}{3D}\left\{H\left[\frac{\alpha^2}{D} - 2\right] - \frac{2\bar{\alpha}H^2}{D} + \frac{4H^3}{3D}\right\}\right] \tag{7.131}$$

The extremal points of Equation (7.131) can be obtained as

$$\bar{\alpha} - 2H - \frac{2s}{3} = 0 \tag{7.132}$$

The equation clearly reveals that D and τ have no influence on the maximum (i.e. the stationary point) of the probability distribution. This result was derived earlier by Ravikumar and Kulkarni (1984) who treated the noise as Gaussian. The equation, however, shows that the Poisson noise can have an effect leading to a shift in the deterministic solution $H = \bar{\alpha}/2$ to a new value $H = \bar{\alpha}/2 - (s/3)$. Clearly no solution exists for $\bar{\alpha} < 2s/3$. While the white and nonwhite Gaussian noises have no effect on the deterministic solution, the Poisson noise alters the solution and for a certain value of s brings about a transition from the one-solution region to the no-solution region. Also, we note that the white and nonwhite noises have identical effects since no participation of τ is involved.

7.4.5 Complete Characterisation of a Nonisothermal CSTR in Presence of Noise: Generalised Stochastic Diagrams

The simple case of nonisothermal continuous stirred tank reactor, where a first-order exothermic reaction occurs, has been extensively analysed by Uppal, Ray and Poore (1974) from a macroscopic viewpoint. They classified the dynamic and steady state behaviour of this system in terms of the system parameters and presented the results in the form of a stationary state diagram indicating the rich variety of behaviour that is possible for this system. The analysis shows the existence of six different

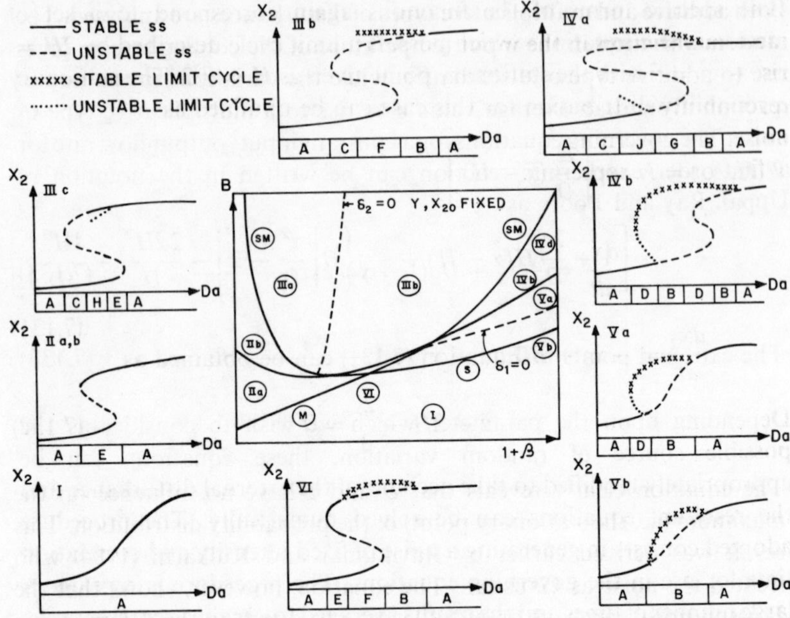

FIGURE 7.76 Permissible stationary states using a macroscopic model.

regions over which a system undergoes drastic stability changes and exhibits features such as monostability, bistability, and oscillatory behaviour. The macroscopic results are summarised in Figure 7.76. In view of the stability considerations involved, the system provides a good case for study from a stochastic viewpoint.

In fact, Ligon and Amundson (1981) have noted the desirability of including fluctuations in the parameter values and, employing a hybrid computer technique, studied the dynamic response and stability characteristics of this system. More specifically, they considered the random fluctuations in the input flow rate and temperature of the feed stream and brought forth a number of results which significantly differed from the results of the corresponding macroscopic model.

In the present section, we shall also investigate this problem in a more systematic way. We begin with the classified diagram of Uppal, Ray and Poore (1974) and study the influence of intensity of noise and its time-correlation effects in each one of the regions of the macroscopic analysis.

Both additive and multiplicative types of disturbances are analysed. The random variation in the input temperature of the cooling stream gives rise to additive type of disturbance, while the variations in the rate of reaction around an average value give rise to a multiplicative type of noise. The governing equations for a constant input–output flow rate for a first-order exothermic reaction can be written in the notation of Uppal, Ray and Poore as

$$\frac{dx_1}{dt} = -x_1 + Da(1 - x_1)\exp\frac{x_2}{1 + x_2/\gamma} \tag{7.133}$$

$$\frac{dx_2}{dt} = -x_2 + BDa(1 - x_1)\exp\frac{x_2}{1 + x_2/\gamma} - \beta(x_2 - x_{2c}) \tag{7.134}$$

Depending upon the parameter which we wish to consider as the possible source of random variation, these equations can be appropriately modified to take account of the external disturbance and the resultant equations can be solved numerically. The procedure adopted consists in generating a noise of fixed intensity and correlation time for use in the governing equations. The procedure is repeated a large number of times and the results averaged to obtain a representative sample function.

Using this procedure the concentration (x_1) and temperature (x_2) profiles for varying intensities of noise (D) and time-correlation effects (τ) have been generated for each one of the regions corresponding to the regions known from the macroscopic analysis. The enormous amount of data (about a 100 profiles) obtained using this procedure has been classified in a way similar to that for the macroscopic analysis. While the detailed results for the case of additive and multiplicative disturbances are presented elsewhere (Kulkarni and Doraiswamy, 1986; Gaikwad *et al.*, 1986), for the present purpose we shall provide the final results in the form of generalised stochastic diagrams. The results for the additive and multiplicative types of noises are presented separately in Figures 7.77 and 7.78.

Comparison of the macroscopic results (Figure 7.76) with those obtained in presence of noise (Figures 7.77 and 7.78) clearly reveals that the system now permits a lesser number of solutions. Also, in all the regions, especially for lower values of Da, the stochastic solutions seem to approach the corresponding macroscopic solutions without actually stabilising. The extent of deviations around the stationary solution

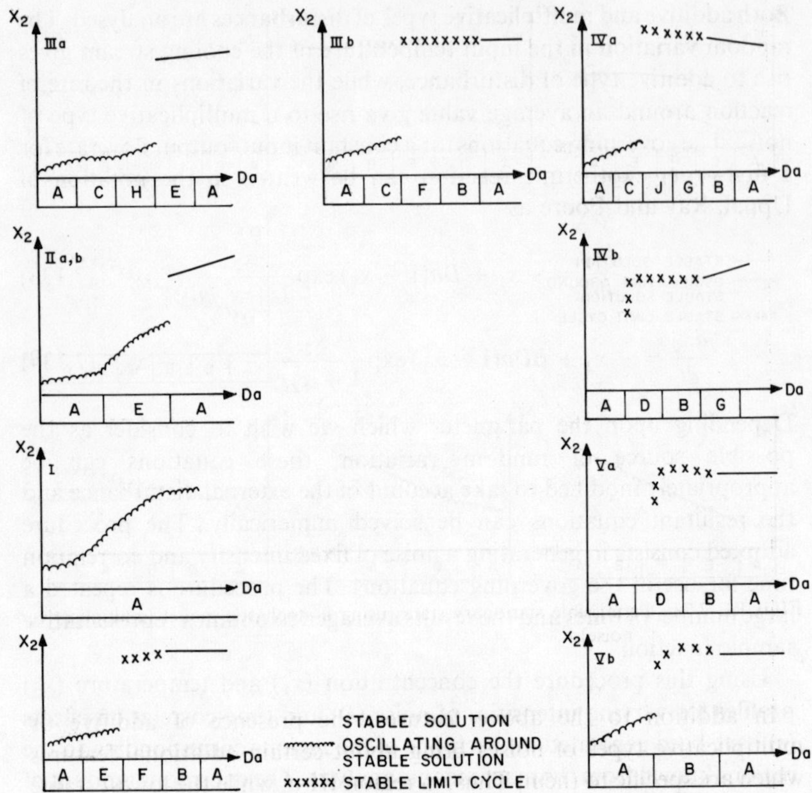

FIGURE 7.77 Permissible stationary states using a stochastic model (additive noise).

depends on the extent of noise and actually increases with it. A comparative study of the effects of additive (Figure 7.77) and multiplicative (Figure 7.78) noise also indicates some vital differences. Thus, while the additive type of noise changes the possible number of solutions, it does not affect the stability properties of the remaining solutions. Also, no new states generated due to noise are evident. The effect of multiplicative noise, on the other hand, indicates a drastic variation and even new solutions not permitted in the macroscopic limit are evident. We shall now highlight the key differences for these types of noises in relation to the results for the macroscopic case.

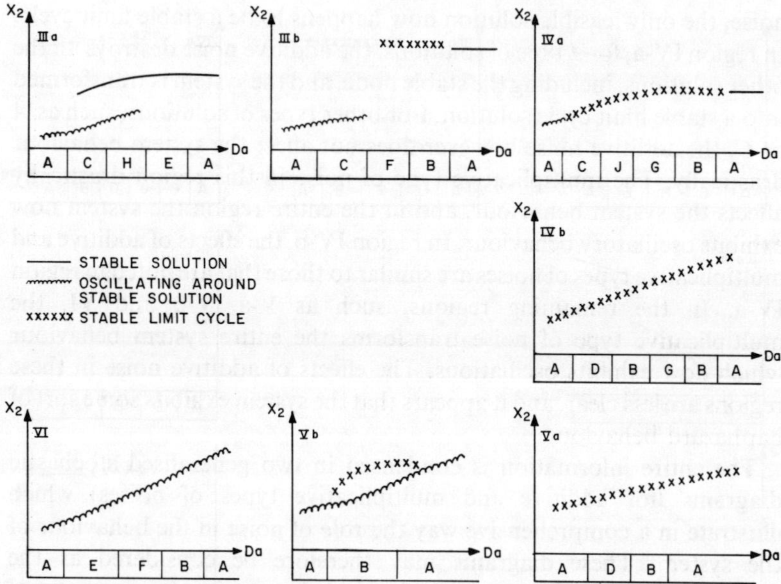

FIGURE 7.78 Permissible stationary states using a stochastic model (multiplicative noise).

In addition to the above features, the presence of additive or multiplicative types of noises bring about certain additional features which are specific to them. Thus, in region III-a, while the lowest stable solutions of the macroscopic analysis are retained, the remaining solutions, especially in the regions marked *C* and *H*, are totally destroyed in presence of additive noise. The corresponding effects for the multiplicative type of noise are different. In presence of this type of noise, the unstable steady state and saddle solution corresponding to *C* in the III-a region are transformed into a new stable solution. This stable solution branch continues with increase in the parameter value *Da*, and eventually merges with the upper stable solution of regions *E* and *A* in region III-a. This entire branch of solutions represents a set of newly generated noise-induced solutions.

In region III-b for *F* type of solutions, the effect of additive noise varies with its extent. Thus, for lower intensities of noise the lower stable solution is realised, while for a higher time-correlation effect the system trips to a stable limit cycle solution. In the case of multiplicative type of noise, the only feasible solution now happens to be a stable limit cycle. In region IV-a, for J type of solutions, the additive noise destroys all the other solutions, including the stable node, and the system is transformed into a stable limit cycle solution. For other types of solutions, such as A or C, the additive noise however does not affect the system behaviour drastically. The multiplicative type of noise in this region drastically affects the system behaviour, and in the entire region the system now exhibits oscillatory behaviour. In region IV-b, the effects of additive and multiplicative types of noises are similar to those that are noted in region IV-a. In the remaining regions, such as V-a, V-b and VI, the multiplicative type of noise transforms the entire system behaviour which now exhibits oscillations. The effects of additive noise in these regions are less clear, and it appears that the system exhibits some sort of haphazard behaviour.

The entire information is condensed in two generalised stochastic diagrams (for additive and multiplicative types of noises) which illustrate in a comprehensive way the role of noise in the behaviour of the system. These diagrams may therefore be considered as the equivalent (or the counterparts) of what is known for the macroscopic case.

7.5 General Comments

In this chapter the theoretical methods developed in Chapter 6 have been applied to model systems. The systems studied range from fluid-bed reactors to complex model systems such as the Brusselator, Oreganator, and Schlogl's model. The developments are largely based on the role of Gaussian noise, but a brief reference has also been made to the behaviour of systems in presence of Poisson noise.

In Section 7.4.2, an observation was made which seems to question the legitimacy of established methods of analysis, in which the occurrence of oscillations (and other unique features) is attributed to certain mechanistic steps in a reaction scheme, such as autocatalysis. The inclusion of the mandatory conditions of detailed balancing and rigorous conservation laws seems to destroy this nonunique feature, thus questioning the role of reaction mechanism in inducing oscillations.

It does not seem possible, or even reasonable, now to reassess all model systems that have been analysed under various conditions and the results reported in the literature. Indeed, the bulk of the material presented in this book is drawn from these publications. Our studies suggest that oscillations can be explained through the presence of noise, but whether noise alone can be regarded as the causative factor is still open to question. There can certainly be other reasons for the occurrence of oscillations or bistabilities.

In the absence of any definitive conclusions on the source of oscillations, the analysis presented in this chapter is based on accepted theories which trace oscillations to reaction mechanism. This would not be true in the thermodynamic limit when the mandatory conditions would strictly hold, but should be entirely acceptable in the time scales consistent with real situations—with which this book is concerned.

REFERENCES

Bukur, D., Caram, H. S. and Amundson, N. R., *Some Model Studies of Fluidised Bed Reactors in Chemical Reactor Theory: A Review*, ed. by Lapidus, L. and Amundson, N. R., Prentice-Hall, New Jersey, 1977.

Dabke, N. S., Kulkarni, B. D. and Doraiswamy, L. K., *Chem. Eng. Sci.* **41**, 1711 (1986a).

Dabke, N. S., Kulkarni, B. D. and Doraiswamy, L. K., *Chem. Eng. Sci.* (communicated) (1986b).

Dabke, N. S., Kulkarni, B. D. and Doraiswamy, L. K., *AIChE J.* (communicated) (1986c).

Dabke, N. S., Kulkarni, B. D. and Doraiswamy, L. K., *Chem. Eng. Sci.* **41**, 2891 (1986d).

Dabke, N. S., Kulkarni, B. D. and Doraiswamy, L. K., *Chem. Eng. Sci.* (communicated) (1986e).

Ebeling, W. and Engel-Herbert, H., *Physica* **104A**, 378 (1980).

Escher, C., *Z. Physik* **35B**, 351 (1979).

Gaikwad, M. S., Tambe, S. S., Kulkarni, B. D. and Doraiswamy, L. K., submitted to *Int. Chem. React. Engng Conf.*, NCL, Pune, 1986.

Gardiner, C. W., *A Handbook of Stochastic Methods for Physics, Chemistry and Natural Sciences*, Springer-Verlag, Berlin, 1983.

Horthemke, W. and Lefever, R., *Noise Induced Transitions*, Springer Series in Synergetics, Vol. 15, Springer-Verlag, Berlin, 1984.

Horsthemke, W. and Lefever, R., *Z. Physik* **40B**, 241 (1980).

Horsthemke, W. and Lefever, R., *Phys. Lett.* **649**, 19 (1977).

Kato, K. and Wen, C. Y., *Chem. Eng. Sci.* **24**, 1351 (1969).

Kulkarni, B. D., Dabke, N. S. and Doraiswamy, L. K., *Chem. Eng. Sci.* **40**, 2007 (1985).

Kulkarni, B. D. and Doraiswamy, L. K., *Ind. Engng Chem. Fundam.* (Hougen memorial issue) (1986).

Ligon, J. R. and Amundson, N. R., *Chem. Eng. Sci.* **36**, 661 (1981b).

Ligon, J. R. and Amundson, N. R., *Chem. Eng. Sci.* **36**, 653 (1981a).

Lindenberg, K. and West, B. J., *Physica* **119A**, 485 (1983).

McQuarrie, D. A., *Stochastic Approach to Chemical Kinetics*, Vol. 8 of Supplementary Review Series in Applied Probability, Methuen, London, 1968.

Prigogine, I. and Nicolis, G., *J. Chem. Phys.* **46**, 3542 (1967).

Ravikumar, V. and Kulkarni, B. D., *Chem. Eng. Commun.* **39**, 69 (1985).

Ravikumar, V., Kulkarni, B. D. and Doraiswamy, L. K., *AIChE J.* **30**, 649 (1984).
Risken, H., *The Fokker–Planck Equation, Methods of Solution and Applications*, ed. by Haken, H., Springer-Verlag, New York, 1983.
Sagues, F., *Phys. Lett. A* **104**, 1 (1984).
Sancho, J. M., San Miguel, M., Katz, S. L. and Gunton, J. D., *Phys. Rev. A* **26**, 1589 (1982).
Sancho, J. M. and San Miguel, M., *Phys. Lett. A* **90**, 455 (1982).
Uppal, A., Ray, W. H. and Poore, A. B., *Chem. Eng. Sci.* **29**, 967 (1974).
van Kampen, N. G., *Phys. Rep.* **24**, 171 (1976).

PART IV

Role of Internal and External Noises

PART IV

Role of Internal and External Noises

8 Simultaneous Presence of Internal and External Fluctuations

8.1 Introduction

8 Simultaneous Presence of Internal and External Fluctuations

8.1 Introduction

THE TREATMENT presented thus far considered the presence of internal or external noise. It is often the case in practice that both these types of fluctuations are simultaneously present, and the present chapter considers and develops the necessary formalism required to handle such situations.

As can be readily appreciated, the internal fluctuations are self-originated in the system and reflect the underlying statistical nature of the processes. These fluctuations scale inversely with the size of the system and in the thermodynamic limit vanish altogether. The conventional way of accounting for these fluctuations is through the description of a Markovian master equation which reduces to the deterministic macroscopic description for large volume (size) systems. This result is essentially valid at points far removed from the points of macroscopic instability. At the critical points the internal fluctuations cannot be ignored and lead to the establishment of some long-range order in the system. How these internal fluctuations can be accounted for in a variety of situations has already been considered in Chapters 3–5.

In contrast to these internal fluctuations, the external ones are not self-originated and owe their existence to the coupling of the system to a fluctuating environment. These fluctuations therefore reflect the statistical nature or the natural randomness of the environment, and quite clearly bear no relationship to the size or volume of the system; they can thus become important even in the thermodynamic limit. Also, since the fluctuations occur at the macroscopic level, a logical way of incorporating them in the analysis would lead to a stochastic differential equation. Chapters 6 and 7 were concerned with the formulation and evaluation of the role of such external fluctuations on the behaviour of reacting systems.

In many practical applications it is often possible to decide *a priori* which of these two types of noises are important in the analysis. Thus, for instance, in a large volume globally stable system we can afford to neglect, without any serious error, the internal noise. Likewise, for a well controlled and stationary environment, the external noise can be ignored. There would, however, be situations, such as when finite size systems are analysed, where both the types of noises would contribute to the total evolution of the system. It is to these types of situations that the present chapter is devoted.

The simultaneous presence of internal and external fluctuations can be handled in two ways. This essentially arises from the fact that, while conventionally the internal noise is handled via the master equation, its inclusion through the description of an equivalent stochastic differential equation or a Fokker–Planck equation is also possible. In the first approach we begin with the master equation formulation and introduce external noise through the variations in the appropriate external parameter. The reformulated equation will now contain the effects of both internal and external noises and can be analysed. Essentially, in this approach the transition probabilities in the master equation would additionally contain the components of the external noise. The drawback of such an approach is the requirement that the transition probabilities in the master equation shall be positive, leading to the restriction that the external noise shall have bounded realisations. In other words, the approach excludes the consideration of white noise. This is an especially serious drawback in view of the fact that a large number of studies involving external noise have assumed white noise. In the second approach we begin with the Langevin description of internal noise and induct in this equation the external noise through an appropriate variation in the external parameter. The resulting stochastic equation, as we know from the developments presented in Chapter 6, can be solved exactly only if the noise considered is white. In what follows we shall explore methods to avoid the shortcomings of the two approaches mentioned.

The literature dealing with the description of systems with both internal and external fluctuations is relatively scarce. The present chapter is therefore based largely on the information available in the research communications of San Miguel and Sancho (1982), Sancho and San Miguel (1984a,b), Rodrigues *et al.* (1985), Horsthemke and Lefever (1984b) and the book by Horsthemke and Lefever (1984a).

8.2 Unified Treatment

8.2.1 Internal Noise Modelled via Master Equation

In this section a unified treatment is considered for cases where the internal noise is modelled via the master equation formulation. The external noise is then incorporated through an appropriate variation in the external parameter that enters the transition probabilities in the master equation. To state the procedure clearly, we begin with a general master equation

$$\frac{\partial p(N,t)}{\partial t} = \sum_{Z=1}^{\infty} \{W(N, N-Z; t)p(N-Z, t) + W(N, N+Z; t)p(N+Z, t)\}$$

$$- \sum_{Z=1}^{\infty} \{W(N+Z, N; t)p(N,t) + W(N-Z, N; t)p(N,t)\} \tag{8.1}$$

where, as before, $p(N,t)$ represents the probability of finding N particles of a given reactant species at time t, and $W(N, N \pm Z; t)$ represent the transition probabilities to move from state $(N \pm Z)$ to state N at any time t. It is often convenient to restrict the development to a unit-step process, in which case Equation (8.1) reduces to the familiar form

$$\frac{\partial p(N,t)}{\partial t} = Q(N-1,t)p(N-1,t) + R(N+1,t)p(N+1,t)$$

$$- [Q(N,t) + R(N,t)]\,p(N,t) \tag{8.2}$$

We note in this equation that

$$\begin{aligned} W(N+1, N; t) &= Q(N,t) \\ W(N-1, N; t) &= R(N,t) \end{aligned} \tag{8.3}$$

The conventional method that ignores the external fluctuations attempts to solve this equation by defining a generating function $\psi(s,t)$

$$\psi(s,t) = \sum_{N=0}^{\infty} s^N p(N,t)$$

which converts Equation (8.2) into the following partial differential equation:

$$\frac{\partial\psi(s,t)}{\partial t} = \left\{(s-1)Qs\left(\frac{\partial}{\partial s}\right) + \left(\frac{1}{s}-1\right)R\left(s\frac{\partial}{\partial s}\right)\right\}\psi(s,t) \qquad (8.4)$$

with the obvious boundary condition: $\psi(1,t) = 1$. The generating function method, discussed in greater detail in Chapter 3, is then employed to obtain the probability distribution and its moments. For convenience we shall recast these equations as

$$p(N,t) = \frac{1}{N!}\frac{\partial^N}{\partial s^N}\psi(s,t)\bigg|_{s=0}$$

$$\bar{N}^\alpha = \sum_{N=0}^{\infty} N^\alpha p(N,t) = \left(s\frac{\partial}{\partial s}\right)^\alpha \psi(s,t)\bigg|_{s=1} \qquad (8.5)$$

Equation (8.2) or (8.4) corresponds to the macroscopic equation (where $x = N/V$)

$$\frac{dx}{dt} = Q(x) - R(x) \qquad (8.6)$$

and, as we know, yields identical results for $N \to \infty$, $V \to \infty$ and finite N/V. In the event we desire to include the external fluctuations, Equation (8.6) provides a convenient starting point. We can identify the existence of fluctuating parameters in the functions $Q(x)$ and/or $R(x)$. Taking, for example, the simple case where $Q(x)$ alone fluctuates we can write Equation (8.6) as

$$\frac{dx}{dt} = \bar{Q}_0(x) + Q_1(x)\xi(t) - R(x) \qquad (8.7)$$

where $Q_0(x)$ represents the mean or nonfluctuating part and the fluctuations are explicitly stated. An analogue of $Q(x)$ can be written in terms of $Q(N)$ as

$$Q(N) = Q_0(N) + Q_1(N)\xi(t) \qquad (8.8)$$

where the properties of the fluctuations $\xi(t)$ are assumed to be defined. It is often convenient to define $\xi(t)$ as a Gaussian white noise with zero mean and correlation $\langle\xi(t)\xi(t')\rangle = 2D\delta(t-t')$. The parameter D signifying the intensity of the noise is not dependent on the system size and remains a finite quantity in the thermodynamic limit. We now substitute Equation (8.8) in the master equation [Equation (8.2)], which includes the internal fluctuations, to obtain

$$\frac{\partial p(N,t)}{\partial t} = [Q_0(N-1) + Q_1(N-1)\xi(t)]p(N-1,t) + R(N+1)p(N+1,t) - [Q_0(N) + Q_1(N)\xi(t) + R(N)]p(N,t) \tag{8.9}$$

Alternatively, we can substitute Equation (8.8) in (8.4) to obtain the result in terms of the generating function as

$$\frac{\partial \psi(s,t)}{\partial t} = \left[(s-1)Q_0 s\left(\frac{\partial}{\partial s}\right) + (s-1)Q_1\left(s\frac{\partial}{\partial s}\right)\xi(t) + \left(\frac{1}{s}-1\right)R\left(s\frac{\partial}{\partial s}\right)\right]\psi(s,t) \tag{8.10}$$

Note that in Equations (8.9) and (8.10), $p(N,t)$ or $\psi(s,t)$ is a function of $\xi(t)$. Thus, for every realisation of the noise $\xi(t)$ these equations are satisfied. The average properties of the system can be obtained by defining $\bar{p}(N,t) = \langle p(N,t)\rangle_{\xi(t)}$ when we intend to use Equation (8.9), or equivalently by defining the function $F(s,t) = \langle \psi(s,t)\rangle_{\xi(t)}$ when we intend to use Equation (8.10). To evaluate this average we need to evaluate the averages $\langle \xi(t)p(N,t)\rangle_\xi$ or $\langle \xi(t)\psi(s,t)\rangle_\xi$. This can be done by using the functional characterisation of the Gaussian property of noise:

$$\langle \xi(t)p(N,t)\rangle_\xi = D\left\langle \frac{\delta p(N,t)}{\delta \xi(t)}\right\rangle_\xi = DQ_1(N-1)\bar{p}(N-1,t) - DQ_1(N)\bar{p}(N,t) \tag{8.11}$$

and

$$\langle \xi(t)\psi(s,t)\rangle_\xi = D\left\langle \frac{\delta \psi(s,t)}{\partial \xi(t)}\right\rangle_\xi = D(s-1)Q_1\left(s\frac{\partial}{\partial s}\right)F(s,t) \tag{8.12}$$

Substitution of Equation (8.11) in (8.9) and Equation (8.12) in (8.10) leads to the final equations

$$\frac{\partial \bar{p}(N,t)}{\partial t} = [Q_0(N-1) - DQ_1^2(N-1) - DQ_1(N-1)Q_1(N)]\bar{p}(N-1,t) + R(N+1)\bar{p}(N+1,t) + DQ_1(N-2)Q_1(N-1)\bar{p}(N-2,t) - [Q_0(N) + R(N) - DQ_1^2(N)]\bar{p}(N,t) \tag{8.13}$$

$$\frac{\partial F(s,t)}{\partial t} = \left[(s-1)Q_0\left(s\frac{\partial}{\partial s}\right) + \left(\frac{1}{s}-1\right)R\left(s\frac{\partial}{\partial s}\right)\right]F(s,t)$$
$$+ D(s-1)Q_1\left(s\frac{\partial}{\partial s}\right)(s-1)Q_1\left(s\frac{\partial}{\partial s}\right)F(s,t) \qquad (8.14)$$

Comparison of the master Equation (8.13) with the master Equation (8.2) in the absence of external fluctuations shows that the incorporation of external noise affects the transition probabilities, and the general equation with reference to Equation (8.3) can be written as

$$\frac{\partial \bar{p}(N,t)}{\partial t} = W(N,N-1)\bar{p}(N-1,t) + W(N,N+1)\bar{p}(N+1,t)$$
$$+ W(N,N-2)p(N-2,t)$$
$$- [W(N-1,N) + W(N+1,N) - W(N-2,N]\bar{p}(N,t) \qquad (8.15)$$

The various transition probabilities appearing in Equation (8.15) can be identified parametrically by comparison with Equation (8.13). In more explicit form these can be written as

$$\begin{aligned}
W(N,N-1) &= Q_0(N-1) - DQ_1^2(N-1) - DQ_1(N-1)Q_1(N) \\
W(N,N+1) &= R(N+1) \\
W(N,N-2) &= DQ_1(N-2)Q_1(N-1) \\
W(N+1,N) &= Q_0(N) \\
W(N-1,N) &= R(N) \\
W(N-2,N) &= DQ_1^2(N)
\end{aligned} \qquad (8.16)$$

It is clear from Equation (8.14) that even for unit-step processes the consideration of external noise leads to new transitions from state $(N-2)$ to state N and vice versa. The equation additionally shows that the terms involving external noise always enter in the form DQ_1^2 showing that the proportionality with the size or volume of the system now varies as V^2 as against the normal case of no external noise where terms proportional to V alone enter. Consideration of individual transition probabilities defined in Equation (8.16) reveals that terms such as $[W(N,N-1)]$ cannot exist if the noise intensity is far too large. Finally, while for unit-step processes a stationary solution to the master

Equation (8.3) exists and is easy to obtain, the corresponding stationary solution to Equation (8.13) is in general difficult to obtain. It is advantageous therefore to define the corresponding generating function equation [Equation (8.14)] which can be used to get additional information such as the moments, etc. for the system.

The analysis thus far assumed, for reasons of simplicity, that the fluctuations occur in only one term $Q(x)$ defined in Equation (8.6). We also assumed that the external noise considered in the analysis was a Gaussian white noise. Though the analysis gives qualitatively the effects of internal and external noises considered together, it is somewhat restrictive due to these assumptions. For instance, the external fluctuations may occur in both the terms $Q(x)$ and $R(x)$ in Equation (8.6). Also an unbounded nature of white noise may not be consistent with the physical reality. We shall therefore consider a more realistic situation with the macroscopic deterministic equation defined as in Equation (8.6) and identify the individual terms $Q(x)$ and $R(x)$ as consisting of

$$Q(x) = Q_0(x) + Q_1(x)\xi_Q(t) \qquad \text{and} \qquad R(x) = R_0(x) + R_1(x)\xi_R(t) \tag{8.17}$$

Here the subscripts 0 and 1 denote, respectively, the noise independent and noise associated parts of the functions $Q(x)$ and $R(x)$, and ξ_Q and ξ_R are the respective noise contributions. The associated master equation for a unit-step process in the absence of external fluctuations is given by (8.3), which can alternatively be written in terms of a difference operator (E) (see Chapter 3) as

$$\begin{aligned}\frac{\partial p(N,t)}{\partial t} &= [(E^{-1} - 1)W_+(N) + (E - 1)W_-(N)]p(N,t) \\ &= \mathscr{L}p(N,t)\end{aligned} \tag{8.18}$$

We note in this equation that $(E^{\pm}) = \exp(\pm\partial/\partial N)$ and the extensive transition probabilities can be identified as

$$W_+(N) = VQ(x), \qquad W_-(N) = VR(x) \qquad \text{with} \quad x = N/V \tag{8.19}$$

In view of Equation (8.17) the extensive transition probabilities in presence of external noise can be written as

$$\begin{aligned} W_+(N) &= VQ(x) = VQ_0(x) + VQ_1(x)\xi_Q(t) \\ &= W_{+,0}(N) + W_{+,1}(N)\xi_Q(t) \\ W_-(N) &= VR(x) = VR_0(x) + VR_1(x)\xi_R(t) \\ &= W_{-,0}(N) + W_{-,1}(N)\xi_R(t) \end{aligned} \tag{8.20}$$

where again for reasons of consistency the subscripts 0 and 1 refer to the noise independent and noise associated extensive transition probabilities.

We now turn our attention to the types of noise $\xi(t)$. For the noises to be consistent with $Q(x)$ and $R(x)$, we would require them to have bounded realisations and a consistent choice could be to regard them as defining a dichotomous two-step Markov process. Alternatively, one can also assume that they define a Poisson white noise process with zero mean. A general definition can then be given [see also Equation (6.133)]:

$$\begin{aligned} \xi_Q(t) &= \sum_i D_{PQ}\delta(t - t_i) - D_{PQ}\lambda_Q \\ \xi_R(t) &= \sum_i D_{PR}\delta(t - t_i) - D_{PR}\lambda_R \end{aligned} \tag{8.21}$$

where D_{PQ}, D_{PR} represent the intensity of noises entering the functions $Q(x)$ and $R(x)$ and λ_Q, λ_R the appropriate Poisson parameters. The processes ξ_Q and ξ_R are bounded from below, in that, omitting the subscripts for generality, we have $\xi(t) \geqslant -\lambda D_P$ and the stochastic master equation is meaningful.

Following the procedure outlined earlier we insert Equation (8.20) in (8.18), which makes $p(N, t)$ dependent upon the noises $\xi_Q(t)$ and $\xi_R(t)$. We then define $p(N, t)$ as the average of $p(N, t)$ over all realisations of ξ_Q and ξ_R. The resulting master equation can be written as

$$\begin{aligned} \frac{\partial \bar{p}(N,t)}{\partial t} = {} & [(E^{-1} - 1)W_{+,0}(N) + (E - 1)W_{-,0}(N)]\bar{p}(N,t) \\ & + \lambda_Q[\{\exp(D_{PQ}[(E^{-1} - 1)W_{+,1}(N)])\}_{\text{av}} \\ & \qquad - \bar{D}_{PQ}[(E^{-1} - 1)W_{+,1}(N)] - 1]\bar{p}(N,t) \\ & + \lambda_R[\{\exp(D_{PR}[(E - 1)W_{-,1}(N)])\}_{\text{av}} \\ & \qquad - \bar{D}_{PR}[(E - 1)W_{-,1}(N)] - 1]\bar{p}(N,t) \end{aligned} \tag{8.22}$$

where the averages in the second and third terms are evaluated over realisations of D_{PQ} and D_{PR}. As before the reformulated master Equation (8.22) contains additional nonvanishing effective probabilities $[\bar{W}(N \rightarrow N \pm n)]$ for any step size n. General equations for these transitions for the case of a multiplicative linear process where $W_{+,1}(N) = \alpha N$ and $W_{-,1}(N) = \beta N$ [reference Equation (8.20)] can be written as

$$\begin{aligned}\bar{W}&(N+1 \rightarrow N)\\ &= W_{-,0}(N+1) + \lambda_R[\{\exp(-D_{PR}\beta N)[1-\exp(-D_{PR}\beta)]\}_{\text{av}}\\ &\qquad \times (N+1) - \bar{D}_{PR}\beta(N-1)] \qquad (8.23)\end{aligned}$$

$$\begin{aligned}\bar{W}&(N-1 \rightarrow N)\\ &= W_{+,0}(N-1) + \lambda_Q[\{-\exp(-D_{PQ}\alpha N)[1-\exp(D_{PQ}\alpha)]\}_{\text{av}}\\ &\qquad \times (N-1) - \bar{D}_{PQ}\alpha(N-1)] \qquad (8.24)\end{aligned}$$

Equations (8.23) and (8.24) for a general n^{th} step transition become

$$\begin{aligned}\bar{W}(N+n \rightarrow N) &= \lambda_R[\{\exp(-D_{PR}\beta N)[1-\exp(-D_{PR})]^n\}_{\text{av}}\\ &\qquad \times (N+1)\ldots(N+n)/n!], \qquad n>1 \quad (8.25)\end{aligned}$$

$$\begin{aligned}\bar{W}(N-n \rightarrow N) &= \lambda_Q[\{(-1)^n \exp(-D_{PQ}\alpha N)[1-\exp(D_{PQ})]^n\}_{\text{av}}\\ &\qquad \times (N-1)\ldots(N-n)/n!], \qquad n>1 \quad (8.26)\end{aligned}$$

The stationary solution to the master equation [Equation (8.22)] may not always be an easy task. Following the procedure outlined earlier one can, however, obtain the equations for the various moments. The first two moments, being normally more important, are presented here for the case of $\lambda_Q = 0$:

$$\frac{d\langle x\rangle}{dt} = \langle Q_0(x) - R_0(x)\rangle + \lambda_R(\{\exp(-D_{PR}\beta)\}_{\text{av}} + \bar{D}_{PR}\alpha - 1)\langle x\rangle \qquad (8.27)$$

$$\begin{aligned}\frac{d\langle x^2\rangle}{dt} &= 2\langle x(Q_0(x) - R_0(x))\rangle\\ &\quad + \lambda_R(\{\exp(-2D_{PR}\alpha)\}_{\text{av}} - 2D_{PR}\alpha - 1)\langle x^2\rangle\\ &\quad + V^{-1}\langle Q_0(x) + R_0(x)\rangle\\ &\quad - \lambda_R V^{-1}(\{\exp(-2D_{PR}\alpha) - \exp(-D_{PR}\alpha)\}_{\text{av}} + D_{PR}\alpha)\langle x\rangle \qquad (8.28)\end{aligned}$$

Of special importance is the presence of the last term in Equation (8.28) which simultaneously involves the internal (V^{-1}) and external contributions. The equation for the mean does not involve such an interactive term; however, if we consider a process where $Q_1(x)$ and/or $R_1(x)$ depends nonlinearly on x, then such interactive terms can appear even in the equation for mean. On the other hand, if we consider the noise to be an additive linear noise, the interactive term does not enter even in the equation for $\langle x^2 \rangle$. In view of this one can regard the existence of interactive terms in the equations for moments as arising due to multiplicative linear or nonlinear type of external noise in a finite size system.

A more general analysis in which the external noise entering the system has some finite time-correlation effects is clearly more relevant but poses considerable problems in mathematical handling. No such analysis is available as of now. The only effective way to take care of such a situation appears to be to move away from the master equation description of internal noise. We shall do so in the following section.

8.2.2 Internal Noise Modelled via Langevin Equation

To begin with we postulate that the macroscopic equation $dx/dt = f(x) = Q(x) - R(x)$ can be modelled in the form of a Langevin equation to account for the internal fluctuations (see Chapter 4):

$$\frac{dx}{dt} = f(x) + V^{-1/2}\tilde{\xi}(t) \tag{8.29}$$

Since our interest now lies in seeking the effects of coloured noise, we shall assume that $\tilde{\xi}(t)$ is not a white noise but follows an Ornstein–Uhlenbeck process:

$$d\xi(t) = -\gamma\tilde{\xi}\, dt + \tilde{\sigma}\, d\tilde{W} \tag{8.30}$$

In Equation (8.29) we supposed that the function $f(x)$ can be rearranged explicitly showing the external parameter (α) as $f(x) = f_0(x) + \alpha f_1(x)$. The parameter varies according to the relation $\alpha = \bar{\alpha} + \xi(t)$, where $\xi(t)$ again follows an Ornstein–Uhlenbeck process

$$d\xi(t) = -\xi\, dt + \sigma\, dW \tag{8.31}$$

Time in this equation has been scaled such that the correlation time of the noise is unity. This also implies that γ appearing in Equation (8.30) represents the ratio of the correlation times of the internal and external noises. The final equation taking account of these fluctuations is then

$$\frac{dx}{dt} = f_0(x) + \bar{\alpha} f_1(x) + \xi(t) f_1(x) + V^{-1/2}\tilde{\xi}(t) \tag{8.32}$$

In the limiting case when the noises are rapidly varying, it is possible to write the Fokker–Planck equation associated with Equation (8.32) thus:

$$\left(\frac{\mathscr{L}_0}{\varepsilon^2} + \frac{\mathscr{L}_1}{\varepsilon} + \mathscr{L}_2\right) p(x, \xi, \tilde{\xi}, t) = 0 \tag{8.33}$$

where the operators are defined as

$$\mathscr{L}_0 = \frac{\partial}{\partial \xi}\xi + \tfrac{1}{2}\sigma^2 \frac{\partial^2}{\partial \xi^2} + \frac{\partial}{\partial \tilde{\xi}}\gamma\tilde{\xi} + \tfrac{1}{2}\tilde{\sigma}^2 \frac{\partial^2}{\partial \tilde{\xi}^2}$$

$$\mathscr{L}_1 = -\frac{\partial}{\partial x}[f_1(x)\xi + V^{-1/2}\tilde{\xi}] \tag{8.34}$$

$$\mathscr{L}_2 = -\frac{\partial}{\partial x} f_0(x) - \frac{\partial}{\partial t}$$

ε in this equation represents an arbitrary parameter introduced to appropriately scale the time and amplitude of the noises so as to ensure the proper limiting form of the white noise. Equation (8.33) can be scaled in terms of the expansion in powers of ε. Defining

$$p(x, \xi, \tilde{\xi}, t) = p_0(x, \xi, \tilde{\xi}, t) + p_1(x, \xi, \tilde{\xi}, t) + \cdots \tag{8.35}$$

we obtain in the white noise limit the following Fokker–Planck equation:

$$\begin{aligned}\frac{\partial p_0(x,t)}{\partial t} &= -\frac{\partial}{\partial x} f_0(x) + \tfrac{1}{2}\sigma^2 f_1'(x) f_1(x) p_0(x,t) \\ &\quad + \tfrac{1}{2}\sigma^2 \frac{\partial^2}{\partial x^2} f_1^2(x) p_0(x,t) + (\tilde{\sigma}^2/2\gamma V)\frac{\partial^2}{\partial x^2} p_0(x,t)\end{aligned} \tag{8.36}$$

Equation (8.36), valid in the limiting white noise, clearly shows the separate contributions of the two noises. The explicit appearance of γ in the last term in this equation also helps to evaluate the effect of the ratio of the correlation times of the two noises. In the limiting case of $\gamma \to \infty$, i.e. when the internal fluctuations are much faster than the external fluctuations, the last term drops out. The solution to Equation (8.33) to higher order in ε can be obtained as

$$p(x,t) = p_0(x,t) + \varepsilon^2\left[H(x,t) + \tfrac{1}{2}\sigma^2 \frac{\partial}{\partial x} f_1(x)\right] p_0(x,t)$$
$$+ \left(\frac{\tilde{\sigma}^2}{2\gamma^3 V}\right)\frac{\partial^2}{\partial x^2} p_0(x,t) + O(\varepsilon^3) \tag{8.37}$$

where $H(x,t)$ is the solution of

$$\tfrac{1}{2}\sigma^2 \frac{\partial}{\partial x} f_1(x) \frac{\partial}{\partial x} f_1(x) H(x,t) + \left(\frac{\tilde{\sigma}^2}{2\gamma V}\right)\frac{\partial^2}{\partial x^2} H(x,t) = I(x,t) \tag{8.38}$$

and the function $I(x,t)$ is defined by

$$I(x,t) = -(\sigma^2\tilde{\sigma}^2/4\gamma V)\left\{\frac{1}{2\gamma^2}\frac{\partial}{\partial x} f_1(x) \frac{\partial^2}{\partial x^2} + \frac{1}{1+\gamma}\frac{\partial}{\partial x}\right.$$
$$\left.\left[\frac{1}{\gamma}\frac{\partial}{\partial x} f_1(x) \frac{\partial}{\partial x} + \frac{\partial^2}{\partial x^2} f_1(x)\right]\right\} p_0(x,t)$$
$$-[3\sigma^2\tilde{\sigma}^2/4\gamma V(2+\gamma)]\left\{\frac{1}{1+\gamma}\frac{\partial}{\partial x} f_1(x)\left[\frac{1}{\gamma}\frac{\partial}{\partial x} f_1(x)\frac{\partial}{\partial x} + \frac{\partial^2}{\partial x^2} f_1(x\right.\right.$$
$$\left. + \frac{1}{2}\frac{\partial^2}{\partial x^2} f_1(x) \frac{\partial}{\partial x} f_1(x)\right\} p_0(x,t)$$
$$-\tfrac{3}{8}\sigma^4 \frac{\partial}{\partial x} f_1(x) \frac{\partial}{\partial x} f_1(x) \frac{\partial}{\partial x} f_1(x) p_0(x,t) - \frac{3\tilde{\sigma}^4}{8\gamma^4 V^2}\frac{\partial^4}{\partial x^4} p_0(x,t)$$
$$+\tfrac{1}{2}\sigma^2\left\{\frac{\partial}{\partial x} f_1(x)\left[\frac{\partial}{\partial t} + \frac{\partial}{\partial x} f_0(x)\right]\frac{\partial}{\partial x} f_1(x)\right.$$
$$\left. + \frac{1}{2}\left[\frac{\partial}{\partial t} + \frac{\partial}{\partial x} f_0(x)\right]\frac{\partial}{\partial x} f_1(x) \frac{\partial}{\partial x} f_1(x)\right\} p_0(x,t)$$
$$+\left(\frac{\tilde{\sigma}^2}{2\gamma^2 V}\right)\left\{\frac{\partial}{\partial x}\left[\frac{\partial}{\partial t} + \frac{\partial}{\partial x} f_0(x)\right] + \frac{1}{2\gamma}\left[\frac{\partial}{\partial t} + \frac{\partial}{\partial x} f_0(x)\right]\frac{\partial}{\partial x^2}\right\} p_0(x, \tag{8.39}$$

The first two terms in Equation (8.39) clearly bring out the interactive effects of the two noises. It appears from these equations that the importance of internal noise increases when the speed of the external noise with respect to the internal noise increases. Equations (8.36) and (8.37) allow us to study the effects of white or coloured internal and external noises together.

8.3 Application to Model Examples

The previous sections were concerned with the general development of the methodology to ascertain the combined influence of internal and external noises. In the present section we shall apply these techniques to some model examples. As a first example we consider the case of a simple reaction $A + X \rightleftarrows B$ carried out in a finite size system. For simplicity we retain the assumption that the concentrations of the species A and B are maintained constant. If N represents the total number of particles of species X at any time, its rate of change can be modelled by the following macroscopic equation:

$$\frac{dN}{dt} = Vk_{-1}B - Vk_1 \frac{AN}{V} \tag{8.40}$$

In the absence of any external disturbances we can write the master equation corresponding to this situation as per Equation (8.2) where, following Equation (8.3), we note that

$$\begin{aligned} Q(N,t) &= W(N+1, N; t) = Vk_{-1}B \\ R(N,t) &= W(N-1, N; t) = k_1 AN \end{aligned} \tag{8.41}$$

The solution of the resulting master equation can be easily given under stationary conditions as a Poisson distribution:

$$P_{\text{st}}(N) = \left(\frac{k_{-1}BV}{k_1 A}\right)^N \exp\left(-\frac{k_{-1}BV}{k_1 A}\right)\frac{1}{N!} \tag{8.42}$$

Let us incorporate the fluctuations in the parameter $(k_{-1}B)$ which is now regarded as consisting of the mean value $(\overline{k_{-1}B})$ with a fluctuating parameter ξ that is assumed to be a Gaussian white noise. With reference to Equation (8.8) we can then identify $Q_0(N) = (\overline{Vk_{-1}B})$ and $Q_1(N) = V$. With these quantities identified we can define the overall transition probabilities as per Equation (8.16) and use them in the formulated master Equation (8.13). The stationary solution to this equation can be obtained by equating it to zero:

$$\bar{P}_{st}(N) = \left(\frac{k_1 A}{2\pi D V^2}\right)^{1/2} \frac{1}{N!} \exp\left(\frac{DV^2}{2k_1 A} - \frac{\overline{k_{-1} B V}}{k_1 A}\right) \times \int_{-\infty}^{\infty} d\alpha\, \alpha^N \exp\left\{-\left(\frac{k_1 A}{2DV^2}\right)\left[\alpha - \frac{V}{k_1 A}(\overline{k_{-1} B} - DV)\right]^2\right\} \tag{8.43}$$

The first two moments can also be obtained using Equation (8.43):

$$\langle N \rangle = \sum_{N=0}^{\infty} N \bar{p}_{st}(N) = \frac{\overline{k_{-1} B V}}{k_1 A} \tag{8.44}$$

$$\langle N(N-1) \rangle = \sum_{N=0}^{\infty} N(N-1) \bar{p}_{st}(N) = \left(\frac{\overline{k_{-1} B V}}{k_1 A}\right)^2 + \frac{DV^2}{k_1 A} \tag{8.45}$$

The mean density $\langle x \rangle = \langle N \rangle / V$ can be used to calculate the normalised mean square fluctuations:

$$\frac{\langle (n - \langle n \rangle)^2 \rangle}{\langle n \rangle^2} = \frac{D k_1 A}{(\overline{k_{-1} B})^2} + \frac{k_1 A}{V \overline{k_{-1} B}} \tag{8.46}$$

The second term in Equation (8.46) gives the additional contribution due to the finite size of the system. Note that in the example considered above external fluctuations alone occur in the function $Q(N)$.

As a second example we consider a simple model $A + X \rightleftarrows C$, $B + X \rightarrow 2X$, and assume that the concentrations of all but the species X are kept constant. The macroscopic model for the system can be written as

$$\frac{dN}{dt} = \alpha V + \gamma N - \beta N \tag{8.47}$$

where $\alpha = k_{-1} C$, $\beta = k_1 A$ and $\gamma = k_2 B$. Let us now assume that the external fluctuations exist in the parameter β such that it is defined as $\beta = \bar{\beta} + \xi_{PR}$ where ξ_{PR} represents the white Poisson noise. Note that the external noise is of the multiplicative linear form so that Equation (8.22) for the probability and the moments Equations (8.27) and (8.28) are applicable. These equations for the particular case can be formally written as

$$\langle x \rangle_{\rm st} = \frac{\alpha}{\beta - \gamma - \lambda_R \bar{D}_{PR} + \lambda_R D_{PR} \,|\, (1 + \bar{D}_{PR})} \tag{8.48}$$

$$\frac{\langle x^2 \rangle_{\rm st} - \langle x \rangle_{\rm st}^2}{\langle x \rangle_{\rm st}^2} = \frac{\bar{\beta}}{\alpha V} + \frac{\lambda_R \bar{D}_{PR}^2}{(1 + \bar{D}_{PR})[(1 + 2\bar{D}_{PR})(\bar{\beta} - \gamma) - 2\lambda_R \bar{D}_{PR}^2]} + \frac{2\lambda^2 \bar{D}_{PR}^4 - \lambda_R \bar{D}_{PR}^2[(1 + 2\bar{D}_{PR}^2)\bar{\beta} - 2(1 + \bar{D}_{PR})\gamma]}{\alpha V(1 + \bar{D}_{PR})[(1 + 2\bar{D}_{PR})(\bar{\beta} - \gamma) - 2\lambda_R \bar{D}_{PR}^2]} \tag{8.49}$$

We note here that the mean value $\langle x \rangle$ does not depend on V, while Equation (8.49) contains the contributions from the internal and external fluctuations as well as from the combined effect of the two.

REFERENCES

Horsthemke, W. and Lefever, R., *Noise Induced Transitions: Theory and Applications in Physics, Chemistry and Biology*, Springer-Verlag, Berlin, 1984a.

Horsthemke, W. and Lefever, R., *Phys. Lett. A* **106**, 10 (1984b).

Rodriguez, M. A., Pesquera, L., San Miguel, M. and Sancho, J. M., *Phys. Lett. A* **107**, 443 (1985).

San Miguel, M. and Sancho, J. M., *Phys. Lett. A* **90**, 455 (1982).

Sancho, J. M. and San Miguel, M., *Recent Developments in Nonequilibrium Thermodynamics*, Lecture Notes in Physics, Vol. 199, ed. by Casas-Vazquez, J., Lon, D. and Lebon, G., Springer-Verlag, Berlin, 1984a.

Sancho, J. M. and San Miguel, M., *J. Stat. Phys.* **37**, 151 (1984b).

Author Index

Subject Index